Springer Proceedings in Earth and Environmental Sciences

The series Springer Proceedings in Earth and Environmental Sciences publishes proceedings from scholarly meetings and workshops on all topics related to Environmental and Earth Sciences and related sciences. This series constitutes a comprehensive up-to-date source of reference on a field or subfield of relevance in Earth and Environmental Sciences. In addition to an overall evaluation of the interest, scientific quality, and timeliness of each proposal at the hands of the publisher, individual contributions are all refereed to the high quality standards of leading journals in the field. Thus, this series provides the research community with well-edited, authoritative reports on developments in the most exciting areas of environmental sciences, earth sciences and related fields.

Kanchan Deoli Bahukhandi · Nitin Kamboj ·
Vishal Kamboj
Editors

Environmental Pollution and Natural Resource Management

Springer

Editors
Kanchan Deoli Bahukhandi
University of Petroleum and Energy Studies
Dehradun, India

Vishal Kamboj
Department of Zoology and Environmental Science
BFIT Group of Institutions
Dehradun, India

Nitin Kamboj
Department of Zoology and Environmental Science
Gurukul Kangri Vishwavidyalaya
Haridwar, India

ISSN 2524-342X ISSN 2524-3438 (electronic)
Springer Proceedings in Earth and Environmental Sciences
ISBN 978-3-031-05334-4 ISBN 978-3-031-05335-1 (eBook)
https://doi.org/10.1007/978-3-031-05335-1

This Springer imprint is published by the registered company Springer Nature Switzerland AG
The registered company address is: Gewerbestrasse 11, 6330 Cham, Switzerland

Contents

Uptake, Accumulation and Translocation of Heavy Metals in Cauliflower Grown in Integrated Industrial Effluent Irrigated Soil in District Haridwar (Uttarakhand)

Roushan K. Thakur and Vinod Kumar

Abstract The combined effluents from various industries (integrated industrial effluent; IIE) in some locations are used as irrigation water to grow cauliflower because it is readily available and low cost. Contents of cadmium (Cd), chromium (Cr), copper (Cu), iron (Fe), manganese (Mn) and zinc (Zn) in soil, roots, leaves and inflorescence of *B. oleracea* cultivated with the IIE irrigated water at Bhagtanpur Village, Ikkarkala Village and Shubrasha Village were higher than the control site (Bongla Village), where irrigation was with well water. The bioaccumulation factor (BCF) and translocation factors (TF) of metals in *B. oleracea* were in Bhagtanpur village due to higher concentration of metals in the effluent. Use of IIE for irrigation of *B. oleracea* increased concentrations of heavy metals in plant parts which may pose a potential threat to human health due to consumption, and to deterioration of soil fertility due to the long-term irrigation practices.

Keywords *Brassica oleracea* var. *botrytis* · Bioaccumulation factor · Enrichment factor · Integrated industrial effluent · Translocation efficiency

1 Introduction

As urban areas in developing countries increase, and people seek better living standards, larger volumes of freshwater are diverted to domestic, commercial, and industrial sectors, which produce greater volumes of wastewater [1–3]. Increasing demand of water for agricultural irrigation has increased use of treated, or untreated, wastewater [4–7]. The State Infrastructure and Industrial Development Corporation of Uttarakhand Limited occupies an area of about 823.13 ha having nearly 700 independent industrial units which are involved in cosmetics, plastic, apparel, agro-food, pharmaceutical products, electronic products, packaging, synthetic fabrics, electroplating, and commercial automotive activities [8, 9]. The areas contain heterogeneous

R. K. Thakur (✉) · V. Kumar
Roorkee Institute of Technology, Roorkee, Uttarakhand 247667, India
e-mail: drroushanthakur@gmail.com

K. D. Bahukhandi et al. (eds.), *Environmental Pollution and Natural Resource Management*, Springer Proceedings in Earth and Environmental Sciences,
https://doi.org/10.1007/978-3-031-05335-1_1

industries with varying pollution loads and concentrations but the industries lack individual effluent treatment systems. The industries use a large quantity of fresh water and generate huge quantities of effluent. The industrial effluents are partially treated in a common effluent treatment plant and most of the partially treated effluent is disposed through various channels near the treatment plant. Unprocessed wastewater flows through channels into rivers where it is diverted by subsistence farmers to small vegetable plots grown for nearby municipal markets [10–12].

There are public health risks of using adulterated streams for irrigation [10]. Disposal of wastewater is a major problem for industries, due to generation of high volume of effluent, and limited space, for land-based treatment and disposal [11, 13–15]. Wastewater contains nutrients that can be used for cultivation of agricultural crops [15–17]. Irrigation with waste water contributes to heavy metal contents of soil and crops [15].

Trace elements play roles in chemical, biological, metabolic, and enzymatic responses in living cells of plants and humans [16]. Large deposits of heavy metals in agricultural soils through wastewater irrigation may result in soil contamination and affect food quality and safety [17–19]. Vegetables plants take up heavy metals and accumulate them in tissues in quantities high enough to cause clinical problems to animals and human beings that consume them [15, 20].

Heavy metal pollution in agriculture can come from atmospheric fall-out, pesticides, and contamination by chemical fertilizers and irrigation water [20]. Heavy metals rank high among toxins of leafy vegetables [21]. Uptake of trace elements by plants varies, and depends on soil pH, soil organic matter content and other factors [22, 23]. Plant uptake is a major route of food chain exposure to trace elements from the soil.

Effluents from industries contain measurable amounts of nitrogen (N), phosphorous (P), sodium (Na), potassium (K), calcium (Ca), magnesium (Mg), zinc (Zn), copper (Cu), iron (Fe), manganese (Mn), lead (Pb), nickel (Ni), and cadmium (Cd), and their disposal causes contamination of soil and water [21, 22, 24, 25]. Pollutants from industry include suspended solids, compounds colored by lignin, dissolved inorganic salts, chlorinated lignins, and phenolic derivatives. Discharge of untreated effluent can create pollution resulting in deterioration of water quality and toxicity to aquatic life [26–28].

Generally, farmers and consumers do not have information about vegetable physiology and morphology; they only consider undamaged, dark green, big, leaves as characteristics of good quality [29, 30]. External appearance of vegetables can not guarantee safety from contamination especially when farming activities are carried out using industrial effluents for irrigation [31–33]. Growers frequently use industrial effluents due to scarcity of clean irrigation water and to save chemical fertilizer cost as the effluents contain nutrients and toxins [15, 32, 34]. Irrigation with effluents can cause deterioration in soil fertility. Cauliflower (*Brassica oleracea* L. var. *botrytis* L.) has been reported to accumulate heavy metals [34–36], with particular affinity for Cu, Cd, Cr, Pb and Zn.

It may be possible to remediate contaminated agricultural soils using vegetables because they grow rapidly, take up high contents of minerals and metals, and translocate them in their aerial plant parts [32, 37–39], which may make them not safe to eat.

In several countries, and in India, vegetables are irrigated with different types of wastewater, generally from a single source of industrial effluent. The general results are that use of wastewater could expose the consumer to hazards. The State Infrastructure and Industrial Development Corporation of Uttarakhand Limited produced IIE supplies the whole district of Haridwar at low cost. Cauliflower is grown in a large area in the region and is irrigated using IIE, but little is known of the effects on plant development or whether the crop is fit for consumption. The study was conducted to evaluate uptake, accumulation and translocation of heavy metals in cauliflower grown in soil irrigated with IIE and to determine if the plants were safe for human consumption.

2 Materials and Methods

2.1 Study Area

Sampling was collected from Bhagtanpur Village (29°53′36.39″N, 78°4′27.64″E), Ikkarkala Village (29°52′38.85″N, 78°4′36.91″E), and Shubrasha Village (29°52′17.36″N, 78°4′26.57″E) where irrigation was with the IIE. Cauliflower is cultivated in these locations using IIE effluents. In Bongla Village, the control site (29°51′57.81″N, 78°4′41.00″E), bore well water was used for irrigation. The soils of treatment and control sites are sandy loam and slightly alkaline, pH 7.8. Cultivation practices at all sites were similar. Irrigation was twice weekly at all sites.

2.2 Collection of Samples and Analysis

The IIE samples were collected in plastic bottles, immediately acidified with HNO_3, and transported to the laboratory. Samples were filtered through Whatman No. 42 filter paper and stored in a refrigerator at 4 °C. Soil samples from each site were composited separately for analysis. Samples of harvest ready cauliflower were collected from different sites. At each sampling inflorescence, leaves, roots, and soils surrounding roots were collected from 3 or 4 plots of treatment and control sites. Plant samples from all sites were washed with double distilled water and plants separated by hand into different parts which were dried at room temperature for 24 h. Tissues were separately chopped and placed in a hot air oven at 60 °C for 48 h to dry. Samples were ground into powder using a mortar and pestle and stored in plastic bags for further use.

2.3 Analysis of Heavy Metals

For analysis of heavy metals, plant, soil and water samples were digested [17]. A 0.5 g of powdered samples of plant and soil, and a 10 mL sample of water, were placed in a digestion tube and 10 mL of nitric acid (HNO_3) and 5 mL of perchloric acid ($HClO_4$) added and digestions completed on digestion blocks (FOSS, Mumbai, India) following standard procedures [21]. After digestion samples were filtered through Whatman No. 42 filter paper and volume made up to 50 mL. The Cd, Cu, Cr, Fe, Mn and Zn contents in the digested aliquot were determined with an atomic absorption spectrophotometer (model 5000, Perkin-Elmer, GenTech Scientific Inc., Arcade, NY).

2.4 Estimation of Enrichment Factor, Contamination Factor and Bio-concentration Factor

Enrichment factor (EF) of different heavy metals was calculated following the formula of Kim and Kim [27], and used to assess the degree of contamination of heavy metals in the soil. The contamination factor (Cf) of heavy metals in the soil was determined by the method of Håkanson [22]. The bio-concentration factor (BCF) was used to describe the transfer of trace elements from soil to plant tissues. The BCF is calculated as the ratio among the concentration of heavy metals in the plant tissue and in the corresponding soil all based on dry weight for each vegetable [35]. The translocation index was used to determine the ability of plants to translocate heavy metals from roots to harvestable aerial plant parts.

2.5 Statistical Analysis

The data was statistically analyzed using multi-way analysis of variance (ANOVA) to determine the significant difference in the characteristics of IIE effluent irrigated soil and cauliflower crops at different sampling sites using MS Excel 2013.

3 Results and Discussion

The physico-chemical and microbiological components of the IIE and well water varied (Tables 1 and 2). The IIE was had a high pH due to concentrations of alkalis used in the manufacturing processes. The BOD, COD, Ca, TKN, Cd, Fe, Mn, MPN and SPC of the IIE were above recommended limits of the Indian Irrigation Standards [17]. Higher BOD and COD of the IIE were due to existence of high oxidizable

Table 1 Physico-chemical, heavy metals and microbiological characteristics of State Infrastructure and Industrial Development Corporation of Uttarakhand limited effluent and bore well water

Parameter	Water source		BIS[a] for irrigation water
	Effluent	Bore well	
EC (dc·cm^{-1})	2.12 ± 0.15	0.49 ± 0.08	– [b]
TDS (mg·L^{-1})	814.38 ± 143.66	154.16 ± 12.29	1900
pH	8.44 ± 0.44	7.15 ± 0.28	5.5–9.0
DO (mg·L^{-1})	ND[c]	7.27 ± 0.25	–
BOD (mg·L^{-1})	231.89 ± 21.48	3.13 ± 0.16	100
COD (mg·L^{-1})	716 ± 29.92	4.82 ± 0.69	250
Ca (mg·L^{-1})	485 ± 38.87	21.80 ± 2.82	200
Mg (mg·L^{-1})	145.89 ± 9.42	8.71 ± 0.76	–
Na (mg·L^{-1})	103.64 ± 6.01	6.41 ± 0.66	–
K (mg·L^{-1})	60.15 ± 5.75	4.03 ± 0.95	–
TKN[d] (mg·L^{-1})	136.55 ± 10.93	15.02 ± 1.12	100
PO_4^{3-} (mg·L^{-1})	166.49 ± 9.63	0.05 ± 0.01	–
Cd (mg·L^{-1})	3.14 ± 0.27	0.08 ± 0.01	2.00
Cr (mg·L^{-1})	1.75 ± 0.37	0.05 ± 0.01	2.00
Cu (mg·L^{-1})	1.70 ± 0.05	0.08 ± 0.01	3.00
Fe (mg·L^{-1})	3.04 ± 0.07	0.19 ± 0.04	1.00
Mn (mg·L^{-1})	1.32 ± 0.12	0.12 ± 0.03	1.00
Zn (mg·L^{-1})	0.70 ± 0.09	0.32 ± 0.02	15
SPC[f] (SPC·mL^{-1})	86 × 10^6	NA[e]	10,000
MPN[g] (MPN/100 mL)	27 × 10^6	NA	5,000

[a]BIS = Bureau of Indian Standards
[b]– = Not given in standard
[c]ND = Not detectable
[d]TKN = Total Kjeldahl Nitrogen
[e]NA = Not available
[f]PC = Standard plate count
[g]MPN = Most probable number

organic matter and rapid consumption of dissolved inorganic materials. The higher bacterial load (SPC and MPN) in IIE was due to presence of more dissolved solids and organic matter including TS, TDS, TSS, EC, BOD, COD, Cl^-, Na^+, K^+, Ca^{2+}, Mg^{2+}, TKN, PO_4^{3-} and SO_4^{2-} [35]. The IIE effluent contained Cd, Cr, Cu, Fe, Mn and Zn [25]. The IIE contained high levels of chemicals that can be plant nutrients as well as heavy metals, which may enhance yield of agricultural crops while still causing them to be contaminated. The ANOVA data presented in Tables 2 and 3 indicated that IIE was significantly enriched with Cd, Cr, Cu, Fe, Mn and Zn in comparison to control (Bore well water). There was no significant effect of location on content of metals in IIE at different sites except Fe (Tables 2 and 3).

Table 2 ANOVA for effect of location, metal, water type and their interactions on heavy metal concentrations in water

Source	Cd	Cr	Cu	Fe	Mn	Zn
Location (L)	ns	ns	ns	ns	ns	ns
Metal (M)	**	**	**	**	**	**
Water type (W)	*	*	*	*	*	*
Interaction						
L × M	ns	ns	ns	ns	ns	ns
L × W	ns	ns	ns	*	ns	*
M × W	*	*	*	**	*	*
L × M × W	ns	ns	ns	*	ns	ns

ns, *, ** not significant or significant at $P \leq 0.05$ or $P \leq 0.01$, ANOVA

Table 3 Interaction of location and water type on concentrations of heavy metals in water

	Concentration (mg L^{-1})[a]						
Location ×	Water type	Cd	Cr	Cu	Fe	Mn	Zn
Bhagtanpur	IEE	3.02[c]	1.68[b]	1.62[b]	2.95[b]	1.17[b]	0.68[b]
Ikkarkala		2.99[b]	1.68[b]	1.62[b]	2.52[b]	1.17[b]	0.60[b]
Shubrasha		2.86[b]	1.73[c]	1.62[b]	3.68[c]	1.32[c]	0.66[b]
Bongla (Control)	Well	0.08[a]	0.05[a]	0.08[a]	0.19	0.12[a]	0.32[a]

[a]Bureau of Indian standards for inland disposal of treated water: Cd (2.0), Cr (1.5), Cu (1.0), Fe (15.0), Mn (2.0) and Zn (10) mg L^{-1}. Data in the interaction analyzed with Least Squares Means and means separated with LSD

[b]values in a group, in the column, followed by ns, *, ** and *** are not different or significantly different at $P \leq 0.05$, $P \leq 0.01$, $P \leq 0.001$

Concentrations of Cd, Cr, Cu, Fe, Mn and Zn in IIE, and bore well, water at sampling sites varied (Tables 1, 2, 3 and 4). Contents of heavy metals in the IIE were higher for Cu, Fe, Mn and Zn than in bore well water. Contents of Cr and Cd were below detection limits in the bore well water. Concentrations of heavy metals in the IIE were above recommended values [35, 36]. The concentrations in the IIE were higher (except Cr) than values reported by Kumar and Thakur [31] for Cd (2.74–2.84 mg·L^{-1}), Cr (1.61–1.65 mg·L^{-1}), Cu (1.42–1.56 mg·L^{-1}), Fe (2.04–2.34 mg·L^{-1}) Mn (1.02–1.08 mg·L^{-1}) and Zn (0.42–0.49 mg·L^{-1}) at sites near the SIDCUL used for irrigation. Values of heavy metals reported here were higher than previous values since the number of industries in operation in SIDCUL has increased and the amount of materials in the effluent has increased. There were increases in Cd, Cr, Cu, Fe, Mn and Zn contents in soil at all sampling sites irrigated with IIE (Tables 5, 6, 7 and 8). The highest Fe content in the IIE irrigated soil was in Shubrasha village followed by Ikkarkala and Bhagtanpur villages (Tables 5, 6, 7, 8 and 9). The increase in Fe content was likely due to Fe content in the water, or soil, at those sites. The ANOVA data presented in Tables 6, 7 and 8 indicated that location have no

Table 4 ANOVA for effect of location, metal, water type, and their interactions on metals found in soil

Source	Cd	Cr	Cu	Fe	Mn	Zn
Location (L)	ns	ns	ns	ns	ns	ns
Metal (M)	**	**	**	**	**	**
Water type (W)	*	*	*	*	*	*
Interaction						
L × M	ns	ns	ns	ns	ns	ns
L × W	**	**	**	*	**	**
M × W	ns	ns	ns	*	ns	ns
L × M × W	ns	ns	ns	*	ns	ns

ns, *, ** not significant or significant at $P \leq 0.05$ or $P \leq 0.01$, ANOVA

Table 5 ANOVA for effect of metal, water type, and their interactions on enrichment of metals in soil

Source	Cd	Cr	Cu	Fe	Mn	Zn
Location (L)	ns	ns	ns	ns	ns	ns
Metal (M)	**	**	**	**	**	**
Water type (W)	*	*	*	*	*	*
Interaction						
L × M × W	*	ns	ns	ns	*	*

ns, *, ** not significant or significant at $P \leq 0.05$ or $P \leq 0.01$, ANOVA

significant effect on Cd, Cr, Cu, Fe, Mn and Zn contents of the soil at all the sampling sites. IIE irrigation have significant effect on the percent increase in the contents of Cd, Cr, Cu, Fe, Mn and Zn in the soil at all the three sampling sites viz., Bhagtanpur, Ikkarkala and Shubrasha villages in comparison to Bongla (Control site) (Table 9). The enrichment factor of Cd, Cr, Cu, Fe, Mn and Zn in the soil was also increased due to the IIE irrigation at all the sampling sites (Tables 6, 7 and 8).

4 Enrichment Factor

The enrichment factor of different heavy metals was varied. The enrichment factor of heavy metals in soils due to IIE irrigation was in the order: Fe > Zn > Mn > Cu > Cd > Cr for Bhagtanpur village; Fe > Mn > Zn > Cu > Cd > Cr for Ikkarkala Village and Fe > Mn > Zn > Cd > Cu > Cr for Shubrasha village. The EF was greatest for Fe and least for Cr in Shubrasha village. The contamination categories established by Sutherland [40] indicated that the IIE irrigated soil was in the significant enrichment category for Fe, Zn and Mn, while Cu, Cd and Cr were in the moderate enrichment

Table 6 Concentration of interaction of metal in soil and location on metal concentration and enrichment factor

Metal ×	Location	Concentration (μg·g^{-1})	Enrichment factor
Cd[a]	Bhagtanpur	23.95[b]	2.6[c]
	Ikkarkala	25.16[a]	2.8[b]
	Shubrasha	27.26[a]	3.0[a]
	Bongla (control)	9.06[c]	–[c]
Cr	Bhagtanpur	15.71[a]	2.4[a]
	Ikkarkala	16.62[a]	2.6[a]
	Shubrasha	14.43[a]	2.2[a]
	Bongla (control)	6.47[b]	–
Cu	Bhagtanpur	40.58[a]	2.9[a]
	Ikkarkala	40.60[a]	2.9[a]
	Shubrasha	34.93[b]	2.5[a]
	Bongla (control)	13.82[c]	–
Fe	Bhagtanpur	391.87[c]	7.3[a]
	Ikkarkala	415.68[b]	7.8[a]
	Shubrasha	425.65[a]	8.0[a]
	Bongla (control)	53.10[c]	–
Mn	Bhagtanpur	50.03[b]	5.0[b]
	Ikkarkala	53.41[b]	5.3[b]
	Shubrasha	55.61[a]	5.6[a]
	Bongla (control)	9.95[c]	–
Zn	Bhagtanpur	43.15[a]	6.4[a]
	Ikkarkala	34.68[b]	5.2[b]
	Shubrasha	36.05[b]	5.4[b]
	Bongla (control)	6.72[c]	–

Data in the interaction analyzed with Least Squares Means and means separated with LSD
[a]Safe Limit of India (Awashthi, 2000): Cd (3–6 μg·g^{-1}), Cu (135–200 μg·g^{-1}), Fe (75–150 μg·g^{-1}), Zn (300–600 μg·g^{-1})
[b]values in a group, in columns, followed by the same letter are not significantly different, $P \leq 0.05$
[c]"–" = Not determined since the values from the control were used as the base line

category (Tables 6). The interaction of location, metals and water type indicated the significant increase in the contents of Cd, Cr, Cu, Fe, Mn and Zn in the IIE irrigated soil in different villages (Tables 7 and 8). There were positive correlations between Cr and Cu, Fe and Cd, Zn and Mn, Cd and Cr, Cd and Zn, Cr and Fe, Mn and Cu in the soil and IIE (Table 9). Therefore, the IIE irrigation increased Cd, Cr, Cu, Fe, Mn and Zn contents in soils in those villages.

Table 7 ANOVA for effect of location, metal, water type, and their interactions on percent increase of metals in soil

Source	Cd	Cr	Cu	Fe	Mn	Zn
Location (L)	ns	ns	ns	ns	ns	ns
Metal (M)	**	**	**	**	**	**
Water type (W)	*	*	*	*	*	*
Interaction						
L × M × W	*	*	*	*	*	*

ns, *, ** not significant or significant at $P \leq 0.05$ or $P \leq 0.01$, ANOVA

Table 8 Percent increase in metals in soil as affected by interaction of metal and location

Metal ×	Location	Percent increase in soil
Cd	Bhagtanpur	264.35b[a]
	Ikkarkala	277.70[a]
	Shubrasha	159.27[c]
Cr	Bhagtanpur	242.81[b]
	Ikkarkala	256.87[a]
	Shubrasha	223.03[c]
Cu	Bhagtanpur	293.63[a]
	Ikkarkala	293.78[a]
	Shubrasha	152.78[b]
Fe	Bhagtanpur	737.98[c]
	Ikkarkala	782.82[b]
	Shubrasha	801.60[a]
Mn	Bhagtanpur	502.81[c]
	Ikkarkala	536.78[b]
	Shubrasha	558.89[a]
Zn	Bhagtanpur	642.11[a]
	Ikkarkala	516.07[b]
	Shubrasha	336.46[c]

Data in the interaction analyzed with Least Squares Means and means separated with LSD

[a]values in a group, in columns, followed by the same letter are not significantly different, $P \leq 0.05$

There are concerns about accumulation of heavy metals present in IIE used for irrigation, and consequently their transference to plants and their eventual entrance into the food chain. The contents of heavy metals in edible and non-edible parts of cauliflower (*B. oleracea*) grown in the IIE and bore well water irrigated soil varied (Tables 10, 11, 12 and 13). The highest concentration of Fe was in cauliflower roots and the lowest concentration of Mn was in the inflorescence. Levels of Fe in

Table 9 Coefficient correlation (r) of heavy metals in soil at locations

	Cd	Cr	Cu	Fe	Mn
Bhagtanpur					
Cr	−0.61				
Cu	−0.14	0.22			
Fe	−0.14	0.18	0.99		
Mn	−0.13	0.85	−0.03	−0.07	
Zn	−0.40	−0.24	0.63	0.65	−0.69
Ikkarkala					
Cr	0.12				
Cu	0.74	0.27			
Fe	−0.17	−1.00	−0.27		
Mn	0.15	−0.95	−0.15	0.93	
Zn	0.14	0.89	−0.04	−0.91	−0.78
Shubrasha					
Cr	0.23				
Cu	0.43	0.84			
Fe	−0.53	−0.07	−0.59		
Mn	−1.00	−0.25	−0.46	0.57	
Zn	−0.05	−0.90	−0.91	0.32	0.08
Bongla (control)					
Cr	0.47				
Cu	0.61	−0.31			
Fe	−0.89	−0.01	−0.84		
Mn	−0.82	0.02	−0.66	0.95	
Zn	−0.72	−0.66	−0.44	0.46	0.22

Table 10 ANOVA for effect of location, metal and tissue on heavy metals found in plant tissue

Source	Cd	Cr	Cu	Fe	Mn	Zn
Location (L)	ns	ns	ns	ns	ns	ns
Metal (M)	ns	ns	ns	ns	ns	ns
Tissue (T)	ns	ns	ns	ns	ns	ns
Interaction						
L × M	ns	ns	ns	ns	ns	ns
L × T	ns	ns	ns	ns	ns	ns
M × T	ns	ns	ns	ns	ns	ns
L × M × T	ns	ns	ns	ns	ns	ns

ns, *, ** not significant or significant at $P \leq 0.05$ or $P \leq 0.01$, ANOVA

Table 11 ANOVA for effect of location, water type and translocation factor of metal found in the entire plant

Source	Cd	Cr	Cu	Fe	Mn	Zn
Location (L)	ns	ns	ns	ns	ns	ns
Metal (M)	ns	ns	ns	ns	ns	ns
Entire plant (E)	ns	ns	ns	ns	ns	ns
Interaction						
L × M	ns	ns	ns	ns	ns	ns
L × E	ns	ns	ns	ns	ns	ns
M × E	ns	ns	ns	ns	ns	ns
L × M × E	ns	ns	ns	ns	ns	ns

ns, *, ** not significant or significant at $P \leq 0.05$ or $P \leq 0.01$, ANOVA

Table 12 ANOVA for effect of metal and organ on bio-concentration factor of metal found in *B. oleracea* plant

Source	Cd	Cr	Cu	Fe	Mn	Zn
Location (L)	ns	ns	ns	ns	ns	ns
Metal (M)	*	*	*	*	*	*
Organ (O)	*	*	*	*	*	*
Interaction						
L × M × O	*	*	*	*	*	*

ns, *, ** not significant or significant at $P \leq 0.05$ or $P \leq 0.01$, ANOVA

Table 13 Bio-concentration factor for heavy metals in *B. oleracea* as affected by interaction of location and organ

Location ×	Organ	Cd	Cr	Cu	Fe	Mn	Zn
Bhagtanpur Village	Root	$0.067b^{a}$	0.157^{b}	0.130^{a}	0.179^{a}	0.124^{a}	0.154^{a}
	Leaves	0.027^{c}	0.064^{c}	0.057^{c}	0.087^{c}	0.034^{b}	0.061^{c}
	Inflorescence	0.077^{a}	0.183^{a}	0.071^{b}	0.114^{b}	0.020^{b}	0.071^{b}
Ikkarkala Village	Root	0.052^{a}	0.156^{a}	0.145^{a}	0.135^{a}	0.112^{a}	0.172^{a}
	Leaves	0.018^{c}	0.055^{c}	0.058^{b}	0.066^{c}	0.026^{c}	0.063^{b}
	Inflorescence	0.043^{b}	0.093^{b}	0.055^{b}	0.094^{b}	0.080^{b}	0.063^{b}
Shubrasha Village	Root	0.051^{a}	0.176^{a}	0.143^{a}	0.134^{a}	0.127^{a}	0.142^{a}
	Leaves	0.018^{c}	0.071^{c}	0.057^{b}	0.044^{c}	0.028^{c}	0.051^{c}
	Inflorescence	0.042^{b}	0.109^{b}	0.064^{b}	0.101^{b}	0.065^{b}	0.075^{b}
Bongla Village (Control)	Root	0.034^{a}	0.102^{a}	0.064^{a}	0.113^{a}	0.102^{a}	0.126^{a}
	Leaves	0.028^{b}	0.077^{b}	0.048^{a}	0.081^{b}	0.054^{b}	0.067^{b}
	Inflorescence	0.026^{b}	0.102^{a}	0.059^{a}	0.074^{c}	0.030^{c}	0.068^{b}

Data in the interaction analyzed with Least Squares Means and means separated with LSD
[a]values in a group, in columns, followed by the same letter are not significantly different, $P \leq 0.05$.

cauliflower were above safe limits for Cd (3–6 μg·g^{-1}), Cu (135–200 μg·g^{-1}), Fe (75–150 μg·g^{-1}), Zn (300–600 μg·g^{-1}) reported by Awashthi (2000). Contents of Cr, Cd, Cu, Fe, Mn and Zn in the soil increased as the number of effluent irrigation increased. The contamination factor was highest for Fe and lowest for Mn at all sampling sites irrigated with IIE. The contamination factor of heavy metals in cauliflower was on the order: Fe > Zn > Cu > Cd > Cr > Mn in Bhagtanpur and Ikkarkala village due to irrigation with IIE (Tables 10, 11, 12 and 13). The concentration of Cd, Cr, Cu, Fe, Mn and Zn were higher in soil irrigated with effluent than in soil irrigated with well water. Fertigation with IIE increased nutrients as well as metals content in soils used for the cultivation of *B. oleracea.* The ANOVA data indicated that location; metal, tissue and their interaction did not show significant effect on the contents of different metals Cd, Cr, Cu, Fe, Mn and Zn on the tissues of *B. oleracea* (Table 10). The contamination factor of metals in *B. oleracea* inflorescence was ranged for Cd (2.99–3.59), Cr (2.70–3.10), Cu (3.59–3.97), Fe (6.19–6.97), Mn (1.92–3.37) and Zn (5.28–5.84) while the average contamination factor of Cd (3.31), Cr (2.88), Cu (3.76), Fe (6.71), Mn (2.56) and Zn (5.50) in the inflorescence of *B. oleracea* was at these sampling sites due to irrigation of *B. oleracea* with IIE. Similarily, location, water type and translocation factor of different metals in the entire plant of *B. oleracea* did not show significant effect on the contents of different metals Cd, Cr, Cu, Fe, Mn and Zn in the entire plant of *B. oleracea* (Table 11). The translocation factor of different metals in the entire plant of *B. oleracea* was ranged for Cd (34.62–40.63%), Cr (35.52–41.05%), Cu (39.73–43.73%), Fe (32.82–48.77%), Mn (22.14–27.65%) and Zn (34.48–39.40%) while the average translocation factor of Cd (36.90%), Cr (38.91%), Cu (41.18%), Fe (43.40%), Mn (24.39%) and Zn (36.81%) in the entire plant of *B. oleracea* was at these sampling sites after irrigation of *B. oleracea* with IIE.

The average contents of Cd (3.33 μg g^{-1}), Cr (5.52 μg g^{-1}), Cu (10.06 μg g^{-1}), Fe (129.92 μg g^{-1}), Mn (10.98 μg g^{-1}) and Zn (10.80 μg g^{-1}) in entire *B. oleracea* plant at IIE irrigated sites as Bhagtanpur, Ikkarkala and Shubrasha villages in comparison to Cd (1.07 μg g^{-1}), Cr (1.82 μg g^{-1}), Cu (1.70 μg g^{-1}), Fe (19.56 μg g^{-1}), Mn (3.63 μg g^{-1}) and Zn (1.76 μg g^{-1}) in entire plant of *B. oleracea* irrigated with bore well water at Bongla (Control site). The content of Cd, Cr, Cu, Fe, Mn and Zn in different tissues and in the entire plant of *B. oleracea* was varied and it was insignificantly different in IIE irrigated *B. oleracea* (Table 10), while content of Cd, Cr, Cu, Fe, Mn and Zn in *B. oleracea* was significantly different at Bhagtanpur, Ikkarkala and Shubrasha villages in comparison to Bongla village (Control). The translocation factor of Cd, Cr, Cu, Fe, Mn and Zn in different parts of *B. oleracea* were also observed varied in IIE irrigated *B. oleracea* (Table 11).

The average contents of different metals in the various organs as root, leaves and inflorescence of *B. oleracea* were varied and higher at all the IIE irrigated sampling sites (Bhagtanpur, Ikkarkala and Shubrasha village) in comparison to the bore well water irrigated sampling site (Bongla village). The average contents of Cd (1.44 μg g^{-1}), Cr (2.53 μg g^{-1}), Cu (5.38 μg g^{-1}), Fe (60.91 μg g^{-1}), Mn (6.44 μg g^{-1}) and Zn (5.92 μg g^{-1}) in the root, Cd (0.53 μg g^{-1}), Cr (0.98 μg g^{-1}), Cu (2.21 μg g^{-1}), Fe (26.65 μg g^{-1}), Mn (1.57 μg g^{-1}) and Zn (2.21 μg g^{-1}) in the leaves while the contents of Cd (1.36 μg g^{-1}), Cr (2.00 μg g^{-1}), Cu (2.47 μg g^{-1}), Fe (42.35 μg g^{-1}), Mn (2.97 μg g^{-1}) and Zn (2.67 μg g^{-1}) was recorded in the inflorescence of *B. oleracea* at all the IIE irrigated sites. the contents of Cd (0.40 μg g^{-1}), Cr (0.66 μg g^{-1}), Cu (0.88 μg g^{-1}), Fe (8.67 μg g^{-1}), Mn (1.01 μg g^{-1}) and Zn (0.85 μg g^{-1}) in the root, Cd (0.34 μg g^{-1}), Cr (0.50 μg g^{-1}), Cu (0.66 μg g^{-1}), Fe (5.36 μg g^{-1}), Mn (1.33 μg g^{-1}) and Zn (0.45 μg g^{-1}) in the leaves while the contents of Cd (0.33 μg g^{-1}), Cr (0.66 μg g^{-1}), Cu (0.82 μg g^{-1}), Fe (5.53 μg g^{-1}), Mn (1.29 μg g^{-1}) and Zn (0.46 μg g^{-1}) were observed at bore well water irrigated *B. oleracea* site (control).

5 Bio-concentration Factor

The BCFs for transfer of heavy metals from soils to cauliflower varied (Table 12). The BCF for heavy metals due to irrigation with IIE was in the order: Cr > Zn > Cu > Fe > Mn > Cd; for the control, the order was: Fe > Mn > Cr > Zn > Cu > Cd. The BCF values for Mn and Cd in cauliflower plants were comparatively low. The highest BCF values were for Cr, Zn and Fe due to irrigation with IIE (Table 13).

The food chain (soil–plant–human) is recognized as the main route whereby humans are exposed to contaminants in the soil [41]. When the BCF is < 1 or BAF = 1, it indicates the plant absorbs, but does not store heavy metals; when the BCF is >1 the plant accumulates metals. The BCF values of <1 were obtained for Cd, Cr, Cu, Fe, Mn and Zn in cauliflower. The bioavailability of metals was low in the areas of study.

The biological interaction between heavy metals and the plant occurs at the root surface and within cauliflower which ultimately affects uptake and translocation of heavy metals [42]. Accumulation of heavy metals depends on plant age and tissue [43]. Plant species differ in their tolerance and ability to take up and transport Cd within the plant [44]. Differences in heavy metal concentrations in vegetables were due to variations in their ability to absorb and accumulate heavy metals [45, 46]. The soil pH, organic matter content, cation exchange capacity (CEC), redox potential, soil texture, and clay content may affect heavy metal uptake [47]. The BCF indicated that cauliflower contained Cd, Cr, Cu, Fe, Mn and Zn after irrigation with IIE. Regular use of IIE for irrigation of cauliflower may build up levels of metals in soil and affect development and yield of cauliflower.

The translocation index of heavy metals in cauliflower irrigated with IIE, or well water, varied. Concentrations of heavy metals in cauliflower confirmed translocation of metals into different parts of cauliflower (root, leaves and inflorescence) from IIE irrigated soil as is evident from the higher translocation factor (Tf) of Fe followed by Cu and Cr in plants grown in the IIE irrigated soil. Cauliflower efficiently translocates higher contents of Cd, Cr, Cu, Fe, Mn and Zn in to leaves and inflorescence of plants irrigated with IIE. These results agree with Sharma et al. [44, 45] who reported higher heavy metals contents in cauliflower tissues after irrigation with municipal wastewater, and concluded that vegetable crops are capable of taking up, accumulating, and translocating higher concentrations of heavy metals, and can be used for the remediation of contaminated agricultural soils.

6 Conclusions

Those results generally indicated that the plants irrigated with the various wastewaters were not fit for human consumption. Cauliflower accumulated and translocated high contents of heavy metals after irrigation with IIE. The BOD, COD, Ca, TKN, Cd, Fe, Mn, MPN and SPC of IIE were above recommended limits. The accumulation of possibly toxic elements can produce adverse effects on plant morphology, and make them hazardous to eat indicating that use of the IIE for agricultural purposes for consumption of the crop is no better than for the other sources previously tested. Cauliflower takes up heavy metals but the levels do not raise to those which indicate it can be used for bio-remediation.

References

1. Lazarova, V. and A. Bahri. 2005. Water reuse for irrigation: Agriculture, landscapes, and turf grass. CRC Press, Boca Raton, FL.
2. Qadir, M., D. Wichelns, L. Raschid-Sally, P.S. Minhas, P. Drechsel, A. Bahri, and P. McCornick. 2007. Agricultural use of marginal-quality water—opportunities and challenges, pp. 425–457. In: Molden, D. (ed.). Water for food, water for life: A comprehensive assessment of water management in agriculture. Earthscan, London.
3. Asano, T., F. Burton, H. Leverenz, R. Tsuchihashi, and G. Tchobanoglous. 2007. Water reuse: Issues, technologies, and applications. McGraw-Hill, New York.
4. Ruma, M.M. and A.U. Sheikh. 2010. Reuse of wastewater in urban farming and urban planning implications in Katsina metropolis, Nigeria. *African Journal of Environmental Science and Technology* 4:28–33.
5. Chopra, A.K., S. Srivastava, V. Kumar, and C. Pathak. 2013. Agro-potentiality of distillery effluent on soil and agronomical characteristics of *Abelmoschus esculentus* L. (Okra). *Environmental Monitoring and Assessment* 185:6635–6644. (doi: https://doi.org/10.1007/s10661-012-3052-8).
6. Chopra, A.K., P. Temin, S. Srivastava, and V. Kumar. 2017. Effects of integrated nutrient management on agronomical attributes of tomato (*Lycopersicon esculentum* L.) under field conditions. *Archives of Agriculture and Environmental Science* 2:86–91.
7. Raj, R. and Thakur, R. K. (2017). A study on physico-chemical and microbiological parameters of ground water in different locations of Gwalior City (MP), India. *Biomedicine and Nursing, 3*(1), 19–23.
8. Kumar, V. and Thakur, R. K. (2018a). Impact of industrial effluents disposal on soil and cabbage grown in district Haridwar (Uttarakhand), Indiat. *International Journal of Current Research in Life Sciences, 7*(04), 1888–1896.
9. Kumar, V. and Thakur, R. K. (2018b). Health risk assessment of heavy metals via dietary intake of vegetables grown in wastewater irrigated areas of Jagjeetpur, Haridwar India. *Archives of Agriculture and Environmental Science, 3*(1), 73–80.

10. Anonymous. 2004. Guidelines for drinking-water quality: Recommendations, 3rd, vol. 1. World Health Organization, Geneva.
11. Shenbagavalli, S., S. Mahimairaja, and P. Kalaiselvi. 2011. Impact of biomethanated distillery spent wash application on soil and water quality: A field appraisal. *International Journal of Environmental Science* 1:1–7.
12. Pandey S.N., B.D. Nautiyal, and C.P. Sharma. 2008. Pollution level in distillery effluent and its phytotoxic effect on seed germination and early growth of maize and rice. *Journal of Environmental Biology* 29:267–270.
13. Kumar, V. and A.K. Chopra. 2013. Enrichment and translocation of heavy metals in soil and *Vicia faba* L. (Faba bean) after fertigation with distillery effluent. *International Journal of Agriculture Policy and Research* 1:131–141.
14. Hati, K.M., A.K. Biswas, K.K. Bandyopadhyay, and A.K. Misra. 2007. Soil properties and crop yields on a vertisol in India with application of distillery effluent. *Soil Tillage and Research* 92:60–68.
15. Anonymous. 1990. Official methods of analysis. Method No. 975.03 metal in plants, atomic absorption spectroscopymethod, 15th ed., Association of Official Analytical Chemists, Arlington, VA.
16. Anonymous. 1991. Indian standards for drinking water specification (BIS 10500:1991). Bureau of Indian Standards, New Delhi, India (http://www.bis.org.in/).
17. Awashthi, S.K. 2000. Prevention of food adulteration Act No. 37 of 1954. Central and State rules as amended for 1999, 3rd ed. Ashoka Law House, New Delhi.
18. Bharagava, R.N., R. Chandra, and V. Rai. 2008. Phytoextraction of trace elements and physiological changes in Indian mustard plants (*Brassica nigra* L.) grown in post methanated distillery effluent (PMDE) irrigated soil. *Bioresource Technology* 99:8316–8324.
19. Chandra, R., R.N. Bhargava, S. Yadav, and D. Mohan. 2009. Accumulation and distribution of toxic metals in wheat (*Triticum aestivum* L.) and Indian mustard (*Brassica campestris* L.) irrigated with distillery and tannery effluents. *Journal of Hazardous Materials* 162:1514–1521.
20. Chaturvedi, R.K. and K. Sankar. 2006. Laboratory manual for the physico-chemical analysis of soil, water and plant. Wildlife Institute of India, Dehradun, India.
21. Håkanson, L. 1980. An ecological risk index for aquatic pollution control. A sedimentological approach. *Water Research* 14:975–1001.
22. Hashmi, D.S., S. Ismail, and G.H. Shaikh. 2007. Assessment of the level of trace metals in commonly edible vegetables locally available in the markets of Karachi city. *Pakistan Journal of Botany* 39:747–751.
23. Overesch, M., J. Rinklebe, G. Broll, and N.H. Neue. 2007. Metals and arsenic in soils and corresponding vegetation at Central Elbe river floodplains (Germany). *Environmental Pollution* 145:800–812.
24. Khanal, B. R., S.C. Shah, S.K. Sah, C.P. Shriwastav, and, B.S. Acharya. 2014. Heavy metals accumulation in cauliflower (*Brassica oleracea* L. var. *botrytis*) grown in brewery sludge amended sandy loam soil. *International Journal of Agriculture Science and Technology* 02:87–92.
25. Olowoyo, J.O., O.O. Okedeyi, N.M. Mkolo, G.N. Lion, and S.T.R. Mdakane. 2011. Uptake and translocation of heavy metals by medicinal plants growing around a waste dump site in Pretoria, South Africa. *South African Journal of Botany* 78:116–121. doi:https://doi.org/10.1016/j.sajb.2011.05.010
26. Kim, K.H. and S.H. Kim. 1999. Heavy metal pollution of agricultural soils in central regions of Korea. *Water, Air and Soil Pollution* 111:109–122.
27. Kumar, V. and A.K. Chopra. 2014a. Response of French bean to fertigation with wine from molasses distillery effluent in two seasons. *International Journal of Vegetable Science* 20:104–123. (doi: https://doi.org/10.1080/19315260.2012.749970).
28. Kumar, V. and A.K. Chopra. 2014b. Accumulation and translocation of metals in soil and different parts of French bean (*Phaseolus vulgaris* L.) amended with sewage sludge. *Bulletin of Environmental Contamination and Toxicology* 92:103–108.

29. Kumar, V. and R.K. Thakur. 2017. Pollution load of SIDCUL effluent with reference to heavy metals accumulated in sediments using pollution load index (PLI and geoaccumulation index (I-GEO) at Haridwar (Uttarakhand), India. *Journal of Environment and Bio-Sciences* 31:163–168.
30. Muchuweti M., J.W. Birkett, E. Chinyanga, R. Zvauya, M.D. Scrimshaw, and J.N. Lester. 2006. Heavy metal content of vegetables irrigated with mixture of wastewater and sewage sludge in Zimbabwe: Implications for human health. *Agriculture, Ecosystem and Environment* 112:41–48.
31. Kumar, V., A.K. Chopra, S. Srivastava, and R.K. Chauhan. 2015a. Accumulation of heavy metals in vegetables grown in wastewater irrigated soil in Haridwar (Uttarakhand), India. *Agriculture Science Research Journal* 5:146–152.
32. Kumar, V., A.K. Chopra, S. Kumar, J. Singh, and K.R. Thakur. 2015b. Effects of pulp and paper mill effluent disposal on soil characteristics in the vicinity of Uttaranchal Pulp and Paper Mill, Haridwar (Uttarakhand), India. *International Journal of Agriculture Science and Research* 4:117–125.
33. Marcovecchio, J.E., S.E. Botté, and R.H. Freije. 2007. Heavy metals, major metals, trace elements, pp. 275–311. In: Nollet, L.M.L. (ed.). Handbook of water analysis, 2nd ed. CRC Press, Boca Raton, FL.
34. Kumar, V., A.K. Chopra, S. Srivastava, J. Singh, and R.K. Thakur. 2016. Irrigating okra with secondary treated municipal wastewater: Observations regarding plant growth and soil characteristics. *International Journal of Phytoremediation* 19:490–499 (doi: https://doi.org/10.1080/15226514.2016.1244169).
35. Liu, W.X., L.F. Shen, J.W. Liu, Y.W. Wang, and S.R. Li. 2007. Uptake of toxic heavy metals by rice (Oryza sativa L.) cultivated in the agricultural soil near Zhengzhou city, People's Republic of China. *Bulletin of Environmental Contamination and Toxicology* 79:209–213.
36. Logan, T.J., H. Goins, and B.J. Lindsay. 1997. Field assessment of trace element uptake by six vegetables from N-Viro soil. *Water Environment Research* 69:28–33.
37. Kumar, V., S. Srivastava, R.K. Chauhan, R.K. Thakur, and J. Singh. 2017. Heavy metals and microbial contamination of certain leafy vegetables grown in abattoir effluent disposal province of Saharanpur (Uttar Pradesh), India. *Archives of Agriculture and Environmental Science* 2:36–43.
38. Lacorte, S., A. Lattorre, D. Barcelo, A. Rigol, A. Malmqvist, and T. Welander. 2003. Organic compounds in paper mill process waters and effluents. *Trends in Analytical Chemistry,* 22:725–737.
39. Liu, W., X. Li, H. Li, H. Sr., and Y.W. Wang. 2006. Heavy metal accumulation of edible vegetables cultivated in agricultural soil in the Suburb of Zhengzhou City, People's Republic of China. *Bulletin of Environmental Contamination and Toxicology* 76:163–170. doi:https://doi.org/10.1007/s00128-006-0981-3
40. Pescod, M.B. 1992. Wastewater treatment and use in agriculture. Irrigation and drainage paper 47. Food and Agriculture Organization of the United Nations, Rome.
41. Rath, P., G. Pradhan, and M.K. Misra. 2011. Effect of distillery spent wash (DSW) and fertilizer on growth and chlorophyll content of sugarcane (*Saccharum officinarum* L.) plant. *Recent Research in Science and Technology* 3:169–176.
42. Sarker, R., M. Ratna, S. Ray, A.H.F. Fahim, and M.J. Tithi. 2017. Effect of planting method on onion (*Allium cepa* L.) bulb production in Faridpur region of Bangladesh. *Archives of Agriculture and Environmental Science* 2:63–67.
43. Sharma, R.K., M. Agrawal, and F. Marshall. 2006. Heavy metal contamination in vegetables grown in wastewater irrigated areas of Varanasi, India. *Bulletin of Environmental Contamination and Toxicology* 77:312–318.
44. Sharma, R.K., M. Agrawal, and F. Marshall. 2007. Heavy metal contamination of soil and vegetables in suburban areas of Varanasi, India. *Ecotoxicology and Environmental Safety* 66:258–266.
45. Sutherland, R.A. 2000. Bed sediment-associated trace metals in an urban stream, Oahu. Hawaii *Environmental Geology* 39:611–627.

46. Vasiliadou, S. and C. Dordas. 2009. Increased concentration of soil cadmium effects on plant growth, dry matter accumulation, Cd, and Zn uptake of different tobacco cultivars (*Nicotiana tabacum* L.). *International Journal of Phytology* 11:115–130.
47. Zhuang, P., M.B. McBride, H. Xia, N. Li, and Z. Li. 2009. Health risk from heavy metals via consumption of food crops in the vicinity of Dabaoshan mine. South China. *Science of Total Environment* 407:1551–1561.

An Intensive Approach to the Renewable Energy Recovery from Agro Waste—A Review

Aditi Bisht, Nitin Kamboj, Akanksha Bisht, Vishal Kamboj, and Manisha Bharti

Abstract Agricultural residues are wastes generated whilst growing and processing these goods. The residues of agriculture are produced from numerous activities like cultivation, aquaculture, and livestock production. Wastes generated in the field or after processing in industries are multiphase and multicomponent. All three forms, liquid, solid, and gaseous wastes are generated by the agriculture sector which tends to pollute soil, water, and air. Waste causing environment degrading ability needs immediate attention. For ages, food and agro-wastes were either burnt or allowed to decompose in fields, but this can be possibly harmful to the environment. With the progress in technology, new approaches concerning their utilization, reuse, and processing need to be established to permit its sustainable utilization of residues and reduce pollution. Agricultural wastes are potential renewable energy resources. It is important to develop proper waste management strategies as it is the only best solution to have a healthy environment. In this paper, we discussed various strategies and technologies introduced to manage and utilize the waste that is generated in the agricultural field or after processing agriculture-based products. Appropriate management of waste is also beneficial in deriving out the essential useful product from the waste.

Keywords Agricultural wastes · Agro-industry wastes · Biofuel · Biogas · Bioethanol · Industrial processing waste

A. Bisht · N. Kamboj (✉) · V. Kamboj · M. Bharti
Department of Zoology and Environmental Science, Gurukula Kangri Vishwavidyalaya, Haridwar, Uttarakhand 249404, India
e-mail: nkamboj@gkv.ac.in

A. Bisht
Department of Medicinal and Aromatic Plants, Uttarakhand Altitude Plant Physiology Research Centre, Hemvati Nandan Bahuguna Garhwal University, Birla Campus, Srinagar, Uttarakhand 249161, India

K. D. Bahukhandi et al. (eds.), *Environmental Pollution and Natural Resource Management*, Springer Proceedings in Earth and Environmental Sciences,
https://doi.org/10.1007/978-3-031-05335-1_2

1 Introduction

Energy demand in India is rising fast with main insinuations for the worldwide energy market. The Government of India has made notable development in providing access to electricity and clean cooking while applying a variety of energy marketplace modifications and mixing a huge share of renewable energy sources into the grid [1]. India is the third-largest producer and consumer of electricity in the globe. It is a fast-budding economy that requires energy to encounter its development objectives in a sustainable way [2]. The economy faces substantial challenges concerning meeting its energy requirements in the coming decade. India is amongst the five Greenhouse-gas (GHG) emitters internationally [3]. Although India has substantially enhanced its producing capacity, it still has trouble in meeting demand and there are power shortages that compel India's monetary growth. The anthropogenic activities degraded the environmental quality by contaminating the quality of soil, water as well as the environment [4, 5]. With the expansion of the manufacturing and commercial sectors as well as the wider use of electric equipment, electricity demand keeps growing [6]. Moreover, around 30% of India's produced power is lost in communication. Circulation is the frailest link of India's power supply chain as it faces substantial technical losses and marketable losses of electricity [7]. Additionally, the lack of broadcast and distribution of power to fewer densely inhabited areas that are situated far away from the power stations is the reason for not being able to attain 100% electrification in the country [8]. For the past two decades, India has had to face a growing shortage in power supply, both for meeting its normal energy supplies as well as its peak load demand. The problem is serious during peak hours and summers and requires deliberate load shedding by many utilities to sustain the grid in a healthy state [9]. The energy consumption in 2017–18 was 23.355 Giga Joules and net energy import dependency was 40% in 2018–19 [10]. Those times are gone when renewable energy is intended to mean hazy solar lamps. Renewable energy power plants of small scale are cheaper, reliable, and efficient now. For example, In Durbuk, (Ladakh) a solar power plant is providing power to approximately 340 households, a clinic, a school, and few government offices [11]. India is presently the third-largest carbon emitter in the world. According to the Copenhagen Accord, which was signed by India in 2009 to pledge that 80% of the world's non-renewable reserves must persist on the earth to avoid warming of the planet [12]. So, from an environmental point of view, renewable energy must come on a large scale and not as isolated stories of miracles.

Agriculture act as one of the supportive pillars of the Indian economy. Its contribution to the economy is more than any other sector [13]. It provides a source for sustenance to about 60% of the Indian population. India ranks first in the production of pulse crops and jute and second rank in the production of wheat, sugarcane, rice, and cotton. It provides large employment opportunities because of which 50% of the total population has adopted agriculture as their occupation [14].

Agriculture has been a key source of raw resources to several leading industries like cotton, jute textiles, sugar, tobacco, edible and non-edible oils, etc. Agriculture has a significant role to play in import and export. Revenue generated from export and import is utilized in many sectors of the countries but more emphasis is given to transportation sectors to facilitate the movement of agricultural produce in and out of the country [15]. Agriculture systems have been enduring changes to a major extent. They need to adjust further in the developing environment worldwide. The agricultural production has drastically amplified amid the years, 1960–2015. This is due to green revolution technologies and extension inland, marine, and natural resources [16].

In this review articles, we discussed about the various types of agricultural waste, proper management of these waste. In addition, we study and discuss the modern technologies used for making the renewable energy from the agro-waste. This paper provides a brief knowledge, how we used the different agro-waste in renewable energy sectors.

1.1 Agricultural Wastes

The remains obtained from the production and processing of agricultural products such as crops, fruits, vegetables, meat, poultry, and dairy goods are considered Agricultural wastes [17]. It can be divided into four categories, that is, crop residues, agro-industry wastes, livestock wastes, and fruit and vegetable wastes [18] (Fig. 1). All categories provide a different chemical and biochemical composition. The composition of chemical and biochemical used as a renewable energy with help of various microorganism and modern techniques that converts the crop residue into energy form i.e., methane, bioethanol etc.

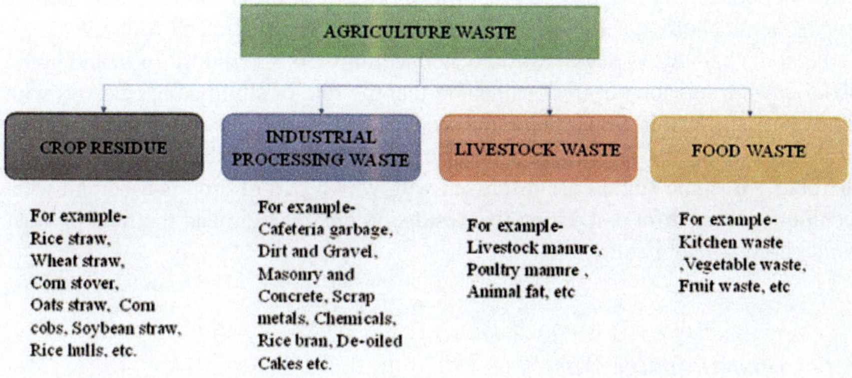

Fig. 1 Basic types of agricultural waste

2 Types of Agriculture Waste

2.1 Crop Residues

The left-over residues produced from direct agricultural production at the field are typically crop residues like leaves, stovers, straws, and seed pods, etc. The global annual predictable production of crop residues is 2802 million tons. Agricultural waste generated from crop residues is the most profuse and cheapest organic waste, which can be easily changed into various value-added stuff [19]. Collectively, three major crop residues are being utilized for bioethanol production, that is, rice straw, wheat straw, and corn stover. These crops are accessible all over the year and a very insignificant portion is used as fodder or biofuel production, with the rest burned and hence triggering critical environmental problems. Rice straw is the utmost promising and abundant biomasses in the world, with a global production of 731 million tons/year, and with Asia being the major producer [20]. The rice straw production in India is expected to be 221.8 million tons/year by 2030 [21]. Corn stover is considered to be one of the most promising crop residues for the production of lignocellulosic ethanol, with an estimated rate of production to be of 4.0 tons/acre and global annual production of 128.02 million tons [22]. The other crop residues obtained from barley, sorghum, and oats also contribute toward agricultural wastes [23].

Plants comprise of cellulose 15–60%, 10–30% of hemicellulose, 5–30% of lignin, 2–15% of protein, and soluble materials, like sugars, amino sugars, amino acids, and organic acids, which might comprise of 10% of dehydrated weight [24]. They also encompass cutin, polyphenols [25], and silica [26]. The degree breakdown of organic material depends on the relative proportions of each of these parts, such as soluble sugars, hemicellulose, cellulose, and lignin [27]. In 1974 Hagin and Amberger testified that the half-life of sugars was 0.6, hemicellulose 6.7, cellulose 14.0, and lignin 364.5 days. It has long been documented that the rate of slight damage drops with time [28–30], and this drop imitates the drop in the quality of the residual substratum. In addition, many researchers used the crop residue for making the renewable energy. In Table 1 in some studies are discussed with which part of crop is used and what techniques is used for converting this residue in energy form and the final product which used as energy alternate.

2.2 Agro-Industry Wastes

The second group of agricultural waste comprises agro-industry processing waste. This consist of by-products produced from food processing industries, such as vegetable and fruit peels after extraction of juice, starch residue from starch-manufacturing industries, sugarcane bagasse, molasses from sugar manufacturing industries, de-oiled seed cake from edible oil manufacturing industries, chicken skin,

Table 1 Types of waste and their products

S. No.	Waste type	Material	Method used	Product	References
1	Crop residues	Rice hull ash	After pulverization it can act as filler in rubber	85–90% amorphous silica, 10–15% carbon	[55]
		Rice hulls and rice straw	Heat treatment at 1400–1600 °C of a homogeneous mixture of calcareous and argillaceous raw materials	Hydraulic cements	[56]
		Inedible crop residues	Physical chemical oxidation and biological oxidation	50–60% nutrients were recovered	[57]
		Corn cobs and corn stover (stalks, leaves and husks)	Fast pyrolysis using a pilot scale fluidized bed reactor	Bio-oil −60% and biochar – 17–19%	[58]
		Wheat straw	Fractionation by steam explosion	Ethanol	[49]
		Oat straw	Sequential pre-treatment, enzymatic pre-treatment	Methane	[59]
		Potato residues	Anaerobic digestion	Replenished inorganic nutrients, recovery of calcium and potassium	[60]
		Glyphosate liquor	Membrane technology	Clean water	[61]
		Straw	Biological processes with microbial degradation and enzymatic hydrolysis	Biofuels/biofertilizers	[39]
		Santalum album leaves	Pyrolysis followed by gas chromatography/ mass spectrometry	Benzene/Ethanol	[52]
		Paper mill residue and rice husk ash	XRF, TG–DTA, XRD and SEM techniques	RPMR bricks	[62]
2	Agro-industry waste	Industrial waste	Pyro/hydrometallurgical processes: thermal treatment followed by leaching, precipitation	Platinum	[63]

(continued)

Table 1 (continued)

S. No.	Waste type	Material	Method used	Product	References
		Coal fly ash and bauxite	Vapor phase reaction followed with co-introduction of MoO_3 and AlF_3	Mullite-whisker-structured ceramic membrane	[64]
		Glycidol	Valorisation	Poly (glycidol)	[65]
		Silicon	Wafers technology	Photoluminescent porous silicon (PSi) nanoparticles	[48]
		Electrical and electronic equipment waste	Hydrometallurgical and pyrometallurgical	Precious metals	[66]
		Industrial effluent	Nanofiltration membranes	Clean water	[67]
		Goethite industrial waste	Nucleation	Glass–ceramic materials	[68]
		Agro-industrial waste	Solid-state fermentation	Amylase	[69]
3	Livestock	Pig manure	Stripping process Anaerobic digestion followed by co-digestion	Energy recovery	[70]
		Liquid swine manure	anaerobic digestion, filtration, flocculation,	Nutrient management	[71]
		Pig manure	Anaerobic digestion followed by co-digestion	Nutrients and energy recovery	[72]
		Cow manure	Anoxic/aerobic-membrane bioreactor	Biogas and biofertilizer	[73]
		Livestock manure	Membrane techniques	Liquid organic fertilizer	[74]
		Animal manure; food processing waste; fats, oils, and grease	Conversion by hydrothermal liquefaction	Biofuel	[42]
4	Food waste	Food waste	Continuous thermophilic composting	Organic fertilizer	[75]
		Kitchen waste and coffee grounds	Vermicomposting	Vermicompost	[76]

egg, meat, and animal fat from slaughterhouses and meat processing industries. The major agro-industrial waste produced after the extraction of juice out of sugarcane in industries is sugarcane bagasse [31].

The global production of sugarcane bagasse is 180.73 million tons. The other agro-industrial wastes comprise orange peel, and other fruit wastes attain from fruit juice, cider, and additional food processing units. Sometimes waste produced from non-food agro-industries like deoiled seed cake obtained from non-edible oil plants like *Jatropha curcans* are also included under agro-industry wastes [32]. The agro-industry waste is best alternate for making the renewable energy with the help of modern techniques. In Table 1, some techniques are mentioned that used by globally with many searchers for making the renewable energy from the residues of agro-industry waste.

2.3 *Livestock Wastes*

The three significant types of livestock wastes comprise liquid manure which mostly urinary waste, solid manure in form of farmyard manure, and wastewater which is generally collected from farms process water, silage juices, liquid manure, and disinfectants. The European agricultural sector yearly produces about 1500 million tons of animal manure. Untreated manure can be the main cause of air and water pollution. The discharge of nutrient-rich liquid manure, wastewater, and pathogen contamination can result in surface water pollution. Solid manure contributes to greenhouse gases [33]. The livestock waste is mainly comprising with high nutrients and the composition of chemical and biochemical properties showed a variety of microorganisms that produces the methane gas. In Table 1, the used of livestock material as alternate of energy is mentioned. Some researchers found that the livestock residues show the higher nutrient values and varieties of microorganism composition that produces the biofuel, biogas and biofertilizers.

2.4 *Fruit and Vegetable Wastes*

Fruits and vegetable wastes comprised of unprocessed fruits and vegetables like mango, pineapple, tomato, banana, and orange, etc. and are also an imperative part of agricultural wastes. The fruits and vegetable waste generation from food processing industries is huge. The landfill dumping of these organic wastes can be problematic and it is a major alarm for environmental pollution due to its extremely perishable nature. The generation of large quantities of waste also increases the operating cost of the markets. The yearly production of FVW in India is approximately 5.6 million tons and presently these wastes are settled by dumping on the peripheries of cities [34].

2.5 *Industrial Waste*

Industrial waste is the waste generated by industrial activity which comprises any material that is left useless through a manufacturing process. Examples of industrial wastes are chemical solvents, sludge, metals, ash, paints, sandpaper, paper products, other industrial by-products, etc., Seepages from industries hold various organic and inorganic waste products [35].

Synthetic organic substances and pesticides can badly affect aquatic ecosystems. Liquid organic seepages change the pH of the water and the toxicity effects on the aquatic plants fluctuate depending on their chemical composition. Insecticides are sprayed on crops to kill the insect that can destroy the crop.

DDT (Dichlorodiphenyltrichloroethane) is a crystalline chemical compound, an organochlorine, originally developed as an insecticide, but now its use has been banned after observing its adverse environmental impacts. Even at lower levels, DDT can distress the growth of babies and has been associated with cancer [36].

3 Management of Agro Waste

India generates more than 620 million tonnes of agricultural waste per annum of which only 25–30% part is utilizing as livestock fodder and energy production. The majority of the Indian farmers are practicing residue burning practices to clear the field for timely sowing of the next crop. The reason being the low nutritive value of waste and more cost of labor to clear the field. The practice of burning crop residue results in the release of harmful gases like CO_2, CH_4, N_2O, H_2S, O_3, and smog. This is the main cause of air pollution which mainly disturb public life and disturb soil physical, biological, and chemical properties by terminating helpful soil microorganism [37]. This problem can be solved by the timely implementation of effective agricultural waste management actions [38]. India produces 284.83 million tonnes of food grains which involves the use of 25.94 million tonnes of chemical fertilizers every year. Agricultural waste holds nitrogen, phosphorus, and potassium as a source of plant nutrients. This can alone substitute the use of 6.5 million tonnes of chemical fertilizer equivalent to 25% of total NPK necessity [18]. The agro-industrial waste comprises high nutritional compositions because of which has drawn the attention of the research community [39].

3.1 *Biofuel*

3.1.1 Bioethanol from Agro-Waste

Bioethanol has received utmost attention as compared to other fuels due to a lesser amount of release of greenhouse gases and its produces from renewable sources. It is a see-through, colorless liquid with a pleasing odor. The taste of ethanol depends on its concentration; when diluted it has a sweet flavor and burning taste when it is concentrated [40]. It is a hydroxyl group-containing and the second member of the alcoholic group. It is chiefly produced from sugar comprising feedstocks. Bioethanol can be best extracted from agriculture waste, which is present in abundance at the global level. Agriculture wastes comprise cellulose, hemicelluloses, and lignin which generates difficulties during conversion to ethanol. Agriculture wastes have to be pre-treated to remove hemicellulose lignin, later it is enzymatically hydrolyzed into sugars. Together pentose and hexose are fermented to bioethanol [41]. Brazil is recognized as the foremost country in production as well as the import of bioethanol in the world. Currently, In India molasses are being typically utilized for ethanol production. Conversion of agriculture waste to ethanol is done in four steps (Fig. 2).

1. Waste is pre-treated
2. Pre-treated waste undergoes hydrolysis of enzymes for fermentable sugar
3. Sugars are fermented to ethanol
4. Ethanol is recovered.

Burning of the bioethanol harvests CO_2 gas which would be adapted again by plants. No net CO_2 is formed by utilizing bioethanol produced from agriculture biomass. It decreases GHGs released. There is no harmful effect of bioethanol even when humans come in exposed to it, mostly by inhalation of vapor. Bioethanol

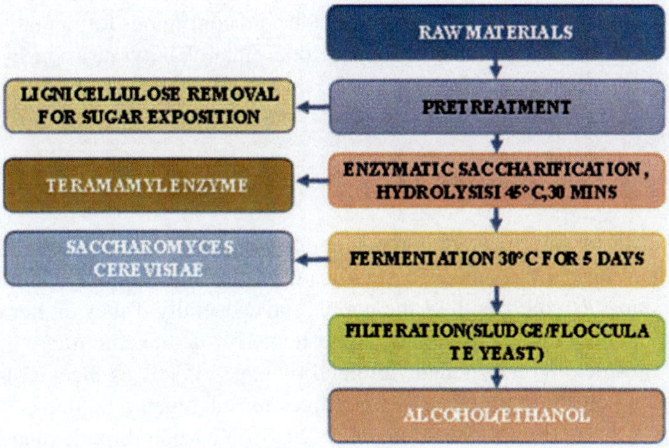

Fig. 2 Process of bio ethanol formation

is capable of degrading in the atmosphere very fast. When ethanol is used along with diesel, a decrease in octane number, heating value is increased, also distillation temperature changes. These properties enable the comprehensive combustion of ethanol and lesser emissions of GHGs [42].

3.1.2 Biobutanol from Agro-Waste

Agriculture residues (or renewable feedstock) are also used for deriving bioenergy which is a green energy technology. It is of biological origin that contributes to a low level of carbon footprints as the biomass of non-edible sources is to dodge the reduction in food security under biorefinery concepts such as biobutanol, biodiesel, biohydrogen, and bioethanol. Comparative to conventional butanol, biobutanol can be generated sustainably via fermentation; it has been tested as a clean fuel and it qualifies under the renewable fuel standard. Recently production of biobutanol through acetone-butanol-ethanol fermentation with the use of some microbial species has grabbed much attention. It is being used as a solvent for manufacturing vitamins, antibiotics, inorganic synthesis, hormones, and also in the production of paint thinner. It can be used as gasoline even without any engine modification [21].

Agricultural and forestry residues signify a wonderful source of readily available biomass for biofuel generation. Such biomasses which are rich in carbohydrate can be utilized for the generation of solvents under severe anaerobic condition using desired microbes. Butanologens are incapable to reduce agricultural biomass to monomeric sugars. To make carbohydrates from biomass available for enzymatic hydrolysis and fermentation, it is important to subject the biomass to pre-treatment. Because of the close association of hemicellulose and cellulose with lignin in the plant cell wall, the agricultural biomass needs to be subjected to pre-treatment and subsequent saccharification for the release of fermentable sugars before acetone-butanol-ethanol (ABE) fermentation. Researchers implicate several modes of ABE fermentation processes such as batch, fed-batch, and continuous fermentation for the production of biobutanol which gives out acetone, ethanol, butyrate as the by-product [43] (Fig. 3).

3.1.3 Biogas from Agro-Waste

Agricultural wastes have become a fundamental source of biogas making especially in areas where agriculture makes a significant part of the economy. These wastes could pose danger to the health of the population especially if they clutter the environment as they are known to be the perfect home for pathogenic microorganisms. They could result in soil loss and farmland damages if left unchecked [44]. The gaseous formation through the anaerobic digestion of organic matter is biogas. It is normally composed of 50–80% methane, 20–40% carbon dioxide, and traces of other gases like CO, H_2S, NH_3, O_2, H_2, N_2, and water vapor. The biogas composition depends on the source of the substrate and the handling of the digestion process [35].

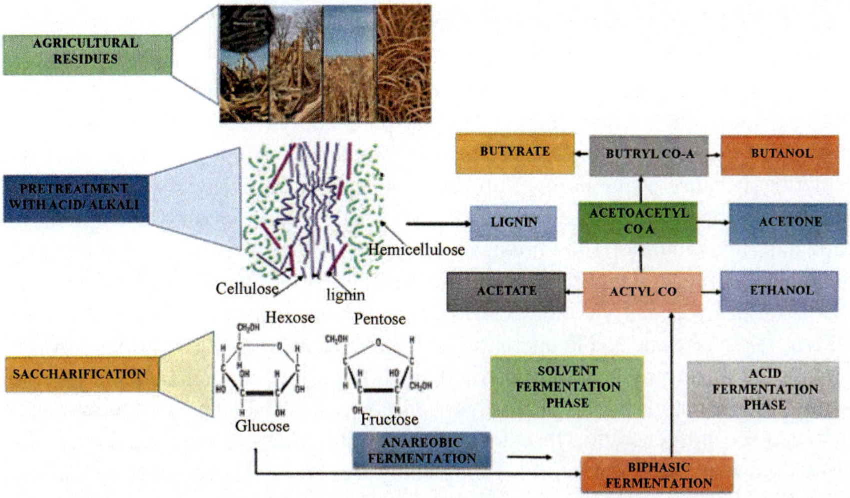

Fig. 3 Process of biobutanol production

The process for wastes transformation into biogas is Anaerobic digestion (AD). It is a natural phenomenon of Earth to re-integrate agriculture wastes into the ecosystem dynamics. This process has been well known and it is an active scientific topic. It is developed by many inter-reliant micro-organism communities, living in an environment free of oxygen, to transform complex substrates in four main stages: hydrolysis, acetogenesis, acidogenesis, and methanogenesis; each stage has definite dynamics and three phenomena are involved that are, physicochemical, hydrodynamic and biological. In optimal conditions, AD produces biogas mainly composed of methane (50–80%) and carbon dioxide (48–18%); depending on the substrate, some other components (NOx, SOx) are also present in low concentration (1–2%). The AD process that uses manure for biogas production, is one of the proficient usages of biomass wastes because it not only provides a source of energy but simultaneously resolves ecological and agrochemical issues. The anaerobic fermentation does not reduce the value of biomass as a fertilizer supplement. The available nitrogen and other substances persist in the treated sludge. This biogas technology is environmentally friendly and inexhaustible [45].

It executes numerous challenges in various aspects of the process; a few of them, which are specially related to the transformation of agriculture wastes, are:

1. Waste's composition
2. Combination of substrates for co-digestion
3. Development of solid-state anaerobic digestion
4. Reduction of inhibitory components.

3.2 Biohydrogen from Agriculture Waste

Hydrogen gas is an energy that can be transformed into electricity and fuel. It is known as a source of clean energy and environmentally friendly. It leaves no toxic waste. Consequently, it does not cause the greenhouse effect, acid rain, or ozone depletion, because of the burning process in the air, it only leaves water vapor and energy. Hydrogen gas has the highest combustion energy out of all types of fuel ever known in the amount of 143 Gj/ton. Combustion of energy hydrogen can be easily channelled in fuel cell technology into electrical energy so that it can be stored. The formation of hydrogen via a process of fermentation is of utmost inexpensive and easy, as it can be done at a temperature and ambient pressure [46]. Various forms of agricultural waste have been verified for their ability to the production of biohydrogen through fermentation. It comprises lignocellulosic materials, that are considered to be called second-generation feedstock. This residue usually comprises barley, rice, straw from wheat, corn stover, etc. Though they are extensively accessible in massive quantities and are a rich source of carbohydrates [47]. The translation processes to produce biofuel have been challenging due to the complex structure, which mainly constitutes cellulose and hemicellulose, and lignin; they are cross-linked with carbohydrates, which decreases the availability of the polysaccharides for microbial attack. Consequently, appropriate pre-treatment is necessary for delignification and saccharification of the polysaccharide's component into monosaccharides, which increases its biodegradability. Agro-based refineries often face difficulties, such as the segregation of dispersed agro-waste and seasonal availability of material [48]. Production of biohydrogen through dark fermentation has numerous benefits over conventional hydrogen production methods, including the ability to practice cheap and renewable substrates.

Biohydrogen production through dark fermentation is a very carbon–neutral process that generates hydrogen and carbon dioxide from residues by facultative and obligate anaerobic microorganisms (Fig. 4). Hydrogen has many valuable features; it

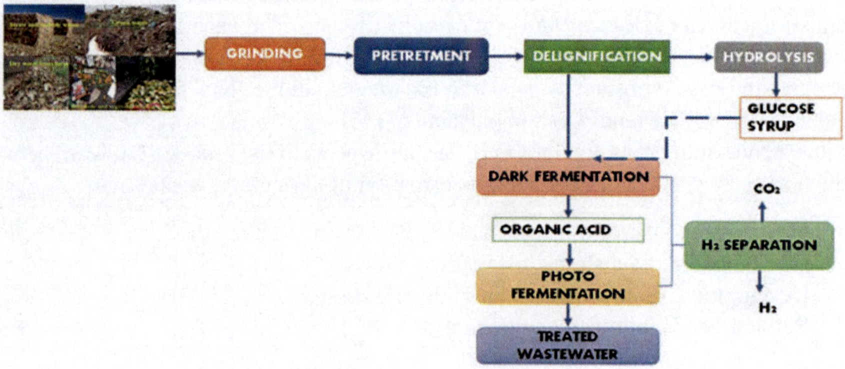

Fig. 4 Hydrogen formation from agro-waste

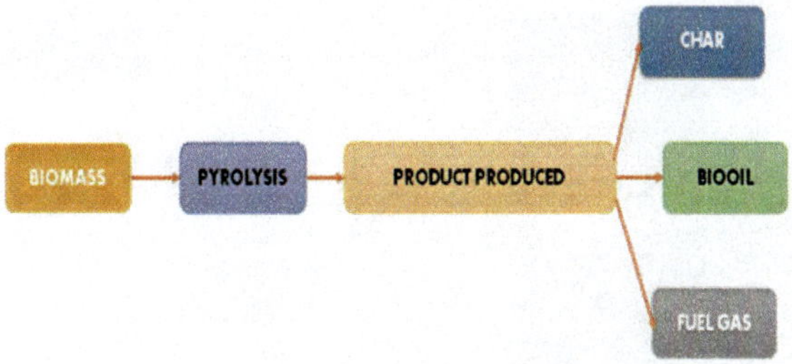

Fig. 5 Bio-oil formation from agro waste

is harmless to animals and the environment. Moreover, it is considered a non-polluting fuel, because its combustion product is water [16].

3.3 *Bio-oil from Agro-Waste*

Bio-oil is a liquid fuel produced from biomass examples, crops, agricultural and forestry by-products, and municipal wastes with help of thermo-chemical processes. It is one of its kind of new, cheap, clean, and green bio-energies. Bio-oil is believed to be a valuable option in place of conventional fuel in the part of reducing environmental pollution. Bio-oil from rice straw contains a high presence of light constituents and a small percentage of volatile and heavy bio-oil [49].

Pyrolysis of agricultural waste can be utilized for the manufacturing of bio-oil (Fig. 5). Few difficulties arise during the conversion process that is needed to be overcome. These problems generally comprise the corrosivity of the oil and poor thermal stability. Advancement done in the process by dropping the oxygen content and eliminating alkalis using catalytic cracking and hydrogenation of the oil could be obligatory for several applications. Energy fuel produced via the pyrolysis process makes it high fuel-to-feed ratios. It is the most efficient process for the conversion of biomass, and also the method which is most capable of competing and substituting non-renewable fossil fuel resources.

Bio-oil thus produced can be utilized in engines and turbines. It is being used as feedstock for refineries. The yield of the bio-oil when produced via pyrolysis of the samples increases with temperature until it reaches 875 K and then decreases with temperature [20].

3.4 *Bio-char from Agro-Waste*

Numerous reports have come out on the production of activated carbon from inexpensive and readily available materials. Activated bio-chars are attained through one of two methods:

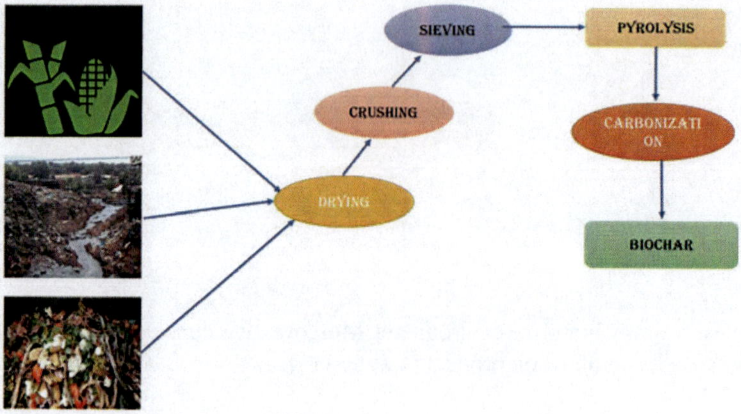

Fig. 6 Process of biochar from agro-waste

1. either by partial gasification of the primary char with use of steam or carbon dioxide or
2. by chemically activating the precursor with a chemical example zinc chloride or phosphoric acid.

Lignin is known to give higher produce of charcoal and tar from wood even though lignin has a greater amount (three-fold) of methoxy content than that of wood. The deeper result of the heating rate on the production of biochar from biomass than from coal may be credited to the cellulose percentage of the biomass [23]. It is known that the heating rate has a very significant impact on the pyrolysis of cellulose. This has a much larger impact on the pyrolysis of biomass than on coal. The rapid devolatilization of the biomass in fast pyrolysis supports the production of char with high porosity and high reactivity (Fig. 6). The reduced formation of char at the higher heating rate was accompanied by an improved formation of tar. Agricultural residues, such as fruit peels, nutshells, and corncobs, are extremely decent precursors for the production of activated carbons. Concerning environmental protection, the use of these wastes has stirred interest in the development of procedures for the production of carbon adsorbents based on agricultural wastes. The nature of the precursors and the production process has a strong effect on the porous structure and adsorption properties of the resulting activated carbons [12].

3.5 Fertilizer Application

The use of animal manure as fertilizer has a certain influence on input energy supplies at the farm level. Manure could supply around 19% of nitrogen, 38% of phosphorus, and 61%, potassium in chemical fertilizer. Many soil mycoflora act as a recovery agent in balancing the nutrients [50]. Fertilizer usage of manures from bulky confinement is related to high energy costs for conveyance, supply, storage facility requirements. It has a bad odor problem and the possibility of groundwater contamination [51]. It has also been reported that poultry manure comprises high phosphorus percentage which has a very positive result on the growth and efficiency of crops. It is extremely effective when combined along with mineral phosphorus fertilizer for farm use. Adding the manure to soil upsurges its fertility because it enhances the nutrient holding capacity or cation exchange capacity, also it can improve the physical condition, the water-holding capacity, and also provide stability to the soil structure [27].

3.6 Animal Feed

In some of the most developed countries, the limited availability of protein-based feedstock for the animals, for which great efforts are being made to provide an alternative supplement. The crop remains have a high fiber percentage, low in protein, starch, and fat [52]. Consequently, the traditional method of increasing livestock production by supplementing pasture and forage with grains and protein concentrate may not meet the future need for meat protein. The utilization of grain and protein for human food will compete with such use for animal feed. These problems may be avoided by using residues to feed domesticated animals [25].

3.7 Direct Combustion

The burning of agricultural waste as fuel is one of the oldest and the simplest biomass conversion processes are known to mankind. Complete burning of agro-waste involves the rapid chemical reaction (oxidation) of biomass and oxygen, the production of energy, and the immediate formation of the ultimate oxidation products of organic matter, CO_2, and water [53]. The energy produced is typically in the form of radiant and thermal energy provided oxidation occurs at an adequate rate; the amount of which is a function of the enthalpy of the burning of the biomass. If agriculture waste is to be used proficiently through the thermal conversion process, there is a need to engineer these biomass wastes into a solid form. It is typically burnt for heating, charcoal production, cooking, and the generation of steam, mechanical and electric power applications. Of all the processes that can be used to convert agricultural waste to energy or fuels, burning is still the leading technology accounting for more than 95% of all biomass energy used today [54].

The table given below shows the agrotypes of waste and its products.

4 Current Scenario and Future Prospects

Global warming and climate change apprehensions, attached with increasing oil charges and growing support of the government, are driving renewable energy regulation, commercialization, and incentives. The spending, regulation, and policies of the government assisted the renewal industry. According to the international energy agency in 2011, estimated that solar power may generate the maximum of electricity in the world within 50 years, dropping the release of greenhouse gases that damage the environment. Approximately 30 nations have renewable energy which contributes more than 20% of energy. The renewable energy market in our country is projected to grow sturdily in the coming years and beyond. Some 120 countries have various policy targets for longer-term shares of renewable energy, including a 20% target of all electricity generated for the European Union by 2020. Some countries have much higher long-term policy targets of up to 100% renewables.

5 Conclusion

Agriculture is considered the backbone of India. Correct and timely utilization of agriculture wastes possibly will lead to sustainable development in India as well as other countries, which require feasible resources of energy.

Agricultural residues can be utilized as a substitute source for the generation of various products, such as biofuel, biogas, and also as the raw material in various researches and industries. The utilization of agricultural wastes as raw materials can help to decrease the production cost and also trim down the pollution burden from the environment. The conversion of the residue for production of enzymes, biofuels, antioxidants, vitamins, animal feed, antibiotics with the help of various processes and techniques. Agriculture waste-based biorefineries can benefit to stabilize the economy of the country by contributing to a clean energy service by the replacement of fossil fuel and generating energy security. Integrated Agricultural Waste Management, should be kept in mind while working in the agriculture sector for proper sustainable management of the waste generated. This system almost covers all parts of the management of the waste generated on-field or generated after processing. The integrated waste management system is in action right from the time waste generation through the collection, transfer, transportation, sorting, treatment, and utilization in various forms. Implementation of IAWM practices can improve the profitability of farms and sustain the sanitation of homes and fields.

Acknowledgements The authors are highly grateful to the Department of Zoology and Environmental Science, Gurukula Kangri Vishwavidyalaya, Haridwar, India for providing infrastructural facilities.

References

1. Kumar, A., Kumar, K., Kaushik, N., Sharma, S., and Mishra, S.: Renewable energy in India: current status and future potentials. Renewable and sustainable energy reviews. Volume 14(8). (2010) 2434–2442
2. Bhattacharyya, S. C.: Shaping a sustainable energy future for India: Management challenges. *Energy policy*. Volume 38(8). (2010) 4173-4185. https://doi.org/10.1016/j.enpol.2010.03.045
3. Garg, A., Bhattacharya, S., Shukla, P. R., and Dadhwal, V. K.: Regional and sectoral assessment of greenhouse gas emissions in India. *Atmospheric Environment*. Volume 35(15). (2001) 2679–2695
4. Bisht, A., Kamboj, N., Kamboj, V. and Bisht, A.: A review on the role of emerging anthropogenic activities in environmental degradation and emphasis on their mitigation. Archives of Agriculture and Environmental Science. Volume 5(3). (2020) 419–425, https://doi.org/10.26832/24566632.2020.0503025
5. Kamboj, N., Bisht, A., Kamboj, V. and Bisht, A.: Leachate disposal induced groundwater pollution: A threat to drinking water scarcity and its management. In: Advances in Environmental Pollution Management: Wastewater Impacts and Treatment Technologies, Volume 1. (2020) 54–76. https://doi.org/10.26832/aesa-2020-aepm-05
6. Tiewsoh, L. S., Jirásek, J., and Sivek, M.: Electricity generation in India: Present state, future outlook and policy implications. *Energies*. Volume 12(7). (2019) 1361. https://doi.org/10.3390/en12071361
7. Green, R.: Electricity transmission pricing: an international comparison. *Utilities Policy*. Volume 6(3). (1997) 177–184
8. India Energy Portal: http://www.indiaenergyportal.org/overview_detail.php
9. Faranda, R., Pievatolo, A., and Tironi, E.: Load shedding: A new proposal. *IEEE Transactions on Power Systems*. Volume 22(4). (2007) 2086–2093
10. Ministry of Power of India 2017. http://www.mop.gov.in
11. Wangchuk, R.B. :23 GW of Potential and Growing: How Ladakh Plans to Lead India's Solar Charge! (2019). https://www.thebetterindia.com/201532/how-much-solar-power-india-rates-cost-ladakh-industry-information/
12. Ramanathan, V., and Xu, Y.: The Copenhagen Accord for limiting global warming: Criteria, constraints, and available avenues. Proceedings of the National Academy of Sciences. Volume 107(18). (2010) 8055–8062. https://doi.org/10.1073/pnas.1002293107
13. Mueller, U. G., Gerardo, N. M., Aanen, D. K., Six, D. L., and Schultz, T. R.: The evolution of agriculture in insects. *Annual Review of Ecology, Evolution, and Systematics*. Volume 36. (2005) 563–595. https://doi.org/10.1146/annurev.ecolsys.36.102003.152626
14. Arjun, K. M.: Indian agriculture-status, importance and role in Indian economy. *International Journal of Agriculture and Food Science Technology*. Volume 4(4). (2013) 343–346.
15. Dandekar, V. M.: The Indian economy. Agriculture. *Sage Publications*.1. (1994) 1947–92
16. Pathak, N.: Contribution of agriculture in the development of Indian economy. The *Journal of Indian Management and Strategy*. Volume 14(1). (2009) 52–57
17. Loehr, R.: Agricultural waste management: problems, processes, and approaches. *Elsevier*. (2012)
18. Litchfield, J. H. *Food Biotechnology*. Volume 1(1). (1987) 29
19. Prasad, R., and Power, J. F.: Crop residue management. *Advances in Soil Science*. Springer, New York. (1991) 205–251
20. Satlewal, A., Agrawal, R., Bhagia, S., Das, P. and Ragauskas, A. J.: Rice straw as a feedstock for biofuels: availability, recalcitrance, and chemical properties. *Biofuels. Bioproducts and Biorefining*. Volume *12*(1). (2018) 83–107. https://doi.org/10.1002/bbb.1818
21. Kumar, S., Paritosh, K., Pareek, N., Chawade, A., Vivekanand, V.: De-construction of major Indian cereal crop residues through chemical pretreatment for improved bio-gas production: an overview. Renewable and Sustainable Energy Reviews. Volume 90. (2018) 160170
22. Pattanaik, L., Pattnaik, F., Saxena, D. K., and Naik, S. N.: Biofuels from agricultural wastes. In *Second and Third Generation of Feedstocks*. Volume (2019) 103–142

23. Dhar, H., Kumar, S., and Kumar, R.: A review on organic waste to energy systems in India. *Bioresource technology*. Volume 245. (2017) 1229–1237. https://doi.org/https://doi.org/10.1016/j.biortech.2017.08.159
24. Paul EA, Clark FE.: Soil microbiology and biochemistry. *Academic Press, San Diego*. (1989)
25. Tian, G., Brussaard, L., & Kang, B. T.: Breakdown of plant residues with contrasting chemical compositions under humid tropical conditions: effects of earthworms and millipedes. *Soil Biology and Biochemistry*. Volume 27(3). (1995) 277–280
26. Goering, H. K., & Van Soest, P. J.: Forage fiber analyses (apparatus, reagents, procedures, and some applications). *US Agricultural Research Service*. (1970)
27. Stout JD, Goh KM and Rafter TA.: Chemistry and turnover of naturally occurring resistant organic compounds in soil. *Soil biochemistry*. Volume 5. (1981) 1–73
28. Minderman, G.: Addition, decomposition and accumulation of organic matter in forests. *The Journal of Ecology*. (1968) 355–362.
29. Jenkinson, D. S.: Studies on the decomposition of plant material in soil. V. The effects of plant cover and soil type on the loss of carbon from14c labelled ryegrass decomposing under field conditions. *Journal of Soil Science*. Volume 28(3). (1977) 424–434
30. Mellilo, J. M., Aber, J. D., Linkins, A. E., Ricca, A., Fry, B., & Nadelhiffer, K. J.: Carbon and nitrogen dynamics along a decay continuum: Plant litter to soil organic matter. *Physio-Chemical Characterization of Plant Residues for Industrial and Feed Use*. (1989)
31. Agamuthu, P.: Challenges and opportunities in agro-waste management: An Asian perspective. In Inaugural meeting of first regional 3R forum in Asia. (2009) 11–12
32. Girotto, F., Alibardi, L., and Cossu, R.: Food waste generation and industrial uses: a review. *Waste management*. Volume 45. (2015) 32–41. https://doi.org/10.1016/j.wasman.2015.06.008
33. Innes, R.: The economics of livestock waste and its regulation. *American Journal of Agricultural Economics*, Volume 82(1). (2000) 97–117
34. Gunaseelan, V.N.: Biochemical methane potential of fruits and vegetable solid waste feedstocks. *Biomass and bioenergy*. Volume 26(4). (2004) 389–399. https://doi.org/10.1016/j.biombioe.2003.08.006
35. Cutler, H. G., and Cutler, S. J. (Eds.).: Biologically active natural products: agrochemicals. *CRC Press*. (1999)
36. Frosch, R. A.Industrial ecology: Minimizing the impact of industrial waste. *Physics Today*, Volume 47(11). (1994) 63–68
37. Bharti, M., Kamboj, N. and Kamboj, V.:Effect of untreated wastewater on soil quality: A case study in Ranipur Rao watershed in Haridwar region (Uttarakhand), India. In: Advances in Environmental Pollution Management: Wastewater Impacts and Treatment Technologies. Volume 1. (2020) 145–157
38. Sadaka, S., and Boateng, A.: A. Pyrolysis and bio-oil. Cooperative Extension Service, University of Arkansas, US Department of Agriculture and county governments cooperating. (2009)
39. Ma, Y., Shen, Y., and Liu, Y.: State of the art of straw treatment technology: challenges and solutions forward. *Bioresource Technology*. (2020) 123656. https://doi.org/10.1016/j.biortech.2020.123656
40. Russ, W. and Meyer-Pittroff, R.: Utilizing waste products from the food production and processing industries. *Critical Reviews in Food Science and Nutrition*. Volume 44(1). (2004) 57–62. https://doi.org/10.1080/10408690490263783
41. Visioli, L. J., Enzweiler, H., Kuhn, R. C., Schwaab, M., and Mazutti, M. A.: Recent advances on biobutanol production. *Sustainable chemical processes*. Volume 2(1). (2014) 15
42. Skaggs, R. L., Coleman, a. M., Seiple, T. E., and Milbrandt, A. R.: Waste-to-energy biofuel production potential for selected feedstocks in the conterminous united states. *Renewable and sustainable energy reviews*. Volume 82. (2018) 2640–2651. https://doi.org/10.1016/j.rser.2017.09.107
43. Rodríguez, L. A., Toro, M. E., Vazquez, F., Correa-Daneri, M. L., Gouiric, S. C., and Vallejo, M. D.: Bioethanol production from grape and sugar beet pomaces by solid-state fermentation. *International Journal of Hydrogen Energy*. Volume 35(11). (2010) 5914–5917

44. Zinoviev, S., Arumugam, S. and Miertus, S.: Biofuel Production Technologies. Dubrovik, Croatia. (2007)
45. Qambrani, N. A., Rahman, M. M., Won, S., Shim, S., and Ra, C.: Biochar properties and eco-friendly applications for climate change mitigation, waste management, and wastewater treatment: A review. *Renewable and Sustainable Energy Reviews.* Volume 79. (2017) 255–273. https://doi.org/10.1016/j.rser.2017.05.057
46. Koukios, E., Koullas, D., Koukios, I. D., and Avgerinos, E.: Critical parameters for optimal biomass refineries: the case of biohydrogen. *Clean Technologies and Environmental Policy.* Volume 12(2). (2010) 147–151. https://doi.org/10.1007/s10098-009-0239-y
47. Alam, M. N. H. Z., Adrus, N., Wahab, M. F. A., Kamaruddin, M. J., and Sani, M. H.:Utilization of Agro-Waste as Carbon Source for Biohydrogen Production: Prospect and Challenges in Malaysia. *In Valorisation of Agro-industrial Residues Biological Approaches.* Volume 1. (2020) 131–147. Springer, Cham
48. Kozlov, N. K., Natashina, U. A., Tamarov, K. P., Gongalsky, M. B., Solovyev, V. V., Kudryavtsev, A. A., ... and Osminkina, L. A.: Recycling of silicon: from industrial waste to biocompatible nanoparticles for nanomedicine. *Materials Research Express.* Volume *4*(9). (2017) 095026
49. Hongzhang, C., and Liying, L.: Unpolluted fractionation of wheat straw by steam explosion and ethanol extraction. *Bioresource technology.* Volume 98(3). (2007) 666–676
50. Bharti, M. and Kamboj, N.: Occurrence and diversity of mycofloral population in soil of two different land use types in Haridwar region (Uttarakhand), India. *Journal of Advance Scientific Research.* Volume 11(1). (2020) 208–214
51. Christy, P. M., Gopinath, L. R., and Divya, D.: A review on anaerobic decomposition and enhancement of biogas production through enzymes and microorganisms. *Renewable and Sustainable Energy Reviews.* Volume 34. (2014) 167–173. https://doi.org/10.1016/j.rser.2014.03.010
52. Zhang, N. N., Xu, D. P., and Li, H.: Analysis on resource recovery of benzene/ethanol extractives of fresh Santalum album leaves in autumn. (2009) In 3rd International Conference on Bioinformatics and Biomedical Engineering. (2009) 1–3. https://doi.org/10.1109/ICBBE.2009.5162362
53. Demirbas, A.: Effects of temperature and particle size on bio-char yield from pyrolysis of agricultural residues. *Journal of analytical and applied pyrolysis.* Volume 72(2). (2004) 243–248. https://doi.org/10.1016/j.jaap.2004.07.003
54. Westendorf, M. L. (Ed.). Food waste to animal feed. John Wiley and Sons. (2000)
55. Haxo, H.E. and Mehta, P.K.:Ground rice hull ash as a filler for rubber. *Rubber chemistry and technology.* Volume 48 (2). (1975) 271-288
56. Mehta, P. K., and Pitt, N.: Energy and industrial materials from crop residues. *Resource Recovery and Conservation,* Volume 2(1). (1976) 23–38
57. Garland, J. L., Mackowiak, C. L., and Sager, J. C.: Hydroponic crop production using recycled nutrients from inedible crop residues. *SAE Transactions.* (1993) 1103–1110
58. Mueller, U. G., Gerardo, N. M., Aanen, D. K., Six, D. L., and Schultz, T. R.: The evolution of agriculture .in insects. Annual Review of Ecology, Evolution, and Systematics. Volume 36. (2005) 563–595. https://doi.org/10.1146/annurev.ecolsys.36.102003.152626
59. Gomez-Tovar, F., Celis, L. B., Razo-Flores, E., and Alatriste-Mondragón, F.: Chemical and enzymatic sequential pretreatment of oat straw for methane production. *Bioresource technology.* Volume 116. (2012) 372–378
60. Strayer, R. F., Finger, B. W., and Alazraki, M. P.: Evaluation of an anaerobic digestion system for processing CELSS crop residues for resource recovery. *Advances in Space Research.* Volume *20*(10). (1997) 2009–2015
61. Shen, J., Huang, J., Ruan, H., Wang, J., and Van der Bruggen, B. :Techno-economic analysis of resource recovery of glyphosate liquor by membrane technology. *Desalination.* Volume *342.* (2014) 118–125. https://doi.org/10.1016/j.desal.2013.11.041
62. Raut, S., Ralegaonkar, R., and Mandavgane, S.: Utilization of recycle paper mill residue and rice husk ash in production of light weight bricks. *Archives of civil and mechanical engineering.* Volume 13. (2013) 269–275. https://doi.org/10.1016/j.acme.2012.12.006

63. Jha, M. K., Lee, J. C., Kim, M. S., Jeong, J., Kim, B. S., and Kumar, V.: Hydrometallurgical recovery/recycling of platinum by the leaching of spent catalysts: A review. *Hydrometallurgy.* Volume 133. (2013) 23–32. https://doi.org/10.1016/j.hydromet.2012.11.012
64. Zhu, L., Dong, Y., Li, L., Liu, J., and You, S. J.: Coal fly ash industrial waste recycling for fabrication of mullite-whisker-structured porous ceramic membrane supports. *RSC advances.* Volume 5(15). (2015) 11163–11174
65. Piana, G., Ricciardi, M., Bella, F., Cucciniello, R., Proto, A., and Gerbaldi, C.: Poly (glycidyl ether) s recycling from industrial waste and feasibility study of reuse as electrolytes in sodium-based batteries. *Chemical Engineering Journal.* Volume *382.* (2020) 122934. https://doi.org/10.1016/j.cej.2019.122934
66. Canda, L., Heput, T., and Ardelean, E.: Methods for recovering precious metals from industrial waste. In IOP Conference Series: Materials Science and Engineering. Volume 106(1). (2016) 12–20
67. Treffry-Goatley, K., and Gilron, J.: The application of nanofiltration membranes to the treatment of industrial effluent and process streams. *Filtration and separation.* Volume *30*(1). (1993) 63–54
68. Pelino, M., Cantalini, C., and Rincon, J. M.: Preparation and properties of glass-ceramic materials obtained by recycling goethite industrial waste. *Journal of Materials Science.* Volume *32*(17). (1997) 4655–4660. https://doi.org/10.1023/A:1018602224392
69. Mojumdar, A., and Deka, J.: Recycling agro-industrial waste to produce amylase and characterizing amylase–gold nanoparticle composite. *International Journal of Recycling of Organic Waste in Agriculture.* Volume *8*(1). (2019) 263–269. https://doi.org/10.1007/s40093-019-00298-4
70. Bonmatí, A., and Flotats, X.: Air stripping of ammonia from pig slurry: characterisation and feasibility as a pre-or post-treatment to mesophilic anaerobic digestion. *Waste management.* Volume 23(3). (2003) 261–272. https://doi.org/10.1016/S0956-053X(02)00144-7
71. Chantigny, M.H., Angers, D.A., Belanger, G., Rochette, P., Eriksen-Hamel, N., Bittman, S.,Buckley, K., Masse, D., Gasser, M.O.: Yield and nutrient export of grain corn fertilized with raw and treated liquid swine manure. *Agronomy Journal.* Volume 100(5). (2008) 1303–1309. https://doi.org/10.2134/agronj2007.0361
72. De Vrieze, J., Colica, G., Pintucci, C., Sarli, J., Pedizzi, C., Willeghems, G., ... and Spiller, M.: Resource recovery from pig manure via an integrated approach: a technical and economic assessment for full-scale applications. *Bioresource technology*, Volume 272. (2019) 582–593. https://doi.org/10.1016/j.biortech.2018.10.024
73. Gong, W., Li, W., and Liang, H. :Application of A/O-MBR for treatment of digestate from anaerobic digestion of cow manure. *Journal of Chemical Technology and Biotechnology*, Volume 85(10). (2010) 1334–1339. https://doi.org/10.1002/jctb.2437
74. Zhang, Z., Xu, Z., song, X., Zhang, B., Li, G., Huda, N., and Luo, W. :Membrane processes for resource recovery from anaerobically digested livestock manure effluent: opportunities and challenges. *Current pollution reports.* (2020) 1–14
75. Waqas, M., Almeelbi, T., and Nizami, A. S.: Resource recovery of food waste through continuous thermophilic in-vessel composting. *Environmental Science and Pollution Research.* Volume 25(6). (2018) 5212–5222. https://doi.org/10.1007/s11356-017-9358-x
76. Adi, A.J., Noor, Z. M.: Waste recycling: Utilization of coffee grounds and kitchen waste in vermicomposting. *Bioresource Technology*, Volume 100(2). (2009) 1027–1030. https://doi.org/10.1016/j.biortech.2008.07.024

Reviewing Medicinal Plants of Treasure Land: The Indian Himalayan Range

Akanksha Bisht, Vishal Kamboj, and Aditi Bisht

Abstract The Himalayan range is home to a diversity of plant due to the altitude, topography gradient and varying climatic conditions. Among the 18,000 already known flowering plants that grow in India, about 2500 of them possess some medicinal property and 25% of these plants reside in Himalayan range at different altitudes. Plants in order to survive and defend themselves in stressful environmental condition adapt and alter their plant physiology leading to the production of secondary metabolites. These medicinal plants have a key role in the lives of the local communities since ages, they rely on them for nutrition and medicinal requirements. Secondary metabolite rich plant is the main source of attraction for researchers as well as the industrialists. Wide Himalayan range not only provide raw material to the pharma industry but also to the cosmetic industry. The vital goal of the current paper is to review the potentially rich medicinal plants, threats and conservation practise of medicinal plants in the Himalayan range.

Keywords Indian Himalaya region · Medicinal plants · Ayurveda · Biodiversity

1 Introduction

Indian Himalayan range is blessed with a diversity of flora. According to the sources out of the 500,000 plant species growing all over the world, only 1% has been phytochemically investigated, which illustrates that these plants have great potential for detecting novel bioactive compounds. The Himalayas stretches cover 3000 km from Northern Pakistan, Nepal, Bhutan and the North-Western and North-Eastern states

A. Bisht
Department of Medicinal and Aromatic Plants, Uttarakhand Altitude Plant Physiology Research Centre, Hemvati Nandan Bahuguna Garhwal University, Birla Campus, Srinagar, Uttarakhand 249161, India

V. Kamboj · A. Bisht (✉)
Department of Zoology and Environmental Science, Gurukula Kangri (Deemed to be University), Haridwar 249404, India
e-mail: minnib18@gmail.com

K. D. Bahukhandi et al. (eds.), *Environmental Pollution and Natural Resource Management*, Springer Proceedings in Earth and Environmental Sciences,
https://doi.org/10.1007/978-3-031-05335-1_3

of India with rich biodiversity due to variations in habitat [17]. The Lower Himalaya promotes the growth of luxuriant trees of *Alnus nitida* and *Pinus roxburghii* on the slopes, while the moist rich slopes are occupied by *Alnus nepalensis*, *Quercus leucotrichophora*, *Rhododendron arboreum*, etc. The forests of *Quercus semecarpifolia*, *Quercus floribunda* and *Pinus wallichiana* are characteristic of the higher altitude zones of the Lower Himalaya. The higher altitudes beyond the tree line (above 3600 m) delivers an ironic assemblage of herbs in the alpine meadows and grasslands. The diverse species of herbal plant belonging to genera *Aconitum*, *Picrorhiza*, *Rheum*, *Meconopsis* and the scrubs *Rhododendron anthopogon* and *Rhododendron lepidotum* are some of the vital plants of these heights. *Rosaceae* and *Umbelliferae* are some of the chief Himalayan families with a high degree of endemism. Total endemic plant species are counted to be 3160 [16]. It is determined that close to 15% of about 70,000 known plant species have therapeutic properties which means that over 10,000 plant species are utilized in medicine at one time or another.

The Indian ancient text provides a reference to the herbs growing in the Himalayan region since ages. Ayurveda is an ancient system of life. It is one of the storehouses of antiquated information, which is still in use [25]. Atharveda has a detailed account of the medicinal properties of countless herbs for curing various ailments.

Treatment provided to a sick person using herbal plants tend to possess no side effects on the body, they are completely safe. Since these herbal plants are in a state of harmony with nature, they hold more prominent favourable position over synthetically treated items and manufactured drugs. Ayurvedic herbs are capable of providing holistic healing to an individual for a long run [18].

Apart from being the main source of raw material for the production of medicines for humans, these herbs can be utilized for the preparation of bio-pesticides, example castor plant, natural dyes example rattan jot, brews example rhododendron and perfumes example Jatamasi [28]. In today's time, every individual has understood the damage one has to face after synthetically prepared medicine. Plants based drugs are a vital element for the current health care system in the world. A total of 85% of world population use the locally available herb as medicine [8]. So, keeping the view of severity of medicinal plants, we reviewed and discussed the description, distribution pattern, phytochemistry, medicinal property of some selected medicinal plant of the treasure land Himalaya and also discussed about the threats and conservation practices.

2 Description of Medicinal Plant

In Himalayan range of India, a lot of medicinal and aromatic plants are present. Here, we discussed about the used five medicinal plants such as *Rhododendron*, *Aconitum*, *Bacopa monnieri*, *Berberis* and *Hedychium*. The distribution pattern of the selected medicinal plants is illustrated in Table 1. However, the phytochemical property and part used for the medicinal is illustrated in Table 2. The detail description of selected medicinal plants is discussed below.

Table 1 Altitude and rainfall of selected medicinal plants of Himalayas

Name of the plant	Altitude	Mean annual rainfall
Rhododendron arboreum	1500–3500 m	200–1800 mm
Bacopa monnieri	1300 m	650–830 mm
Berberis aristata	2000–3000 m	1200–1400 mm
Hedychium spicatum	1500–2500 m	1000–1500 mm
Asparagus racemosus	1500 m	600–1000 mm

Table 2 Chemical constituent and phyto-constituent of parts used in selected medicinal plants of Himalayas

Name of the plant	Chemical constituents	Part used	Phytoconstituent present
Rhododendron arboreum	Quercetin	Flower	1.051–12.5%
Bacopa monnieri	Bacoside	Leaves	50–55%
Berberis aristate	Berberine	Root	1.6–4.3%
Hedychium spicatum	β-pinene	Rhizome	16.7%
Asparagus racemosus	Sapogenin	Tubers	0.7–0.9%

2.1 *Rhododendron*

2.1.1 Descriptions

The presence of Rhododendrons was first verified by Captain Hardwick in Jammu and Kashmir in the year 1776. Rhododendron commonly called as Buras or Laligurans (Nepali name) belongs to the family Ericaceae [10]. There are in total 80 species of Rhododendron but only 12 Species are found in India. Most of these species are reported to be very valuable for the local population. Many species are being widely exploited for fuelwood and incense [4].

It is an evergreen tree with a wide crown, that grows up to 20 m and above in height. The leaves are glossy green from the top and rustic brown at the back, they are crowded towards the base and are lanceolate, narrow in shape. They are 10–20 cm long and 3–4 cm wide. The colour of the flower ranges from pink to red [35]. Fruit type is capsular and has ellipsoid seeds. Last two to three weeks in October are considered suitable for the collection of seeds [39].

2.1.2 Distribution

Rhododendron has been named as the national flower of Nepal and the state tree of Uttarakhand. It is well distributed in Himachal Pradesh, Sikkim, Nagaland, Mizoram, Uttarakhand etc. The moist slopes in the eastern part of the Himalayan region provide the best habitat for well-flourished growth of Rhododendron species. In Uttarakhand, the *R. arboreum* species are confined to an altitude of 1500–3500 m above mean sea level. It grows best on acidic soil where under such pH condition very few plants can survive [45].

2.1.3 Phytochemistry

Numerous phytochemicals have been extracted from different parts of the Rhododendron. 34 compounds were acknowledged i.e. Leaves of Rhododendron is rich in flavonoid. Quercetin is found in flower and leaves. Rutin and sterol are found in leaves. Betulin, lupeol, ursolic acid is present in the bark [27].

2.1.4 Medicinal Uses

The fresh blossoms are likewise utilized in the treatment of dysentery and dyspepsia. The bioactive constituents of the bark of this plant have antifungal properties. The ethanolic extracts from the leaves exhibit antitumor activity.

2.1.5 Other Rhododendron Species

1. *Rhododendron barbatum, Rhododendron falconeri, Rhododendron hodgsonii* are a source of fuel for local people.
2. Leaves and twigs of *Rhododendron anthopogon* are used as a raw material for the preparation of incense stick.
3. Fresh corolla from *Rhododendron arboretum* and *Rhododendron cinnabarium* is used for the preparation of local brews. The tender leaves and woods are used for the production of cups, spoons, boxes and saddle.
4. Leaves and flowers of *Rhododendron anthopogon* are effective in treating indigestion and lung infection [9].

2.2 *Aconitum*

2.2.1 Description

Aconitum which is also known as aconite, wolfbane, queen's poison or monkshood, comprises of 250 species. It belongs to Ranunculaceae family [13]. Some of the species of Aconitum are remarkably deadly, despite this, they possess a lot of therapeutic importance. Proper measures must be taken to extract the essential compound from the plant without the toxic nature [40]. It is one of the most valued drug and treasured components of Ayurvedic system of medicine. Yearly demand of raw drugs is fairly high in the local market, which has generated immense pressure on the genus leading to its threatened status in different states of Himalayan range [11]. They are biennial tubers and are economically important. It grows up to a height of 15–20 cm. Leaves are glabrous, heteromorphous and long petioled. It has violet-blue sepals and black coloured pyramidal seeds.

2.2.2 Distribution

Aconitum occurs in the alpine and sub-alpine regions of Himalayas at an elevation of 2000–5000 m. It is native to parts of Western Himalaya [29].

2.2.3 Phytochemicals

Diterpene alkaloids and flavonoids are found in abundance in Aconitum species. The alkaloid aconitine in the tuber's attributes to the medicinal properties of the plant. The other compounds existing in the plant are tannic acid, starch, fatty acids and their glycerides, carbohydrates, etc. [1].

2.2.4 Medicinal Uses

Aconites are quite beneficial in treating the communicable disease, for example, measles. It fruitfully cures problems like asthma, diabetes, leukoderma, leprosy and convulsions etc. phenolic and flavonoid elements of the plant have antioxidant as well as anti-bacterial effect. It is effective in counteracting cough and congestion. It is a good aphrodisiac and bears the features of being a diuretic [2].

2.2.5 Other Aconitum Species

- *Aconitum heterophyllum* is widely used in Ayurvedic medicines because of its hot potency. The seeds have a laxative property. The roots are effective for health problems like anorexia, arthritis and ascaris [42].
- *Aconitum ferox* is also known as Meetha Vish or Monkshood. It has sedative nature because of which it is also prescribed to people dealing with severe anxiety issues, also it is highly effective in fever, diuretic action, asthma and arthritis.
- *Aconitum heterophylloides* also known as Nepal Monkshood or safed bish is a tall hairless herb with tuberous roots. The root of the herb contains 0.9% total alkaloid content which is used in acute headache and rheumatism [24].

2.3 *Bacopa monnieri*

2.3.1 Description

Bacopa monnieri is commonly known as Brahmi. It is a member of Scrophulariaceae family. This family includes more than 4500 species [47]. It has been effectively used in improving memory and other positive psychological abilities in both normal and mentally retarded kids with promising results. It is a creeping plant with a tender stem which grows up to a height of 10–30 cm in length with 1–2 mm of a thickness [48]. It has succulent and sessile leaves that are oppositely arranged on the main stem. Flower and the fruit emergence take place in the summer season. It bears flowers possessing a colour range from white to slightly violet-blue. The fruit is ovoid, with no discrete odour. It has cooling potency [12].

2.3.2 Distribution

It is found in the tropic and sub-tropics regions. It is found in moist and wet places throughout India and Nepal [37].

2.3.3 Phytochemicals

The therapeutic importance of this plant is because of the presence of several compounds in it such as alkaloids, saponins, glycosides, flavonoids and stigmasterols. The saponins of this plant are called "bacosides". Luteolin and apigenin are the two flavonoids that are found in it. Two alkaloids isolated from *Bacopa monnieri* are herpestine and brahmine, if taken in high dosage may lead to extreme headache [22].

2.3.4 Medicinal Uses

Bacopa monnieri is beneficial for treating skin diseases example eczema, ulceration etc. It is an excellent medicine for Alzheimer and amnesia, thus commonly used as a brain tonic. It is an effective herb for the prevention of hair fall and early greying of hair. It rejuvenates the hair follicle and prevents baldness [3]. *Bacopa monnieri* consists of powerful antioxidants that protect the body against free radical impairment [30].

2.3.5 Other Bacopa Species

- *Bacopa crenata* is a perennial medicinal plant also called moneywort.

- *Bacopa repens* is an annual plant and has a creeping stem.
- *Bacopa rotundifolia*: This plant is not of commercial use [23].

2.4 *Berberis*

2.4.1 Description

Berberis aristate also called as Indian barberry in English or Daru Haldi in Hindi belongs to the family Berberidaceae. Most of the *Berberis* sp. have acquired a significant position in the traditional system of medicine. It is a perennial shrub which grows in a temperate climate [19]. It can grow up to 3 m height. The flower is yellow coloured. The fruit is in the form of an ovoid berry and contains 2–5 seeds, the colour of seed ranging from yellow to pink. The fruit has acidic to sweet taste [38].

2.4.2 Distribution

Berberis is a well know herbal plant of Indian Himalayan Region. It growing in small patches in hilly slopes. It is found at an altitude up to 1800–3000 m (middle altitude areas) in Uttarakhand and Himachal Pradesh [36]. IUCN has kept this plant under endangered status. It is typically cultivated for its edible fruits [34].

2.4.3 Phytochemicals

The most vital constituent of this herbal plant is berberine, a quaternary isoquinoline alkaloid, which is mostly found in the stems and roots. The bark consists of the quaternary ammonium salt of isoquinoline alkaloid. Western Himalayan species of *Berberis* are rich in fibre, protein, fats and some minerals like Calcium and Potassium [41].

2.4.4 Medicinal Use

Berberis is super effective in curing skin disease like melasma. It is effective in safeguarding liver against disorders. It prevents water loss from the body due to loose motion and diarrhoea so utilized as a mild laxative [14]. The roots, stems, leaves and fruits are traditionally utilized to treat wounds, diabetes, inflammations and jaundice. The extracts of this plant have been testified with antibacterial, antiviral, antifungal, anticancer, anti-inflammatory and antidiabetic profiles [33].

2.4.5 Other *Berberis* Species

- *Berberis* vulgaris has anti-inflammatory properties.
- *Berberis heterophylla* has anti-microbial properties [6].
- *Berberis repens*: It has anti-bacterial properties.

2.5 *Hedychium*

2.5.1 Description

Hedychium spicatum belongs to the family Zingiberaceae, commonly known as Kapoor katcheri in Hindi and spiked ginger lily in English and as Shati Ayurveda. It is a small hardy rhizomatous herb [15]. It is a perennial herb which raises to 1–2 m, which consists of orange whitish flowers. Flowering can be seen in July–August and seeding in September–October. This plant has fragrant leaves with a robust stem which grows up to 1.5 m tall. Leaves are 30 cm long and lanceolate in shape. The rhizome has a strong odour and it tastes bitter. It bears black with red aril [32].

2.5.2 Distribution

Hedychium spicatum is an endemic herb species which is found in countries of South-East Asia. This species is also native to a Himalayan region consisting of various habitats in sub-tropical to temperate zones. The species can be commonly found in the of Central Himalayan Region of India from 1200 to 2400 m in moist and rocky habitat near a water body and oak or mixed forest [32]. It is distributed in Andhra Pradesh, Arunachal Pradesh, Darjeeling, Himachal Pradesh, Karnataka, Manipur, Meghalaya, Mizoram, Nagaland, Orissa, and Sikkim. The species is best for the sub-tropical zone which requires sunny weather. The lowest temperature it can tolerate is −2 °C [31].

2.5.3 Phytochemicals

The extract of tubers contains 4% essential oil, starch, resins, organic acids, glycosides, albumen and saccharides. These extracts are good for the treatment of bronchitis, indigestion, eye disease, inflammations, laxative, stomach ache, carminative, stimulant, tonic to the brain and diarrhoea. Numerous chemical ingredients testified in essential oil are α-pinene, β-pinene, limonene and many others [31] used as tranquillizer, CNS depressant, hypotensive, respiratory disorders, antimicrobial, analgesic, anti-spasmodic and antifungal activities [21].

2.5.4 Medicinal Use

Pharmacological activities of *Hedychium spicatum* testified till now are antidiabetic, tranquillizing, pediculicidal, antimicrobial, antioxidant, antimalarial hepatoprotective, cytotoxic activity, hypocholesterolemic, anthelmintic, stomachic and tonic activity. 1, 8 cineole is compound obtained from the rhizome is used for its anti-diabetic property. The anti-microbial property of the essential oil can be seen in against *Staphylococcus aureus*, *Escherichia coli*, *Shigella flexneri*, and *Pasteurella multocida.* It is used for anthelminthic activities also but is not very effective. Most of the essential oils extracted from the tuber are used for the treatment of various problems [5].

2.5.5 Other *Hedychium* Species

- *Hedychium acuminatum*—It best used for liver disease.
- *Hedychium coronarium*—It is super effective in curing cataract.
- *Hedychium marginatum*—Best used against the problem of indigestion [7].

3 Threats and Conservation Practices for Medicinal Plants in Indian Himalayan Region

Loss in Genetic Resources of Medicinal Plants
The demand for raw medicines is pretty high per annum in the domestic marketplace [46]. Major medicinal species of Himalayas are susceptible to extinction due to overharvesting [44]. Result of expansion in urban areas, increasing tourism and other progressive activities like hydroelectric schemes and highway projects cause excessive damage to genetic diversity [43].

Effect of Overgrazing
Grazing animal causes a serious threat to and the dispersal of medicinal plants. The area protected from nomadic grazing shows healthier flora. The diversity of medicinal plants had decreased in the overgrazed locations by nearly 90%. It is also observed that the accessibility of economically and therapeutically important plant species is declining.

Disproportionate Collection of Medicinal Plants
The indiscriminate harvesting of medicinal plant from wild become a source of revenue when sold to traders in informal market. The risk of over-harvesting leading to loss of susceptible plant population has been recognized for years. It is necessary to introduce good harvesting practices to the collectors to decrease the avoidable losses.

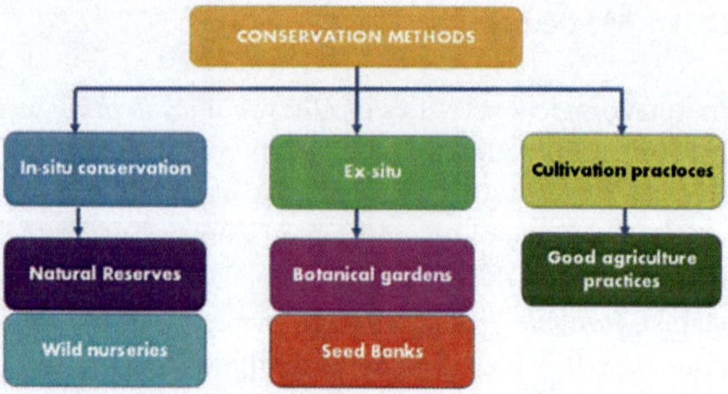

Fig. 1 Common conservation practices

Climate Change

There are indications that climate change is producing visible effects on life cycles and distribution of species of the plants. Although, recently this viewpoint has been given renewed attention.

4 Conservation and Sustainable Development

Conservation of medicinal plant species is done for maintaining gene bank of the vulnerable species. On the other hand, sustainable harvesting decreases the degeneration of renewable non-timber plant resources. Conservation of medicinal plants and their genetic resources can be done through in-situ and ex-situ practices. Ex-situ conservation includes conservation of average plants outside their natural habitat used to protect them from destruction, replacement or deterioration. It includes procedure like seed storage, DNA storage, field gene banks and botanical garden. In-situ is opposite of ex-situ, the conservation of a particular species is done in its habitat (Fig. 1).

5 Organizations Involved the Conservation of Medicinal Plants

The Government of India (GOI) is thoughtful about the maintenance of priceless herbal wealth. The banking segment has also been established for providing economic aids to numerous plans involved in medicinal plant conservation and cultivation plans. The National Bank for Agriculture and Rural Development is supportive in the development of medicinal plant sector in synchronization with NMPB. It is providing bankable models for medicinal and aromatic crops [20] and is also supporting in the capacity building curricula in this direction [26].

6 Conclusion

The Himalaya enjoys the benefit of diverse medicinal and aromatic species, which occupies a very important place in the herbal pharmaceutical sector. Constant research will help in further promoting the rich and rare biodiversity of the range. The medicinal plants of the Himalayan range have a great diversity of medicinal properties including antioxidant, neuroprotective, antidiabetic, hepatoprotective, anticancer, antimicrobial, and immunomodulatory functions. Harvesting these herbs directly from wild has been in practice for a long time, and rising curiosity in herbal medicines has led to the bigger demand of these plants at national as well as global level, which has put this treasure under pressure. Preparing suitable agro-techniques for the domestication of wild species under threat and finding out other ways with which we can conserve them has become the need of the hour.

Acknowledgements The authors highly thanks to the department of zoology and environmental science, Gurukula Kangri (Deemed to be University), Haridwar (Uttarakhand) for providing the facility and also thank the Department of Medicinal and Aromatic Plants, HNB Garhwal University.

References

1. Bahuguna R., Purohit M.C., Rawat M.S.M., Purohit A.N.: Qualitative and quantitative variations in alkaloids of Aconitum species from Garhwal Himalaya. *Journal of Plant Biology*. Volume 27(2). (2000) 179–183
2. Beigh S.Y., Nawchoo I.A., Iqbal M.: Cultivation and conservation of Aconitum heterophyllum: A critically endangered medicinal herb of the northwest Himalayas. *Journal of Herbs, Spices & Medicinal Plants*. Volume 11(4). (2006).47–56
3. Bhandari P., Kumar N., Gupta A.P., Singh B, Kaul VK.: Micro-LC determination of swertiamarin in Swertia species and bacoside-A in Bacopa monnieri. *Journal of Chromatography*. Volume 64(9–10). (2006) 599–602
4. Bhattacharyya D.: Rhododendron species and their uses with special reference to Himalayas–a review. Volume 7(1). (2011) 161–167
5. Chaudhary A.A., Dhar U., Ahmad A., Bhatt I.D., Jugran A., Kaur G.: Analysis of genetic diversity in Hedychium spicatum Buch. Ham. ex. D. Don in West Himalaya, India. *Journal of Medicinal Plants Research*. Volume 6(23). (2012) 3984–3991
6. Das S., Das M.K., Mazumder P.M., Das S., Basu S.P.: Cytotoxic activity of methanolic extract of Berberis aristata DC on colon cancer. *Global Journal of Pharmaceutical Sciences*. Volume 3(3). (2009) 137–140
7. Devi N.B., Singh P.K., Das A.K..: Ethnomedicinal utilization of Zingiberaceae in the valley districts of Manipur. *Journal of Environmental Science Toxic Food Technology*. Volume 8(2). (2014) 21–23
8. Dubey N.K., Kumar R, Tripathi P.: Global promotion of herbal medicine: India's opportunity. *Current Science*. Volume 86(1). (2004) 37–41
9. Gaira K.S., Rawal R.S., Rawat B., Bhatt I.D.: Impact of climate change on the flowering of Rhododendron arboreum in central Himalaya, India. *Current Science*. (2014) .1735–1738
10. Iqbal K., Negi A.K.: Rhododendrons in Uttarakhand: Diversity and conservation. *International Journal of Environmental Science and Technology*. Volume 6(1). (2017) 31–45

11. Jabbour F., Renner S.S.: A phylogeny of Delphinieae (Ranunculaceae) shows that Aconitum is nested within Delphinium and that Late Miocene transitions to long life cycles in the Himalayas and Southwest China coincide with bursts in diversification. *Molecular Phylogenetics and Evolution.* Volume 62(3). (2012) 928–942
12. Jana J C.: Use of traditional and underutilized leafy vegetables of Sub-Himalayan Terai region of West Bengal. *International Conference on Indigenous Vegetables and Legumes. Prospectus for Fighting Poverty, Hunger and Malnutrition.* Volume 752. (2006) 571–575
13. Jeelani SM, Siddique MAA, Rani S.: Variations of morphology, ecology and chromosomes of Aconitum heterophyllum Wall., an endangered Alpine medicinal plant in Himalayas. *Caryologia.* Volume 68(4). (2015) 294–305
14. Joshi P.V., Shirkhedkar A.A., Prakash K., Maheshwari V.L.: Antidiarrheal activity, chemical and toxicity profile of Berberis aristata. *Pharmaceutical Biology.* Volume 49(1). (2011) 94–100
15. Jugran A., Bhatt I.D., Rawat S., Giri L., Rawal R.S., Dhar U.: Genetic diversity and differentiation in Hedychium spicatum, a valuable medicinal plant of Indian Himalaya. *Biochemical genetics.* Volume 49(11–12). (2011) 806–818
16. Kala C.P.: Status and conservation of rare and endangered medicinal plants in the Indian trans-Himalaya. *Biological Conservation.* Volume 93(3). (2000) 371–379
17. Kamboj N., Bisht A., Kamboj V., Pandey N., Bisht A.: Role of natural disasters in environmental degradation: An overview. *Environmental Degradation: Causes and Remediation Strategies.* Volume 1 (2020) 21
18. Khalsa K.P.S.: The practitioner's perspective: Introduction to Ayurvedic herbalism. *Journal of Herbal Pharmacology.* Volume 7(3–4). (2008) 129–142
19. Komal S., Ranjan B., Neelam C., Birendra S., Kumar S.N.: Berberis aristata: A review. Inter. *International Journal of Research in Ayurveda and Pharmacy.* Volume 2(2). (2011) 383–388
20. Kumar R. Medicinal aromatic and herbal crops. *Financial Agriculture.* Volume 36. (2004) 3–5
21. Mishra T., Pal M., Meena S., Datta D., Dixit P., Kumar A, et al. Composition and in vitro cytotoxic activities of essential oil of Hedychium spicatum from different geographical regions of western Himalaya by principal components analysis. *Natural Product Research. Volume* 30(10). (2016) 1224–1227
22. Mitra P.K, Mitra P., Das A.P., Ghosh C., Sarkar A., Chowdhury D.: Screening the efficacy of some East Himalayan medicinal plants on ethanol induced gastric ulcer in Albino rats. *Pleione.* Volume 4(1). (2010) 69–75
23. Mukherjee S., Dugad S., Bhandare R., Pawar N., Jagtap S., Pawar P.K., Kulkarni O.: Evaluation of comparative free-radical quenching potential of Brahmi (Bacopa monnieri) and Mandookparni (Centella asiatica). *An International Quarterly Journal of Research in Ayurveda.* Volume 32(2). (2011) 258
24. Nandi S.K., Pandey H., Nadeem M., Palni L.M.S.: Chemical stimulation of seed germination in Aconitum heterophyllum Wall and A. Balfourii Stapf.: important Himalayan species of medicinal value. *Seed Sci Tech (Switzerland).* Volume 28(1). (2000) 39–48
25. Patil D.P., Urunkar N., Patil S.S.: Tomorrow-Ayurveda in a Daily Routine. *Paripex - Indian Journal of Research.* Volume 8(11). (2019)
26. Prahalathan S.: Export potential of Indian medicinal plants and products. *Financial Agriculture.* Volume 36. (2004) 33–36
27. Prakash D., Upadhyay G., Singh B.N., Dhakarey R., Kumar S., Singh K.K.: Free-radical scavenging activities of Himalayan rhododendrons. *Current Science.* (2007) 526–532
28. Rahman H., Shaik H.A., Madhavi P., Eswaraiah M.C.: A review: pharmacognostics and pharmacological profiles of Nardostachys jatamansi DC. *Elixir pharmacy.* Volume 39. (2011) 5017–5020
29. Raina R., Chauhan R., Naithani R.: A Note on the Floral Morphology of Aconitum heterophyllum Wall. (Ranunculaceae)-An Important Temperate Medicinal Plant. *Journal of Applied Research on Medicinal and Aromatic Plants.* 2(2). (2011) 1
30. Rakhi R., Vashistha D.P:. Shortlisting of cultivable herbal plants in the Bhabhar region of the Garhwal Himalaya, Uttarakhand, India. *International Journal of Medicinal and Aromatic Plants.* Volume 1(1). (2011) 23–27

31. Rasool S., Maqbool M.: An overview about Hedychium spicatum: a review. *Journal of Drug Delivery and Therapeutics*. Volume 9(1-s). (2019) 476–480
32. Rawat S., Bhatt I.D., Rawal R.S.: Total phenolic compounds and antioxidant potential of Hedychium spicatum Buch. Ham. ex D. Don in west Himalaya, India. *Journal of Food Composition and Analysis*. Volume 24(4–5). (2011) 574–579
33. Ray R., Gururaja K.V., Ramchandra T.V. Predictive distribution modeling for rare Himalayan medicinal plant Berberis aristata DC. *Journal of Environmental Biology*. Volume 32(6). (2011) 725
34. Samant Pant S., Singh M., Lal M., Singh A., Sharma A., Bhandari S.: Medicinal plants in Himachal Pradesh, north western Himalaya, India. *International journal of biodiversity science, ecosystem services & management*. Volume 3(4). (2007) 234–251
35. Sekar K.C., Srivastava S.K.: Rhododendrons in Indian Himalayan region: diversity and conservation. *American Journal of Plant Sciences*. Volume 1(02). (2010) 131
36. Shahid M., Rahim T., Shahzad A., Latif T.A., Fatma T., Rashid M. et al.: Ethnobotanical studies on Berberis aristata DC. root extracts. *African Journal of Biotechnology*. Volume 8(4). (2009)
37. Sharma J., Gairola S., Gaur R.D., Painuli R.M., Siddiqi T.O.: Ethnomedicinal plants used for treating epilepsy by indigenous communities of sub-Himalayan region of Uttarakhand, India. *Journal of Ethnopharmacology*. Volume 150(1). (2013) 353–370
38. Singh J., Kakkar P.: Antihyperglycemic and antioxidant effect of Berberis aristata root extract and its role in regulating carbohydrate metabolism in diabetic rats. *Journal of Ethnopharmacology*. Volume 123(1). (2009) 22–26
39. Singh K.K., Kumar S., Rai L.K., Krishna A.P.: Rhododendrons conservation in the Sikkim Himalaya. *Current Science* .85(5). (2003) 602–606
40. Singhuber J., Zhu M., Prinz S., Kopp B.: Aconitum in traditional Chinese medicine-a valuable drug or an unpredictable risk? *Journal of Ethnopharmacology*. Volume 126(1). (2009) 18–30
41. Srivastava S.K., Khatoon S., Rawat A.K.S., Mehrotra S., Pushpangadan P.: Pharmacognostic evaluation of the root of Berberis aristata DC. *Natural Product Sciences*. Volume 7(4). (2001) 102–106
42. Srivastava N., Sharma V., Kamal B., Dobriyal A.K., Jadon V.S.: Advancement in research on Aconitum species (Ranunculaceae) under different area: a review. *Biotechnology*. Volume 9(4). (2010) 411–427
43. Tandon P., Kumaria S., Nongrum I.: Conservation and management of plant genetic resources of Northeast India. *Indian Journal of Traditional Knowledge*. Volume 8(1). (2009) 29–34
44. Ticktin T.: The ecological implications of harvesting non-timber forest products. *Journal of Applied Ecology*. Volume 41(1). (2004) 11–21
45. Tiwari O.N., Chauhan U.K.: Rhododendron conservation in Sikkim Himalaya. *Current Science*. (2006) 532–541
46. Ved D.K., Goraya G.S.: Demand and supply of medicinal Plants in India. *NMPB, New Delhi; FRLHT, Bangalore*. (2007)
47. Yadav K.D., Reddy K.R., Agarwal A.: Preliminary physico-chemical profile of Brahmi Ghrita. *An International Quarterly Journal of Research in Ayurveda*. Volume 34(3). (2013a) 294
48. Yadav K.D., Reddy K.R.C., Kumar V.: Encouraging effect of Brahmi Ghrita in amnesia. *International Journal of Green Pharmacy*. Volume 7(2). (2013b)

Anthropogenic Impact on Fish Faunal Diversity and Their Habitat Ecology in the Ganga River and Its Tributaries, Uttarakhand

D. S. Malik, Arvind K. Sharma, Sunil Kumar, Rakesh Kumar, Ranjit Kumar, Vishal Kamboj, and Amit K. Sharma

Abstract The Ganges is one of the largest riverine ecosystem of India which supports diverse aquatic biodiversity in comparison to other aquatic ecosystems. The endemic fish species richness has significantly contributed to the eco-biocoenosis of Ganges riverine ecosystem. The main purpose of the current research was to assess the assemblage structure of endemic fish fauna in the river Ganges and its correlation towards the major water quality parameters. Collection of endemic fish species and important water quality parameters from eight sampling zones of river Ganges were done for the period of two years from September 2017 to August 2019. Fish diversity status was assessed by software PAST (version 3.0). The results of the current experimental research work evidently showed that Ganges riverine system offers the natural habitat of 31 fish species. Cyprinidae was the most dominant family followed by Balitoridae, Sisoridae, Schilbidae, Mastacembelidae, Salmonidae. Several physico-chemical parameters i.e. Water velocity, depth, water temperature, turbidity and dissolved oxygen subsidized as key significant environmental factors which directly and indirectly influence the diversity, species richness and distribution of fish population in the Ganges riverine ecosystem.

Keywords Fish fauna · Habitat ecology · Anthropogenic factors · Ganges · Uttarakhand

1 Introduction

Biodiversity is fundamental for change of an organic framework ensuring of general condition quality and for understanding the common worth of all species on earth. The fish faunal diversity has its significance like other aquatic and terrestrial animals.

D. S. Malik · A. K. Sharma · V. Kamboj
Department of Zoology and Environmental Science, Gurukula Kangri (Deemed to be University), Haridwar, Uttarakhand, India

S. Kumar · R. Kumar · R. Kumar · A. K. Sharma (✉)
Department of Animal Science, Central University of Himachal Pradesh, Dharamsala, India
e-mail: dr.amitgkv@gmail.com

K. D. Bahukhandi et al. (eds.), *Environmental Pollution and Natural Resource Management*, Springer Proceedings in Earth and Environmental Sciences,
https://doi.org/10.1007/978-3-031-05335-1_4

Fishes involve all the conceivable natural habitats of the aquatic ecosystem. The river Ganga and its tributaries form the largest riverine ecosystem in India. The river Ganga is divided into three major stretches due to the variance in geomorphology, ecology and rheology [1]. i.e. Gangotri to Haridwar (upper Ganga river), Haridwar to Varanasi (middle Ganga river) and Varanasi to Bay of Bengal (lower Ganga river). Upper Ganga basin (Gangotri to Haridwar) lies in the Garhwal region of Uttarakhand (Latitude 29° 26′ to 31° 28′N) and (Longitude 77° 49′ to 86° 06′E) with about 39,090 km^2 geographical area [2]. The Alaknanda and Bhagirathi river are sister streams joins at Devprayag and below at this confluence point, the joint stream is known as the Ganga river. The upper basin of the Ganga river flows on the steep and narrow bed and generally has the substrate in form of boulders, cobbles and rocks carried the very cold water and is less exposed to anthropogenic pollution. The upper Ganga basin is mainly characterized by very Coldwater, high water velocity, average productivity and diverse aquatic biodiversity with abundant rapids, runs, riffles and few deep pools [3]. The Upper Ganga basin supports the persistence of a million population with an average density of about 1000 populaces per square km. Primarily the water of the upper Ganga river is being utilized for numerous purposes i.e. aquaculture, agriculture, transportation and domestic usage. These types of events are essential for the dietary prerequisite and upliftment of the financial grade of millions of domiciliary [4, 5]. The fish species in the riverine network is very much dynamic in tempo-spatial scales due to intra-annual ecological modifications and several short-term changes i.e. day/night cycle, interaction amongst the spatiotemporal distribution and abundance of these communities. The upper Ganga and its tributaries are mostly controlled by barrages for various purposes as a result fish catches have been declined, and thereafter, loss of species diversity [6, 7]. Fish diversity and distribution are partially depending several environmental variables which directly influence the competing population. Several research studies regarding the distribution of fish fauna in the river Ganga and its tributaries have been made by various fisheries workers [7–11]. However, considering the above-mentioned reason, the present experimental research aimed to inspect the anthropogenic influence on fish faunal diversity and their habitat ecology.

2 Materials and Methods

2.1 Study Area

The current experimental research work was mainly focused on the Ganges riverine system and its tributaries at Uttarakhand for the period of two years from 2017 to 2019. An extensive ground investigation was conducted for the collection of primary data of endemic fish species and their habitat ecological parameters. The whole study area was divided into eight sampling zones (Fig. 1) for the collection of endemic fish

Fig. 1 Geo-coordinates of selected sampling zones. *Source* [12]

species along with water quality parameters. Monthly sampling was done for two years in all eight-sampling zone.

Experimental fishing was done with the help of previously contacted native fishermen's and also from the local landing centres. Fishes sample were collected by using different fishing nets i.e. gill nets (mesh size 1.2 × 1.2 cm. L × B = 12 × 1.5 m), cast net (dia. 2.5 m, mesh size 4.5 × 5.5 cm) and several indigenous nets. Catch and release fishing technique was done. All the fish species were released in the river however few samples were taken for laboratorial analysis. In laboratory, fish sample were arranged and identified to species level [6, 13, 14]. The temperature, pH, TDS, Conductivity were determined with the help of digital type instruments (Table 1).

Table 1 Instruments used for analysing habitat ecological characteristics

Parameters	Instruments used
Water temperature	Thermometers **(Testo EN50081)**
pH	Digital pH meter **(HI96107) (HANNA)**
Conductivity	Conductivity meter **(HI98303P, DIST3) (HANNA)**
TDS	Water analyzing kit **(S-967)**, digital TDS meter **(HI98303P, DIST3) (HANNA)**
Turbidity	Digital nephelometer **(Systonic, S-967)**
DO, BOD	Water analyzing kit **(S-967)**, titrimetric methods

2.2 Data Analysis

The fish diversity analysis and statistical comparison was done with the help of Paleontological Statistics (PAST version 3.1) by P. D. Ryan, D. A. T. Harper and J. S. Whalley. Diversity among the endemic fish fauna was evaluated by different diversity indices i.e. Shannon–Wiener diversity index [15]; Simpson Index; Evenness and Margalef index. Correlation between endemic fish fauna and important habitat ecological factor was done by Past and MS Excel (Figs. 2 and 3).

3 Result and Discussion

Physicochemical parameters: The results of the important water quality parameters assessed in all eight-sampling zones of the upper Ganga and its tributaries. Maximum water temperature was ranged from 23.2 to 25.0 °C at zone 7 and Zone 8 (Fig. 4) during summer (in June and July) where the minimum water temperature was ranged from 12.9 to 13.1 °C at Zone 1 and zone 2 during winter (during January and February). The maximum water temperature during summer is due to straight rapport amongst clear daylight, its time interval and high air temperature. Maximum and minimum variation in water temperature at Zone 1 and Zone 8 may be due to seasonal and altitudinal variation. The seasonal variation in water temperature was also reported by Refs. [9, 16] in the Ramganga river. In the Bhagirathi river seasonal disparity of water temperature vacillated between 8.0 and 18.0 °C was observed by Ref. [17]. The pH values ranged from 6.23 to 7.63 in the upper Ganga basin and were within the acceptable limit i.e. 6.5–8.5. The pH amongst all the sampling zone in

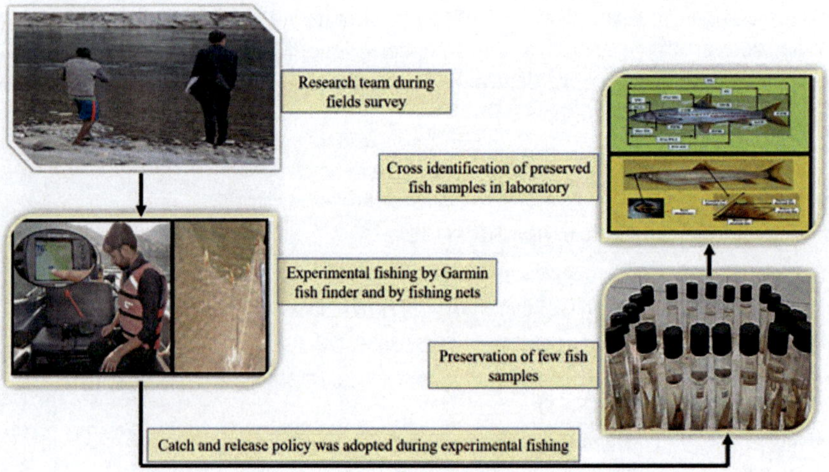

Fig. 2 Overview of method used during experimental fishing

Fig. 3 Various views of selected sampling zone on Ganges riverine system

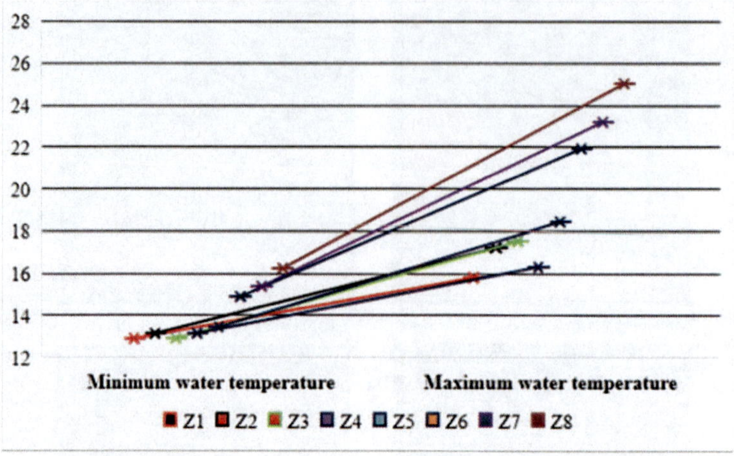

Fig. 4 Mean variation of water temperature °C in the Ganga river and its tributaries

Ganga riverine system was recorded alkaline accept little acidic (6.23) in the sampling zone 8 because of the accessibility of several types of carbonates and bicarbonates in the river which increase dissolve carbon dioxide level by severance and acts as a raw material for photosynthesis. References [3, 18] reported the alkaline nature of Ganga water and its tributaries at Uttarakhand. During the present study, turbidity in the upper Ganga riverine system and its tributaries ranged from 25.50 to 73.29 NTU. The Maximum (73.29 NTU) turbidity was recorded in sampling zone 8. During the monsoon season, turbidity surpassed in all the sampling zones from the standard limit i.e. 5 NTU posed by Ref. [19]. The upsurge in the amount of turbidity value results in dissolved oxygen stratification [20]. The same trend of turbidity value was reported by Refs. [21, 22] in the river Ganga. The high turbidity value in the aquatic ecosystem directly and indirectly disturbs the aquatic organism [23, 24]. Dissolved oxygen is a key indicator for the growth and development of aquatic organisms, water quality assessment and also a significant aspect for regulation of metabolic progressions of aquatic organisms and their population dynamics. The dissolved oxygen in the upper Ganga river and its tributaries ranged between 7.20 and 8.69 mg/L in all eight sampling zones. The DO at all the sampling zones was within the acceptable limit by Ref. [19] i.e. >6 mg/L. A similar trend of DO was reported in the Ganga river by Ref. [21]. References [12, 25] reported that DO is a key factor for freshwater ecosystem and above 5 mg/L dissolved oxygen is important for the aquatic ecosystem and also for sustenance of diverse aquatic dynamics and virtuous fish productivity. The lessening of DO might be due to organic load through the various anthropogenic factors like upsurge in municipal, domestic sewage and nutrients. Some of the factors distressing the DO content are mostly air and water temperature, rate of photosynthesis, respiration and decomposition processes. Biological oxygen demand is a vibrant prerequisite to all aquatic organisms for their metabolic activities in an aquatic ecosystem. BOD increases with an increase in the biodegradable organic content in a river. The BOD of the upper Ganga river and its tributaries ranged from 1.20 to 3.05 mg/L. All the values of BOD were within the acceptable limits accept in sampling zone 8 where the value exceeded the acceptable limit i.e. 3 mg/L mostly. A significant difference was observed in sampling zone 8 and sampling zone 1, 2 and 3 ($P < 0.05$). References [26, 27] reported a similar trend of BOD in the Ganga river. BOD is an important pollution indicator that helps to indicates the level of pollution triggered by the organic pollutant in any riverine ecosystem [28]. The potassium in upper Ganga was within the acceptable limit and was vacillated from 0.05 mg/L at sampling zone 2 and sampling zone 4 to 0.2 mg/L at sampling station 8. Similarly, the values of sodium were also within the acceptable at all the sampling zones. Average mean values of calcium were within the acceptable limit and were ranged from 19.3 mg/L at sampling zone 2 to 35.1 mg/L at Sampling zone 8. The calcium values at sampling zones 8 were significantly higher in contrast to other sampling zones. The mean magnesium values ranged from 6.2 mg/L at sampling zone 2 to 16.5 mg/L at sampling zone 8. All the sampling zones have an acceptable limit of Magnesium values except sampling zone 8. Magnesium in sampling zone 8 was significantly higher in comparison to other sampling zones. All the cations like Calcium, Magnesium, Potassium and Sodium were recorded within the acceptable limits. The reason beyond the low values of all the cations may be due to the geology and morphology of the river. Reference [25] recorded little value of cations in the Bhagirathi river. The water quality of selected sampling zones in upper Ganga river and its tributaries was observed good during the present study except in sampling zones 7 and 8 was inclining in the direction of slight polluted class (Fig. 5).

Fish species diversity (Species abundance and distribution): A total of 1640 individuals were reckoned which comprises 31 fish species belonging to the 6 families. The maximum number of fish species was counted for the Cyprinidae family followed by Balitoridae, Sisoridae, Schilbidae, Mastacembelidae, and Salmonidae The highest number of 520 individuals was counted in zone 4 throughout the study period where the lowest number of individuals (391) was found in zone 3 due to relatively low human interference and optimum environmental condition and on the other hand, the lowest number of individuals observed at zone 3 due to an extreme environmental condition. The family Cyprinidae was recorded higher in the upper Alaknanda river (Z1) and lower in the lower Ganga river (Z8) and lower Bhagirathi river (Z4). Lower to medium diversity of family Cyprinidae was recorded from the upper Bhilangana river (Z3) and Alaknanda-Bhagirathi confluence point (Z5). Medium to higher in the middle (Z7) and upper Bhagirathi river (Z6) (Fig. 6). The family Balitoridae was recorded higher in Z7 and Z8 and lower in Z1, Z2, Z3, and Z4. Medium diversity of family Balitoridae was recorded from Z5 and Z6 (Fig. 7). The family Sisoridae was recorded higher in Z5, Z6 and Z7 and lower in Z1, Z2, Z3, Z4 and Z8 (Fig. 8). The family Schilbidae was recorded higher in Z1, Z5, Z6 and Z7 and lower in Z2, Z4, and Z8 (Fig. 9). The family Mastacembelidae was recorded higher in Z5, Z6 and Z7 and lower in Z1, Z2, Z3, Z4 and Z8 (Fig. 10). The family Salmonidae was recorded higher in Z5, Z6 and Z7 and lower in Z1, Z2, and Z3. Medium diversity of family Salmonidae was recorded from Z4 and Z8 (Fig. 11). The highest number (249) of fish was recorded in January. Significant fluctuations were observed for

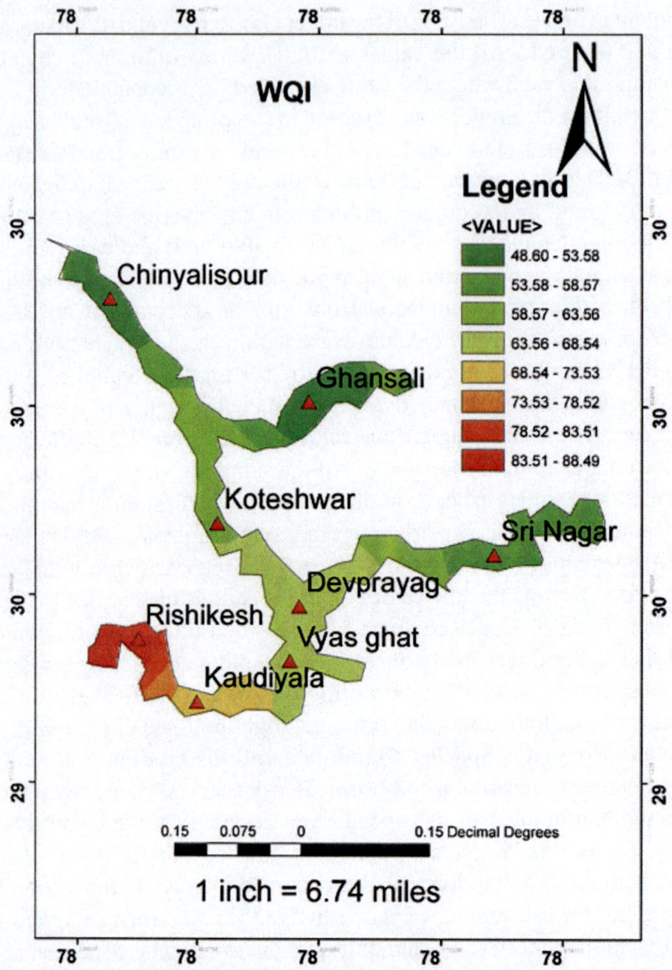

Fig. 5 Mean variation of WQI in the Ganga river and its tributaries

endemic fish families at selected sampling zones (Fig. 12). The seasonal abundance of all the sampling zones abruptly reduced from June to September. The population dynamics of endemic fish species and their relative abundance depend upon several factors i.e. water temperature, water velocity, nature of substratum, depth of water, accessibility of food, water quality and river length. Dissolved oxygen showed the positive correlation and turbidity showed negative correlation with endemic fish species in upper Ganges riverine ecosystem (Fig. 13). The declining trend of diversity and distribution of endemic fish species is due to many factors i.e. diverting the river in the tunnel at hydropower project, damming the river, chocking the migratory routes etc. During the present experimental study, a total of 31 endemic fish species was recorded in the upper Ganges riverine system and its tributaries. Among them

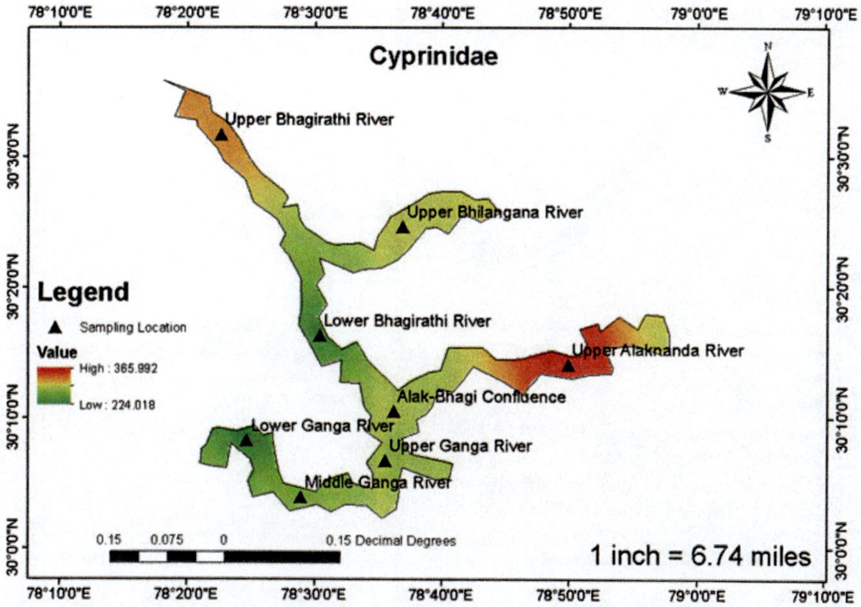

Fig. 6 Mean variation of Cyprinidae family in the Ganga river and its tributaries

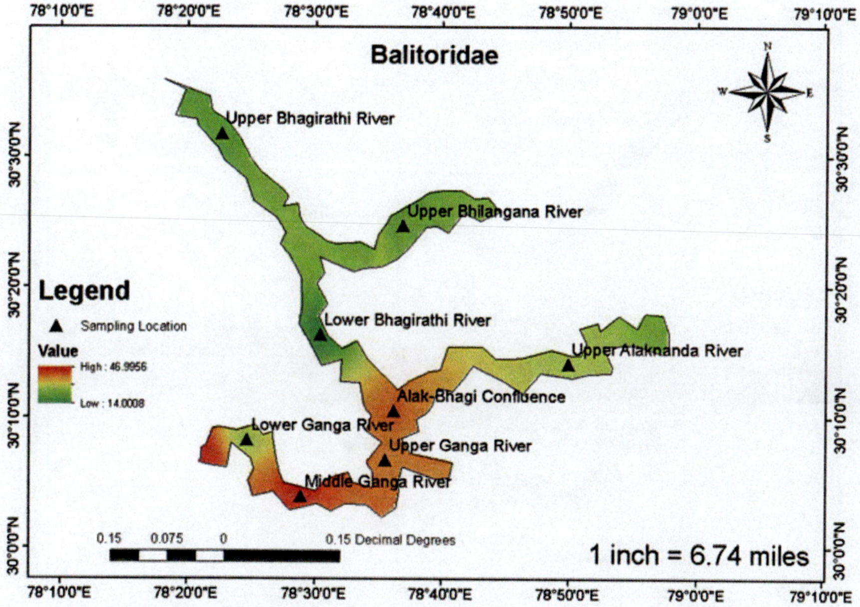

Fig. 7 Mean variation of Balitoridae family in the Ganga river and its tributaries

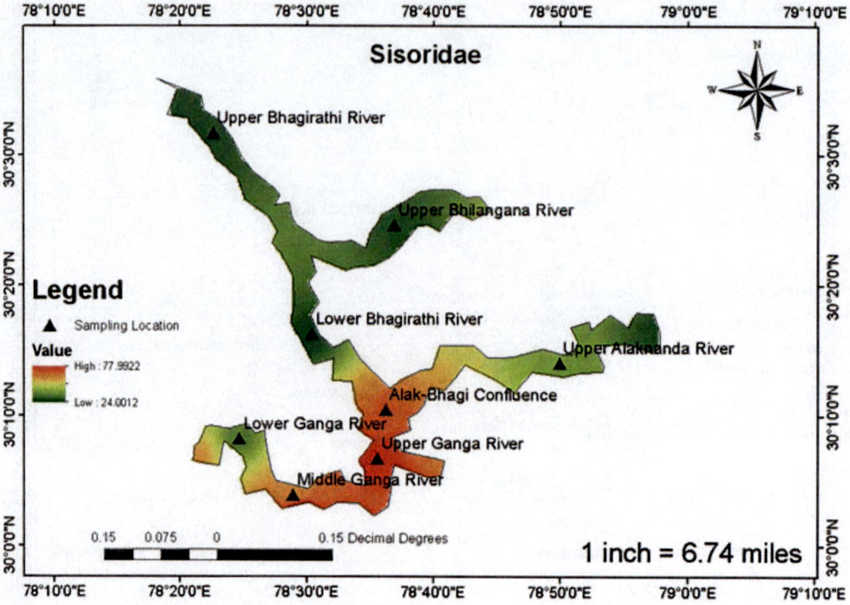

Fig. 8 Mean variation of Sisoridae family in the Ganga river and its tributaries

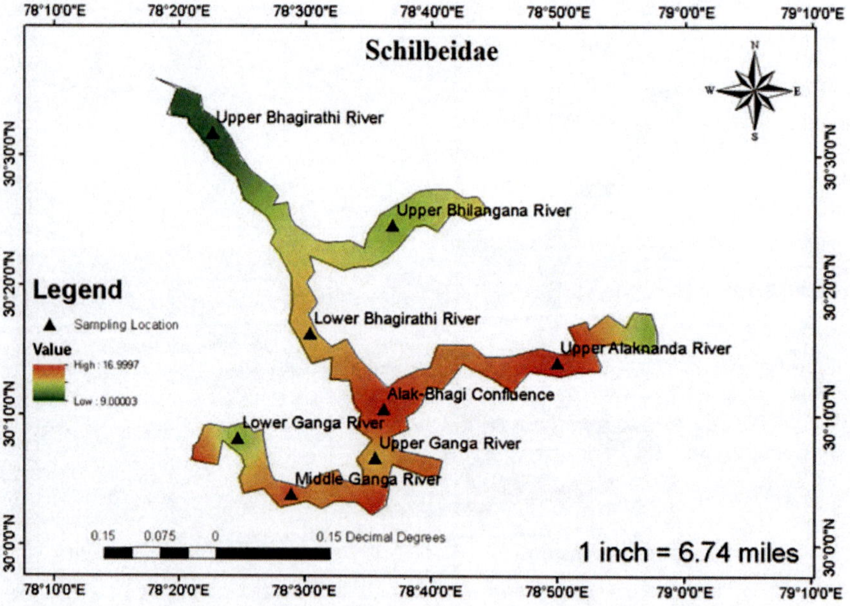

Fig. 9 Mean variation of Schilbeidae family in the Ganga river and its tributaries

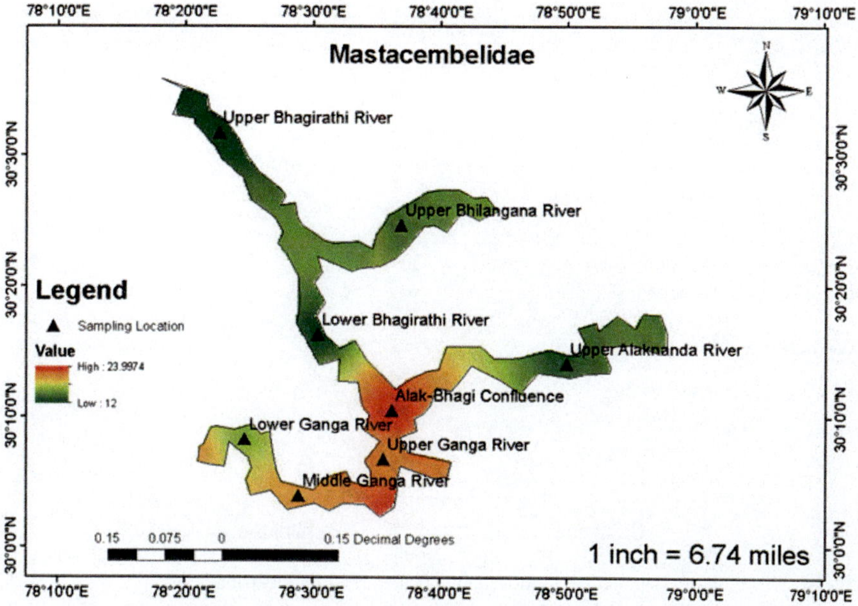

Fig. 10 Mean variation of Mastacembelidae family in the Ganga river and its tributaries

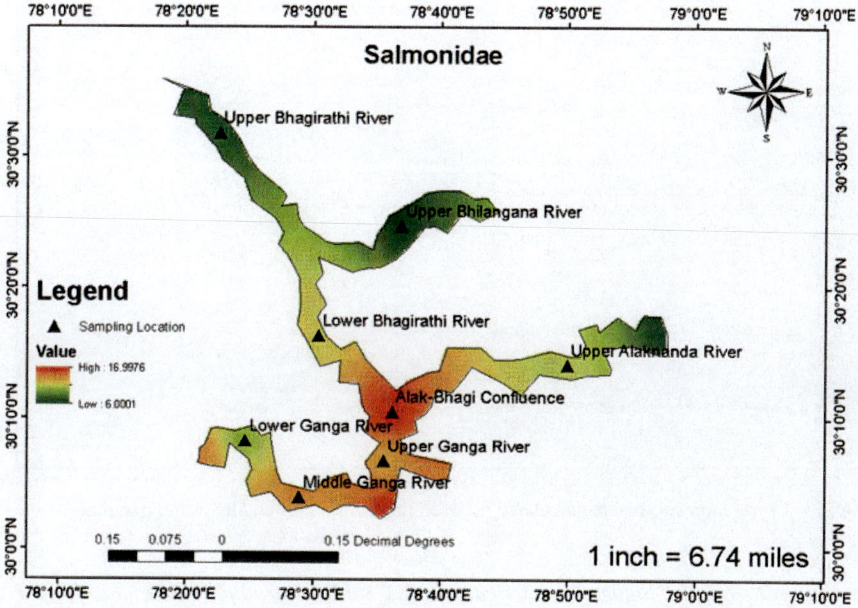

Fig. 11 Mean variation of Salmonidae family in the Ganga river and its tributaries

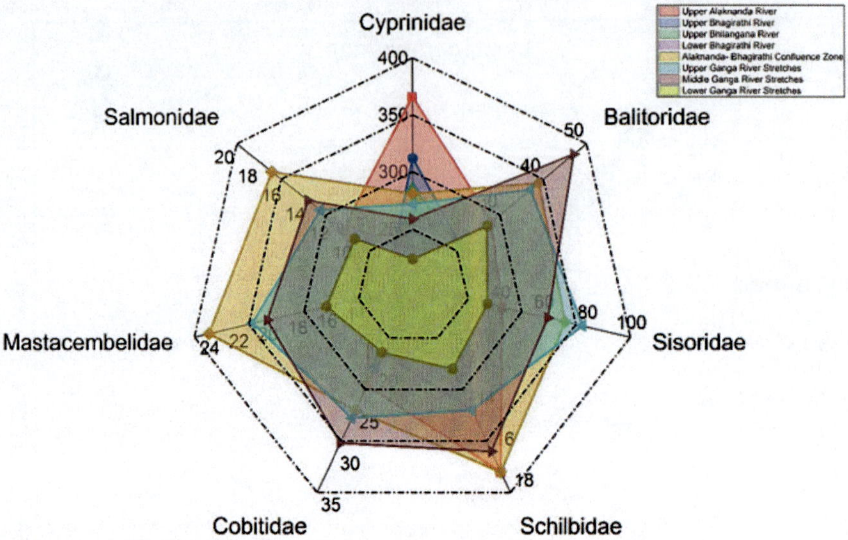

Fig. 12 Variation of endemic fish families in the Ganga river and its tributaries

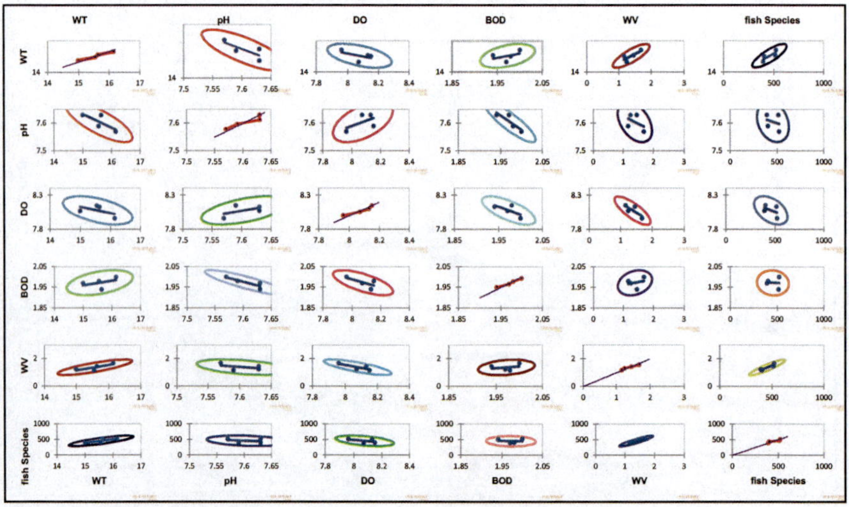

Fig. 13 Correlation coefficient calculated between the fish species and the water parameters

is *Cyripnius carpio*, *Schizothorax richardsonii*, *Schizothorax plagiostomus*, *Tor tor*, *Tor putitora* and *Barilius bendelisis* are dominant species. Sharma et al. [25] reported 20 endemic fish species from the Bhagirathi river, where Khanna et al. [18] detected 53 endemic fish species from the Ganga river. Reference [29] reported 22 endemic fish species from the river Bhilangana (a major tributary of the Bhagirathi river). An unscientific fishing method used by local fishermen is also an important reason for the loss of biodiversity (Table 2). The main reason for the reduction in endemic fish species diversity depends upon various natural and anthropogenic factors (Table 3). The highest number of individuals was observed at Zone 1 and Zone 2 because of less human interference and optimum ecological condition and on the other hand lowest number of individuals observed at Zone 4 due to extreme environmental interference. Three endemic fish species were observed dominantly in the upper Ganga and its tributaries which is similar to numerous studies [29, 30]. Shannon–Wiener diversity index ponders the lushness and proportion of species while evenness and dominance indices denote the relative number of individuals in the sample and the fraction of common species respectively. The biodiversity index values (H^0) obtained from the present study are not so very high according to Shannon–Weaver biodiversity index values and they do not exactly show the differences occurring among the stations. According to Ref. [31] the purpose for lower fish biodiversity is due to the fishing gears used. Different conservation strategy for endemic fisheries developments are presented in Table 4.

3.1 Diversity Status

The Highest Simpson index (1-D) (0.9587) was recorded at sampling zone 4 and the lowest (0.9120) at Z1 and Z2. Higher Simpson index (1-D) values were recorded in the month of April (0.9527) while lower in the month of December (0.8785). The Highest Shannon index (H) (3.181) was recorded at Z4 and Z5 and the lowest (2.8689) at Z1 and Z2. Higher Shannon index (H) values were recorded in the month of April (3.110) while lower in the month of December (2.535). The highest evenness value (0.8825) was recorded at Z4, Z5, Z6 and the lowest (0.6561) were recorded at Z1 Z2 and Z8. The highest evenness value (0.8303) was recorded in the month of April, May and the lowest (0.6086) in the month observed in December. The Highest Margalef index (4.362) was recorded at Z3, Z6 and lowest (4.121) at Z1, Z2, Z8. Higher Margalef index values were recorded in the months of April and May (5.214) while low in the month of December (3.910).

4 Conclusion

The water quality of the upper Ganga riverine system and its tributaries are very much capable and supportive to sustain the diversity and distribution of aquatic biodiversity. From the present experimental research study, it can be concluded that the water quality of upper Ganges in all eight sampling zones was good except in sampling zones 7 and 8 was inclining to slightly polluted class. The endemic fish diversity of the upper Ganges and its tributaries show seasonal and altitudinal variation in selected sampling zones due to various factors mainly damming the upheld water

Table 2 Fish and fishery survey observed along the upper Ganga river and its tributaries

Parameters	Alaknanda river Basin	Bhagirathi river Basin	Bhilangana river Basin	Ganga river Basin
Average length (km)	35	36	21	44
Fishing spots in the area	Ramol Gaon, Jogiyara Goan, Bandra Koti, Baldogi Gaon	Nakot, Jhinwali, Godi siran, Panch Koti, Padiyar Gaon	Bhatkot, Fafaruwan Gaon, Pendhars, Ali Tali	Gwalna, Siraun, Talya Gaon, Jakher, Gusai Gaon
Fishing practices	a. Cast net b. Gill net c. Hammering d. Electric shock e. Single hand operation	a. Cast net b. Gill net c. Hook and lines d. Ichthyotoxic plants e. Electric shock	a. Cast net b. Gill net c. Hook and lines d. Bleaching powder e. Ichthyotoxic plants	a. Cast net b. Gill net c. Electric shock d. Single hand operation e. Electric shock
Main fishing season	Winter to pre-monsoon	Whole year except breeding season	Winter to pre-monsoon	Winter to pre-monsoon
Fishing composition	*Schizothorax*, *Mahseer*, *Garra*, *Glyptothorax*, *Pseudchenies* etc	*Schizothorax*, *Mahseer*, *Labeo*, *Glyptothorax*, *Catla Cyprinius carpio*(in impoundemt segment) etc.	*Schizothorax*, *Mahseer*, *Labeo*, *Glyptothorax*, *Cyprinius carpio*(in impoundemt segment) etc.	*Mahseer*, *Garra*, *Barilius*, *Namachelius*, *Cyprinius carpio* etc.
Number of fishermen	9–12	15–20	15–18	10–15
Fishery community	Local	Local	Local	Local

Table 3 Various categories of causative and unscientific fishing practices observed along upper Ganges and its tributaries

Causes of depletion of fish fauna in Bhagirathi river basin								
Over exploitation of fish fauna	Lack of policy/law implementation	Deforestation	Introduction of exotic fish species (in reservoir)	Weed infestation	Construction activities	The fishing festival (Maund)	Pollution	Unawareness among masses
Unscientific methods of fishing i.e • Dynamiting • Hammering • Bleaching powder • Ichthyotoxic plants	• Law was not strictly followed • No check on daily catch, size of fish and mesh size of nets • No licenses were issued to the villagers	• Excessive deforestation adversely affects the fish population	• Exotic fish species grow faster than native species and adversely affect their population and growth	• The weed infestation increases the turbidity and decreases the oxygen percentage of the river, resulting in the fish mass mortality	• The construction of dam and road adversely affect the migration route of the fishes which result in destruction in breeding and spawning	• Maund-fishing festival celebrated in Jaunsar and Garhwal Himalaya. Due to this a lot of endemic fish species die which is excessive risk to the fish fauna	• Domestic garbage, • Solid waste, • Sewage disposal • Pesticides and • Insecticides	• Unplanned manner of fishing • Importance of conservation

Table 4 Different conservation strategy for endemic fisheries developments

Action oriented plan	1. Migratory channels in between upstream and downstream
	2. Maintain minimum quantity of water for fishery in downstream
	3. Construct separate breeding channels along with the main downstream river
	4. Restoration of river system habitat, feeding and breeding spawning grounds
	5. Seed stock enhancement
	6. Bio monitoring of the water quality
	7. Afforestation on river basin catchments
	8. Formation of threatened fish species conservation youth groups in local area
	9. Stop illegal fish catching practice
Future govt. oriented plan	1. Strictly implement modified fishery act
	2. Ban in fishing in pre-breeding and breeding season
	3. Declare fish sanctuaries as Reserved Protected Area Network
	4. Declare new fish conservation policy for threatened species
	5. Co-ordination committee of Central, state different departments
	6. Action oriented groups from Central, State Universities and Govt. Research Institutes
	7. Mass conservation action groups of local peoples and NGO
Status of research, gaps and identify the areas of research	1. Establishment of habitat-based brood bank for captive breeding of different Mahseer species
	2. Studies on fish behavior, migratory pattern of different species
	3. Development of appropriate protocol for nutrition and disease control for hatchery systems, fry and fingerling production units
	4. Ascertain the viability of fish ladders/fish passages for Mahseer migration
	5. Assessment of environmental flow requirements
	6. Identification of ecotourism centers, national parks and sanctuaries in potential river systems for the Mahseer
	7. Developing site-specific in-situ and ex-situ knowledge-based Mahseer conservation strategies in potential river systems
	8. Conducting studies on Mahseer population dynamics in important river basin stretches
	9. Developing a national angling protocol for game fish on existing population trends
Monitoring at national, state and river basin levels	1. Development of a database along river basins at different altitudes for the purpose of long-term conservation
	2. Methodology to be devised for proper monitoring of ranching programmes
	3. Capacity and funding needs for dedicated research
	4. Training and awareness programmes for different stakeholders at the state and national levels

for hydropower purposes and various other anthropogenic factors. The water quality of the upper Ganges is still very proficient for the sustenance of diverse aquatic biodiversity. Although, some water quality parameters exhibited upsurge maximum physio-chemical was within the acceptable limits. All the selected water quality parameters were recorded slightly higher in sampling zones 7 and 8 in contrast to others. The current research has also revealed that various developmental happenings including edifice of dam and barrages, fluctuations in water discharge, choking of migratory routes, river bed mining, etc. possessed direct impact on the water quality and population dynamics of endemic fish species especially at sampling zone 4 (Lower Bhagirathi river). Yet, these sorts of happenings will be observed habitually and must be controlled according to a systematic and scientific manner.

Acknowledgements The authors are thankful to Department of Zoology and Environmental Science, Gurukula Kangri (Deemed to be University) Haridwar" for providing the laboratory facilities. The authors also acknowledge to the local fishermen for their helps in fish sampling during the study.

References

1. Tare, V., Yadav, A. V. S and Bose, P.: Analysis of Photosynthetic Activity in the Most Polluted Stretch of River Ganga. Water Research, 37(2003) 67–77.
2. Nautiyal, P., Agarwal, N. K. and Singh, H. R.: Certain abiotic components of the fluvial system and their interrelationships as evidenced by statistical evaluation. In: Advances in Limnology, edited by HR Singh, Narendra Publishing House, Delhi. (1993). pp. 181–186.
3. Malik, D.S., Sharma, A.K. and Bargali, H.: Status of phytoplankton diversity in relation to water quality of Bhagirathi riverine system in Garhwal Himalaya. International Journal of Advanced Science and Research, 3(1) (2018). 30–37.
4. Das, M. K.: Environment and fish health: a holistic assessment of Inland fisheries in India. In: Goswami UC (ed) Natural and anthropogenic hazards on fish and fisheries. Narendra Publishing House, Delhi, (2007) pp 137–151
5. Lakra, W. S., Sarkar, U. K., Gopalakrishnan, A. and Pandian, A. K.: Threatened freshwater fishes of India. NBFGR publication, Lucknow. (2010).
6. Hossain, M. S. Das, N. G. and Chowdhury, M. S. N. Fisheries Management of the Naaf River Chittagong, Coastal and Ocean Research Group of Bangladesh, (2007) 257 pp.
7. Payne, A. I., Sinha, R. K., Singh, H. R. and Haq, S.: A review of the Ganges Basin: its fish and fisheries. In: Welcomme RL, Peter T (eds) Proceedings of the second international symposium on the management of large rivers for Fisheries, vol 1, FAO Regional Office for Asia and the Pacific, Bangkok, Thailand, (2004) pp 229–251
8. Hamilton, F.: An account of the fishes found in the river Ganges and its branches. Edinburgh and London. (1822).
9. Malik, D. S.: Population dynamics and conservation management of Himalayan mahseer (*Tor species*) in riverine aquatic ecosystem in Garhwal region of Uttarakhand (India). Journal of Applied and Natural Science, 3(1) (2011) 9–101.
10. Sarkar, U. K., Gupta, B. K. and Lakra, W. S.: Biodiversity, ecohydrology, threat status and conservation priority of the freshwater fishes of river Gomti, a tributary of river Ganga(India). Environmentalist, 30(1) (2010) 3–17.
11. Sinha, R. K.: The Ganges river dolphin *Platanista gangetica*. Journal of Bombay Natural History Society, 103(2006) 254–263.

12. Malik, D. S., Sharma, M. K., Sharma, A. K., Kamboj, V. and Sharma, A. K.: Anthropogenic influence on water quality and phytoplankton diversity of upper Ganga basin: A case study of Ganga River and its major tributaries. World Water Policy, 07 (2021) 88–111.
13. Day, F.: The Fishes of India: Being a Natural History of the Fishes known to inhabit the Seas and Fresh water of India, Burma and Ceylon. William Dawson & Sons Ltd., London. (1876) 778p.
14. Talwar, P. K. and Jhingran, A. G. Inland Fishes of India and Adjacent Countries (Vol. 1 & 2). Oxford and IBH Publishing Co. Pvt. Ltd., Calcutta. (1991) 1158p.
15. Shannon, C. E. and Weaver, W. The mathematical Theory of Communication, University of Illinois Press, Urbana, USA. (1949).
16. Tyagi, D. and Malik, D. S. Status of phytoplankton diversity and biomass in relation to productivity of Ramganga reservoir at Kalagarh (Uttarakhand). International Journal of Fisheries and Aquatic Studies, 5(3) (2017) 430–434.
17. Khanna, D. R., Bhutiani, R., Matta, G., Singh, V. and Ishaq, F.: Physico-chemical and microbiological status of River Asan in Dehradun Uttarakhand. Environment Conservation Journal, 13(1&2) (2012) 145–150.
18. Khanna, D. R., Bhutiani, R. and Mukesh, R.: Fish Diversity and Their Limnological Status of Ganga River System in Foothills of Garhwal Himalaya, Uttarakhand, India. Journal of Environmental Research and Development, 7(4) (2013) 1374–1380.
19. FMEnv.: National Environmental (Surface and Groundwater Quality Control) Regulations, S.I. No. 22, Gazette no. 49, vol. 98. Abuja: Federal Ministry of Environment. (2011).
20. Halim, M. A., Sharmin, S., Rahman, H. M. H., Haque, M. M., Rahman, M. S. and Islam, M. S.: Assessment of water quality parameters in baor environment, Bangladesh: A review. International Journal of Fisheries and Aquatic Studies, 6(2) (2018) 263–269.
21. Kamboj, V. and Kamboj, N.: Spatial and temporal variation of zooplankton assemblage in the mining-impacted stretch of Ganga River, Uttarakhand, India. Environmental Science and Pollution Research, 27(2020) 27135–27146.
22. Malik, D. S., Sharma, A. K. and Sharma, A. K. Current Status of Microbenthic Diversity And their Habitat Ecology in River Ganga and its Tributaries, Uttarakhand. Journal of Experimental Zoology, India, 23(2) (2020) 1413–1420.
23. Kale, V. S.: Consequence of temperature, pH, turbidity and dissolved oxygen water quality parameters. International Advanced Research Journal in Science, Engineering and Technology, 3(8) (2016) 186–190.
24. Malik, D. S., Sharma, A. K. and Sharma, A. K.: Status of Zooplankton Diversity, Abundance in Relation to Hydrography of Ganga River and its Tributaries at Rishikesh. Indian Journal of Ecology, 47(3) (2020) 739–745.
25. Sharma, A. K., Malik, D. S. and Bargali, H.: Present status of fish diversity and population abundance of selected fish species in Bhagirathi river at Uttarakhand. International journal of creative research thoughts, 6 (1) (2018) 432–438.
26. Kamboj, V., Kamboj, N. and Sharma, A. K.: A review on general characteristics, classification and degradation of river systems. In: Environmental Degradation: Causes and Remediation Strategies, Volume 1, Eds. Kumar, V., Singh, J. and Kumar, P., (2020). pp. 47–62.
27. Kumar, V, Malik, D. S. and Sharma, A. K.: Study on Distribution Pattern of Benthic Diversity in Relation to Water Quality of River Ganga And Its Tributaries., Journal of Emerging Technologies and Innovative Research, 5(5) (2018) 88–103.
28. Nwankwo, C., Mohammed, A., Ikyereve, R. E. and Dawari, B. K.: The Impact of Human Water Exploitation on Physico-Chemical Characteristic of Mmubete River in the Niger Delta, Nigeria. International Journal of Science and Technology, 3(5) (2014) 292–297.
29. Agarwal, N. K, Singh, G., Singh, H.: Present status of Ichthyofaunal diversity of Garhwal Himalayan river Bhilangana and its tributaries with reference to changing environment. Environment Conservation Journal, 12(3) (2011) 101–108.

30. Sharma, A. K., Malik, D. S. and Tyagi, D.: Food and feeding habits of *Tor tor* (Hamilton, 1822) and *Schizothorox richordsonii* (Gray, 1832) inhibiting Bhagirathi river, Tehri Garhwal, India. International journal of Science and Nature, 10(1) (2019) 97–103.
31. Keskin, C. and Unsal, N.: The Fish Fauna of Gokceada Island, NE Aegean Sea, Turkey. Italian Journal of Zoology, 65(1998) 299–302.

Assessment of Land Use Change and Climate Change Impact on Biodiversity and Environment

Arvind K. Sharma, Amit K. Sharma, Munit Sharma, and Munish Sharma

Abstract *Aim* Climate change and land use land cover change LULC, are the supreme pressures on biodiversity that can straightly impact each other. The land use land cover change has strong impact on local climatic conditions such as anthropogenic altered regions leads to more warmer and drier as compared to natural habitats. The several impacts on biodiversity due to anthropogenic factors can be developing area of research. The present state of approach is on the impact of local climatic changes on biodiversity that responds to land-use land cover changes. *Location* Global. *Methodology* We review published research studies by searching procedure on google using keywords on different databases. The strength of this methodology is that it delivers a rapid synopsis of intensity on Climate change and land use change in relation to biodiversity trends across the globe. *Results* A total of 1223 published research items from 2000 to 2022 were extracted from searching on online databases based on scenarios of biodiversity due to climate change and land use changes. Out of 1223 published research items we randomly sampled research items within the list (sample size N = 250). We collected information critically from each of the above selected studies on spatial and temporal scales on different geographical locations. It revealed that natural and anthropogenic factors resulted into loss of biodiversity and its services. *Conclusion* Climate change and Land use change are both major contributing factor and are a main mode in which the impacts of climate change are expressed. As a contributing factor, land use influences the change in land cover patterns. Projected climate variations resulted in changes land

A. K. Sharma
Department of Zoology and Environment Science, Gurukula Kangri (Deemed to be University), Haridwar 249404, India

A. K. Sharma
Department of Animal Science, Central University of Himachal Pradesh, Dharamsala, India

M. Sharma
Department of Botany and Microbiology, Gurukula Kangri (Deemed to be University), Haridwar 249404, India

M. Sharma (✉)
Department of Plant Sciences, Central University of Himachal Pradesh, Dharamsala, India
e-mail: rs.munish.cuhp@gmail.com

K. D. Bahukhandi et al. (eds.), *Environmental Pollution and Natural Resource Management*, Springer Proceedings in Earth and Environmental Sciences,
https://doi.org/10.1007/978-3-031-05335-1_5

cover patterns at a variety of spatial and temporal scales. Therefore, it is required to achieve ecological functions efficiently on local, regional and global scales.

Keywords Biodiversity · Climate change · Land use change · Environment · Anthropogenic factors

1 Introduction

Biodiversity globally faces increasing pressures from climate change, invasive species, habitat loss, and environmental pollution [1, 2]. Ecosystem services reflect and describe complex relationships between human society and the environment have become an important subject of research in recent years. Global ecological changes are predicted to have a major effect on human civilization and culture which include ecological, economic, social as well as political aspect in the coming next 10–20 years. The ecological impacts are migration, productivity, biodiversity change and sustainability. In the near coming future, it is predicted that land use and climate change are among the two-main global ecological change. Human induced climate change and land use activities causes a large impact on biodiversity [3]. Rapidly increasing growth in human population has led to increasing exploitation and conversion of agricultural and forest land for human use and similarly environment is being degraded [4]. Sampaio et al. [5] reported that the conversion of primary natural and undisturbed habitats to agriculture and urban areas dominated land uses by human resulted to changes in vegetation structure and land cover which can impact directly on local climatic conditions of the region [6]. Human inhabited areas of converted land for human use are warmer and drier as compared to natural habitats. Biodiversity is affected by land use change caused by climate change, either it affects directly by change in precipitation, temperature, moisture levels and indirectly through changes in vegetation structure, resource availability resulted into alterations in community composition [7, 8]. The earth is undergoing extensive biodiversity losses, although the richness of species or abundant communities size is increasing or declining at regional and local scales [9, 10]. Climatic conditions are affecting wide landscapes due to rise or fall of average annual temperature and precipitation that led to their extreme values may become stronger and their frequency of extreme climatic events may increase [11]. Land use change have a significant role in the climate system ranging from the local regional scale to the global scale. Land use change causes biodiversity loss by disturbing every single biome on planet Earth [12]. It has been found that land use change either directly or indirectly are major pressures impacting biodiversity negatively through biofuel production mostly when natural grasslands, tropical and subtropical forests ecosystems with high biodiversity values are changed into biofuel plantations [13]. As the impacts of climate change and land use change intensify resulted into amplified the major threats to biodiversity [14, 15]. In this chapter, we have highlighted on the two major global change drivers such as climate change and land use change and make an effort to estimate their interactive effects on biodiversity. The impacts on biodiversity caused by the climatic change and habitats/land uses changes with land conversions set to continue making influence on native climatic conditions that ultimately effect on biodiversity and therefore, in order to predict forthcoming biodiversity changes and set up suitable management and conservation plans [16, 17].

2 Methodology

2.1 Data Collection

The data had been collected by finding the best available understanding and knowledge reported by the scientific community across the globe. We had screened and selected the academic peer reviewed published literature from 2000 to 2020. These publications were identified by using keywords like climate change, Land Use Change, Land Use Land Cover (LULC), Biodiversity, Environment, species richness or abundance, diversity indices, habitat loss, etc. from renowned databases such as Scopus, ISI, ResearchGate, Google Scholar, Web of Science. Original research article, review article and book chapters had been collected and selected for studying abstract and methodology and duplicate copies were excluded from study to avoid double counting and errors.

A total of 1223 published research items from 2000 to 2022 were extracted from searching on online databases based on scenarios of biodiversity due to climate change and land use changes. Out of 1223 published research items we randomly sampled research items within the list (sample size N = 250). After carefully checking the title and abstracts we had assign each of them into one of the following categories:

1. Published research items reporting scenarios of biodiversity based only on climate change predictions.
2. Published research items reporting scenarios of biodiversity based only on land use/land cover changes (LULC) predictions.
3. Published research items reporting scenarios of biodiversity based on the use of both climate and land use/cover change (LULC) predictions.
4. Published research items reporting scenarios of biodiversity based on the other types of environmental change predictions.

Available predictions on Land Use Land Cover (LULC) and Climate Changes were used to estimate and assess the future environmental conditions for biological diversity across species rich variety of spatial and temporal scales. Some reported study had predicted the consequences of expected Climate changes on biological diversity [18, 19]. We collected information critically from each of the above selected studies on spatial and temporal scales on different geographical locations. For specificity, ecological effects were grouped into 4 different temporal zones of five years gap from in 20 years; 2000–2005, 2205–2010, 2010–2015, 2015–2020 and similarly at the spatial scale, they were grouped into 4 regions: patches (1–100 ha), landscape (101–100,000 ha), regions (>100,000 ha) and a large global.

2.2 *Data Analysis*

We had a critical analysis of published research items derived from online databases on climate change and land use change. Firstly, we had assessed ecological responses to land use change and climate change in relation to biodiversity. Secondly, to understand the interactive effect, quantitative analysis based on interaction mechanism between both Climate change and land use change and land cover change.

3 Results

Biodiversity has a significant role in providing ecosystem functions and services. It is experiencing a substantial decline globally due to anthropogenic factors induced environmental changes. We had compiled the result on reviewing the published literature from year 2000 to 2020. We had calculated for each year between 2000 and 2020 by random sampling among published articles, the amounts of studies assigned to each of the 4 categories (Fig. 1). We selected and estimated these

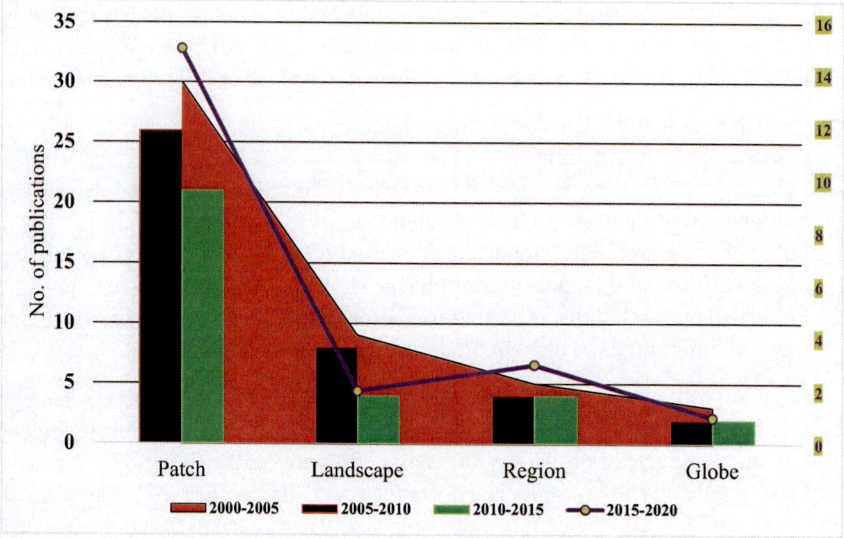

Fig. 1 Temporal and spatial scale investigation after observation of peer review research items reported on effect of climate change and land use change on biodiversity

round proportions from these samples of published articles and the total number of 1223 published research items extracted from the google searching methodology to estimate the yearly numbers of published articles during 2000–2020 that had been primarily reported category 1—was on development of biodiversity situations and using climate change predictions; category 2—was on land use land cover change (LULC) predictions and category 3—was on both types of environmental change predictions. Our extensive survey reported that the number of studies included on expected influences of impending land use land cover change (LULC) on biodiversity falls behind in comparison to studies included great focus on the influences of future climate change. Among all the published research studies during the past period 2000–2020; those depicted upon one of these two driving forces to affect biodiversity scenarios we evaluated that 87% focuses on the use of climate change projections alone and that 5% used only projections of land use land cover changes (LULC). Both combined climate change and land use land cover changes (LULC) projections resulted 8% in the studies.

3.1 Ecological Responses

Biological Diversity is greatly altered by climatic variations that includes changes in evaporation, temperature, radiant flux, precipitation, carbon emissions, and the increasing frequent extreme climatic events. The responses of living organism in relation to climate change have gained immense attention at local, regional, and global scales [20]. The increasing rate of climate change continues to alteration at an increasing rate and in relation to it living organisms and ecosystems respond to it accordingly that depends on numerous factors from species, community and ecosystem levels [21]. Therefore, it resulted into alteration of distribution and abundance of species, range shifts and interspecific relations. Most clearly, distribution and ranges of all living organisms are potentially get affected by climate change, in context to it, current geographic distributions are helping in prediction of shifting ranges of species [22–24]. When the changing ecological conditions are suitable for species they are able to adapted to climate change. It has been confirmed along altitudinal and latitudinal gradients where shifts are allowed due to dispersal mechanism and availability of natural resource [25–27]. It has been also reported that the shift is not fast enough to equals the pace with climate change in plants whose ability to move seems too conservative as compared to moveable animals. The new communities established due to shifting in response to climate change are novel and dynamic until the climate returns to stability because some species vary in their capacity to movement [28, 29]. Impacts on biological parameters of living organisms brings out major changes in composition and structure of community [30]. Biological Diversity have a vital role in the establishment of various ecosystem functions and its services [31–33]. The result compiled is supported by a substantial literature reported positive correlation between biodiversity and ecosystem services [33]. It had been found that from the year 2010 and onwards the ecosystem services related to biodiversity has grown rapidly. Approximately 80% of studies were conducted in North America, Europe, and the remaining 20% of studies conducted in Asia, Africa and Oceania. The functional significances of loss of species can be mitigated or intensified by the instantaneous effect of species richness and their relative abundance [34]. It has been reported that due to rapid climate change colder regions become warmer and arid regions obtain increased precipitation support a greater diversity of species. It has been seen in maximum ecosystem processes, the impact of biodiversity is non-linear, and variation may accelerate as increases in biodiversity loss take place [35, 36].

When impact of climate change is compared with land use land cover change (LULC). It has a massive impact on urban green spaces, recreation, greenhouse gases, agricultural production species and genetic diversity [37]. The adaptive responses shown by species show complex scenarios of land use change in response to climate change. Land use land cover change has been used in forecasting future changes in habitats, detecting invasive species, estimating potential extinctions of species in biodiversity hotspots, and stress on the restricted potential of expansion of protected areas [38]. The most common response at the species level to LUCC is increasing extinction rates of species. Species with less ability to adapted to an area are more expected to become threatened or extinct when the available area is reduced due to habitat destruction and fragmentations (Fig. 2). Species placed at higher trophic levels are more vulnerable to extinction due to from habitat fragmentation [39].

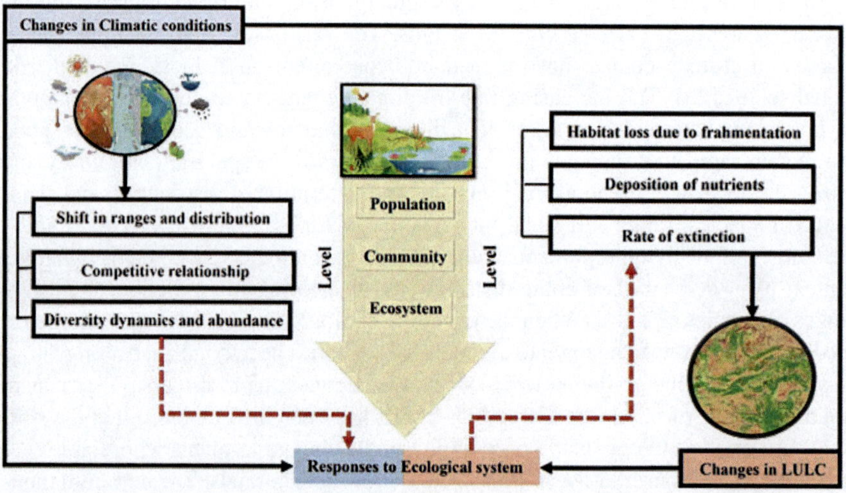

Fig. 2 Impact of climate change and land use land cover change on ecological responses

3.2 Climate Change and LULC Interaction Mechanisms

Climate change and land use land cover change (LULC) both together interact make huge effect on biodiversity through many pathways. Climate change can hamper the ability of living species to deal with land use land cover change and it can reduce resilience to climate change [40]. A meta-analysis study resulted that on interactions between effects due to habitat loss and climate change effects concluded that the effects of habitat loss were seen greatest in those regions with higher mean temperatures and decreased mean precipitation [41] and therefore land use land cover change must be a major concern on climate change Some of the interactive mechanisms that reported showed interactions between climate change land use land cover change (LULC) in several illustrative pathways and eventually affecting the biodiversity (Fig. 2). In future extreme climatic events such as droughts, floods, and hurricanes, are expected more frequent and with high intensity [42, 43]. The potential of living organisms to adaptation to a particular habitat is influenced by both the severity and frequent extreme climatic events [44]. It has been reported a process which is intensified by climate changing under extreme weather conditions may push a plant species closer to the edge of its fundamental niche, ultimately reducing its competitive ability and allowing other plant species to become dominant in the space. In some places cold living species being replaced by warm living species due to increase and frequent incidence of extreme weather events. But most research carried out on climate change effects has focused mostly on the significances of increased average temperatures as compared to the impacts of extreme climatic events. The responses to increasing in extreme climatic events impact functioning of plant species across spatial and temporal scales and also due to these extreme events plant water relations will be very vulnerable. Plant physiological processes may get strongly affected by flooding and changing mean climate [45].

3.3 Effects of Rising CO_2 Concentrations Mechanisms

12% of anthropogenic carbon emissions is contributed by alone land use land cover change (LULC) [46]. Land-use related variables such as changing vegetation types, extensive losses of forest, green cover vegetation and higher rates of decomposition are closely associated to rising CO_2 concentrations. Complex deviations linked to carbon emissions have involved wide changes in the structure, quantity, and land-use types spatial pattern in the soil food web. The substantial land use land cover change (LULC) employs a direct significant impact on soils, and indirect impacts arising from anthropogenic activities such as heavy metal pollution, acid deposition, and industrialization [47, 48]. In contrast, many plants species are getting benefit from increased in CO_2 concentration through increasing photosynthesis and decreasing water stress [49]. Early life stages of marine organisms in coastal marine systems are potentially susceptible to the stresses caused with global climate change. It has

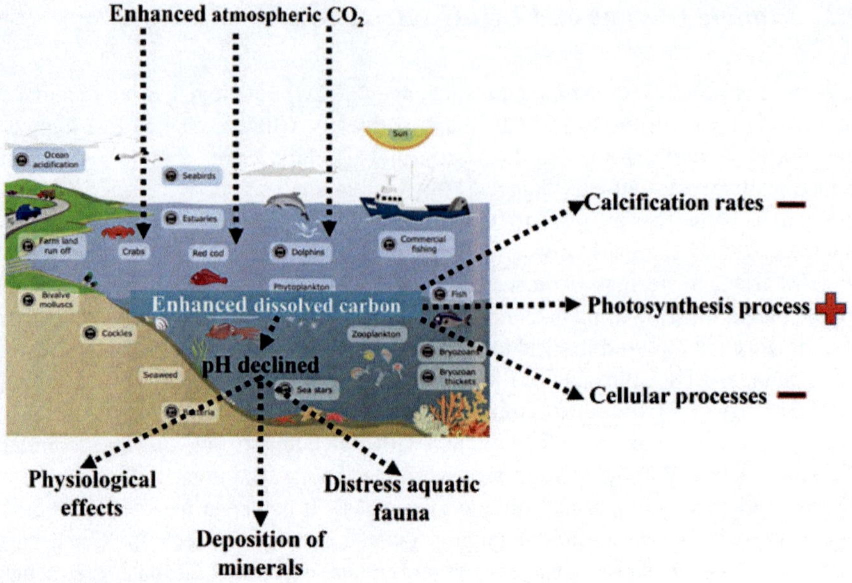

Fig. 3 Impact of CO_2 enrichment on aquatic biodiversity in marine ecosystem

been noticed under enhanced CO_2 rates of calcification in coralline red algae and corals ecosystems are likely to drop significantly. Due to increase dissolved CO_2 concentrations physical and chemical changes cascade is appeared (Fig. 3). Some experimental output showed that increasing CO_2 concentrations can resulted into decreases protein synthesis and ion exchange subcellular processes. Also, acidification of the marine ecosystem could be anticipated to have severe negative impacts on marine invertebrates and algae [50, 51].

3.4 Anthropogenic Factors with Potential Impact on Biodiversity

3.4.1 Land Use/Land Cover Change

Land is the platform on which every anthropogenic activity is being done and also it is the only source of materials for the conduct of these activities. Land Use is resulted from extensive human use of land resources for their own purposes and survival such as conversion into agriculture land for food production, road constructions, construction of houses for shelter, railway construction, dams and another project construction, etc. Land cover in a particular area refers to the natural vegetation cover and are reflection of the local climate and land forms. Some of the wide land cover categories include forest, tundra, savannah, desert or steppe that can be subdivided into categories representing specific plant communities such as oak-pine scrublands, mangroves, seasonally flooded grassland, etc. Anthropogenic factors are found increasingly as a dominant factor in global environmental change. Land use Land cover change, specifically when the forested areas are converted into other human uses, has been found to be major contributing factor that led to global climate change, that accounts for about 33% increase in atmospheric carbon dioxide (CO_2) and a main factor in the loss of biological diversity. Agricultural practices, forest fire, soil erosion and overgrazing are factors causing land degradation and desertification. Land use land cover change drivers are classified into two major categories: Direct and Indirect drivers. The main direct human drivers include: land use and land cover change due to global increased inlands to agriculture and grazing, Invasive species introductions/removals. Use of fertilizers and pesticides in agriculture, environmental pollution and climate change. The main indirect human drivers include: increased demographic change economic, scientific and technological, sociopolitical, cultural and religious factors. Land use land cover change is one of the significant changes that have a distinct relevance to biodiversity maintenance in the particular landscape. Land use land cover changes occurring at various spatial levels and within various time periods are resulted from environmental and human dynamics mediated on land. Land use land cover changes (LULC) can have both positive and negative impact on natural biodiversity. Negative outcomes come from the most intensive land use, grazing a wide cropping area of natural vegetation. It had found that the impact on natural biodiversity of a land use change is mainly related to alteration the extent, pattern and quality of native vegetation (Fig. 4).

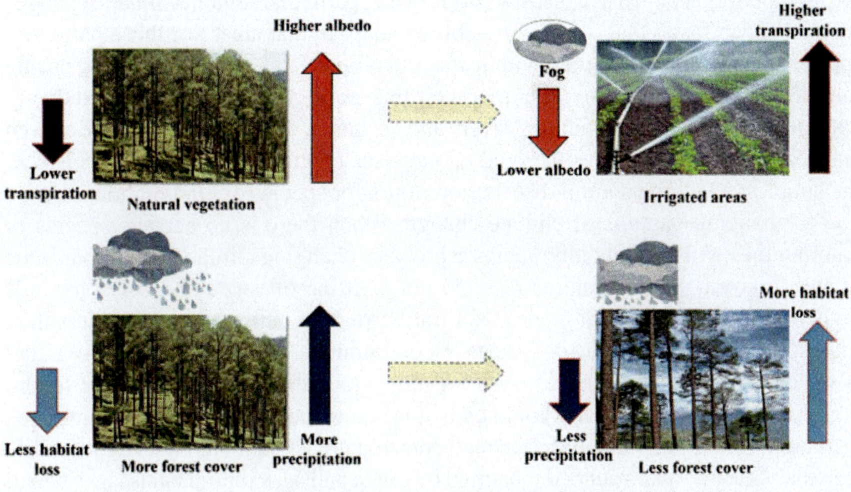

Fig. 4 Changes in climatic conditions due to anthropogenic influence

3.4.2 Climate Change

Climate change is found to be caused by induced human activities that led to increase in the amount of greenhouse gases in the atmosphere that ultimately enhancing the natural greenhouse effect. Of Land use changes, fossil fuels combustion and cement production and have led to increased CO_2 concentration by 30%. Greenhouse gases are also increasing in concentration in the atmosphere due to anthropogenic activities such as use of landfills, agriculture practices and rice/paddy cultivation, animal rearing, use of fossil fuel burning and industrial production. These greenhouse gases include methane (CH_4), which is found to be more than doubled amount since pre-industrial times and similar trend is followed by nitrous oxide, Sulphur dioxide and ozone led to increase in global mean surface temperature and global warming. The effects that the global mean change in temperature may have effect on local and regional scales that are extremely variable and uncertain due to the influence of atmospheric circulation and ocean bodies. In addition to these changes there is possibilities in the increase in the frequencies of irregular seasons and extreme events that includes forest fires, hurricanes and droughts, mean climatic conditions such as temperature and rainfall, are likely to change and in some regions. The introduction of invasive species, land conversion, and other related threats to biodiversity will affect even remote wilderness areas that have experienced a slight anthropogenic change. There are major issues that likely to impact on biodiversity such as land use land cover and climate change. It has been found that as there is change in global climate, ecosystems are responding to changes by enhancing temperature, precipitation and CO_2 concentrations in the atmosphere. In respond to these changes that can be positively favour some of the species and can also negatively affect other species result into competition that led to cause invasions in that particular region. Climatic changes would happen faster than species are able to adapt in that area. For those particular species that are able to migrate with respect to climate change, there is a risk during migration from one place to other that their migration routes will be blocked due to natural barriers such as mountains, rivers and oceans as well as anthropogenic altered landscapes. The ultimate result could be large scale extinctions of these species. At the simplest level, the natural distribution limits for species or communities change due to changing patterns of climate change. When there is no barriers species or communities will possibly migrate in response to changing climatic conditions. Due to shift in average temperatures resulted into shifting of vegetation zones towards higher altitudes or higher latitudes. Natural barriers or anthropogenic barriers may impact the natural migration of species or communities. Due to human land use land cover change pattern natural ecosystems may squeezed and such as Arctic tundra and alpine meadows may have confined by the natural configuration of the landscape or region. It had been found that many protected areas, national parks and wildlife sanctuaries are now surrounded inhabited by urban and agricultural landscapes which will ultimately prevent the simple migration of species beyond their boundaries. It has been observed in many temperate regions that changes in seasons being takes place. In tropical and sub-tropical zones drought and desertification may increase. Negative impacts may also include increased ranges of insect pests infestation and diseases. Rainfall and drought will also have equally importance as crops failure in some regions from drought or flooding is being observed.

3.4.3 Environmental Pollution

Environmental pollution is the addition of any undesirable substance or any form of energy such as heat, sound, radioactivity or nutrients/metals to the natural environment at a much faster rate than the environment that can tolerate it by various processes through dispersion, breakdown, recycling and storage into harmless form for living organisms. A pollutant must need not to be harmful in itself, for example, Carbon dioxide (CO_2), is a normal component of the atmosphere and a by-product of respiration that is found in all animal tissues. But in concentrated form it can kill animals. Human sewage can be a useful as fertilizer, but when it is in concentrated too highly it becomes a serious pollutant by causing the depletion of oxygen in water bodies. Radioactivity is in small quantity is harmful for life. Pollution has go together with humans ever since groups of people persisted for a long time in any particular place. With the permanent human settlements establishment by large numbers of people, still, pollution became a problem in cities made poisonous places, contaminated by human wastes and debris resulted into unsanitary conditions in urban areas which favored the outbreaks of epidemic diseases. Also, air, water pollution and solid wastes accumulation were mostly the major problems of large cities. Increase of advanced technology as well as rapid spread of industrialization and the rapid increase in human populace to extraordinary levels. Hence, pollution has become a major universal problem.

3.4.4 Desertification

Climatic changes can cause the desertification process in which anthropogenic activities commonly are the prime cause. The agriculture land has suffered soil erosion and soil degradation resulted to people leave their farms. Desertification takes place in dryland regions because in that areas earth is delicate and friable, rainfall is almost zero and the climate is very harsh. It resulted in the destruction of topsoil leading to the loss of land ability to sustain human activity, agriculture and livestock farming. Deforestation eliminates green cover and trees that hold the soil to the land. Animals overgrazing removes the green cover of grasses on land. It leads to create situations that largely intensify wildfires, forest fires and winds that make huge pressure to most valuable natural resources of earth.

3.4.5 Persistent Organic Pollutants

Anthropogenic activity resulted into release of all the primary and secondary pollutants into the environment among them most dangerous are Persistent Organic Pollutants (POPs) because of highly toxic nature causing adverse effects like deaths, epidemic disease, and defective births among humans as well as animals. Some of the specific adverse effects can be disruption of the immune system, cancer, reproductive disorders, allergies, hypersensitivity, damage to the central and peripheral nervous systems. Persistent Organic Pollutants are highly stable organic compounds before breaking down lasts in the environment for decades. By the process of evaporation and deposition these POPs released in one place can be transferred to another place during the seasonal process in atmosphere. Another process called bioaccumulation these POPs can concentrate in the living organisms because these are not soluble in water can readily absorbed in fatty tissues where the concentration magnified very much large as compared to the basic levels. For example, Fishes, humans, predatory birds and mammals are placed in higher level of food chain can absorb greater concentrations. DDT (Dichlorodiphenyltrichloroethane), PCBs (Polychlorinated biphenyls), and TBTs (Tributyltin chloride) are the organic secondary pollutants can have the capacity to alter the reproductive and endocrine systems of living organisms making an impact on human population to serious extent.

3.4.6 Pesticides

Pesticides such as dieldrin and toxaphene are added into environment through runoff from agriculture farming practices. It had been reported that these chemicals led to disruption of endocrine system. It has also found that higher concentrations of the chemicals/pesticides led to testicular cancer in men.

4 Discussions

Biodiversity refers to variety of all kinds of living organisms such as plants, animals, fungi and organisms living in a particular area. Biological diversity is the network of life that make planet earth different from other lifeless planets. It is measured as number of species inhabiting in an ecosystem. Land use can be managed by execution of humans on a site through plantations or agroforestry and land cover determine status of the green vegetation on a site such as forest or crop. Land-use land cover change make huge impact that alter global climate change by atmospheric flux of CO_2. The negative effects of alteration in climate change due to land use change can be mitigated through management practices. The spatial scales and temporal stresses the impact on regional and global landscapes at which land use and climate-change interactions occur. The process of land use land cover conversion is go together with habitat fragmentation, which may ultimately make effect on ecosystem functions by rapidly decreasing cover size or extend distances between cover [52]. It has been reported at the ecosystem scale that increasing the size or number of covers may expand flexibility by decreasing edge effects and increasing the current size of populations in the landscapes. In ecosystems with widespread fragmented areas, cover size and segregation effects will increase the effect of habitat loss [53]. The survival of a species in a land cover mainly depends on the cover size and connectivity between different land covers in adjacent habitat. The effects on biodiversity due to habitat fragmentation are diverse and significant with different ways of measuring the magnitude and direction of effects [54–57]. It is good to avoid or banned anthropogenic activities to stop habitat degradation and fragmentations in intact ecosystems region to uphold its intrinsic stability of ecosystem. Species or individual living organisms migrate and drift with global warming changes that helps to increase the capacity to recover the fragmented landscapes due to climate change that ultimately helps in habitat restoration process. It has been reported that as the effort increases in reducing the severity of extreme climate changes resulted into decline in the loss of natural vegetative cover that caused directly from land use land cover changes. Thus, execution of actions to lessen climate change could at the same time prevent habitat loss and fragmentation driven by land use land cover changes in biodiversity hotspots. It has been seen that in ecological freshwater communities such as buffer strips whose mobility is limited can be ability to minimize damage caused from agricultural runoff [58]. Besides this setting up of climate protections laws and protection [59] changing the sites of endangered species [60] and or restoration and construction of riparian habitats [61]. Global biodiversity is greatly changing with land use land cover changes (LULC) and climate changes gradient and also lead to modify microclimatic gradients. The spatial habitats arrangement at any given time may influence landscape processes such as edge effects, functional connectivity and ecological integrity [62]. In general, increasing land use land cover heterogeneity has limited potential to counterbalance negative effects of intensification. Greater variation in spatial heterogeneity tends to a wider environmental tolerance which may improve flexibility to climate change. Mountainous regions including great topographic and altitudinal diversity have a tendency to be more heterogeneous so that species may be better able to escape the effects of climate change [63]. However, with low heterogeneity in landscape ecosystem, species or living beings are forced to search for adaptive climatic conditions for stretched distance movement. Such as, tropical ecosystems with narrow latitudinal gradient in temperature that makes it harder for living organisms to shift ranges poleward compared with temperate organisms [64]. By achieving the goal of maintaining spatial heterogeneity at the landscape ecosystem scale, biodiversity conservation strategies become more effective [65–73].

5 Conclusion

Climate change is disturbing ecosystem at numerous scales and is an unescapable and increasing major global threat to disturb natural biodiversity and ecosystems.

Hence, at present we should have to plan future action that must to be executed to the fundamental ideas of native biodiversity conservation and climate mitigation that must should reflect habitat protection and conservations as a dual conservation priority. Species inhabiting within anthropogenic altered land uses are facing global climate change pressures and is budding area requiring extensive further research.

Acknowledgements I am highly thankful to Central University of Himachal Pradesh for support and facilities.

Conflict of Interest The authors declare no conflict of interest.

Funding No funding.

References

1. Tittensor, D. P., Walpole, M., Hill, S. L., Boyce, D. G., Britten, G. L., Burgess, N. D., Butchart, S. H., Leadley, P. W., Regan, E. C., Alkemade, R. A.: Mid-term analysis of progress toward international biodiversity targets. *Science.* 346 (2014) 241–244.
2. Gallardo, B., Clavero, M., Sanchez, M. I., Vila, M.: Global ecological impacts of invasive species in aquatic ecosystems. *Global Change Biol.* 22 (2016) 151–163.
3. Foley, J. A., DeFries, R., Asner, G. P., Barford, C., Bonan, G., Carpenter, S.R., Snyder, P. K.: Global consequences of land use. *Science,* 309 (2005) 570–574. https://doi.org/10.1126/science.1111772.
4. Laurance, W. F., Sayer, J., Cassman, K. G. Agricultural expansion and its impacts on tropical nature. *Trends in Ecology and Evolution.* 29 (2014) 107–116. https://doi.org/10.1016/j.tree.2013.12.001.
5. Sampaio, G., Nobre, C., Costa, M. H., Satyamurty, P., Soares-Filho, B. S., Cardoso, M. Regional climate change over eastern Amazonia caused by pasture and soybean cropland expansion. *Geophysical Research Letters.* 34 (17) (2007) 1–7. https://doi.org/10.1029/2007GL030612.
6. Senior, R. A., Hill, J. K., González del Pliego, P., Goode, L. K., Edwards, D. P. A.: pantropical analysis of the impacts of forest degradation and conversion on local temperature. *Ecology and Evolution.* 7(19) (2017) 7897–7908. https://doi.org/10.1002/ece3.3262.
7. Barnagaud, J. Y., Barbaro, L., Hampe, A., Jiguet, F., Archaux, F.: Species thermal preferences affect forest bird communities along landscape and local scale habitat gradients. *Ecography,* 36 (2013) 1218–1226. https://doi.org/10.1111/j.1600-0587.2012.00227x.
8. Frishkoff, L. O., Hadly, E. A., Daily, G. C.: Thermal niche predicts tolerance to habitat conversion in tropical amphibians and reptiles. *Global Change Biology.* 21(11) (2015) 3901–3916. https://doi.org/10.1111/gcb.13016.
9. Wardle, D. A.: Disturbance and landscape dynamics in a changing world. *Science.* 91(2011) 2833–2849.
10. Yoan, P., Laurent, B., Joakim, H., Peter, O., Catherine, A., Markus, B., Rienkjan, B., Luc, D. B. F., Marc, G. U.: Biodiversity Differences between Managed and Unmanaged Forests: Meta-Analysis of Species Richness in Europe. Conservation Biol. 24 (2010) 101–112.
11. Garcia, R. A., Cabeza, M., Rahbek, C., Araujo, M. B.: Multiple Dimensions of Climate Change and Their Implications for Biodiversity. *Science.* 344(2014) 1247579.
12. Mahmood, R., Pielke, R. A., Hubbard, K. G., Niyogi, D., Dirmeyer, P. A., Mcalpine, C., Carleton, A. M., Hale, R., Gameda, S., Beltrán-Przekurat, A. Land cover changes and their biogeophysical effects on climate. *Int. J. Climatol.* 34 (2014) 929–953.

13. Correa, D. F., Beyer, H. L., Possingham, H. P.: Thomas-Hall, S. R., Schenk, P. M. Biodiversity impacts of bioenergy production: Microalgae vs. first generation biofuels. *Renewable Sustainable Energy Rev.* 74 (2017) 1131–1146.
14. Payne, J. L., Bush, A. M., Heim, N. A., Knope, M. L., Mccauley, D. J.: Ecological selectivity of the emerging mass extinction in the oceans. *Science.* 353(2016) 1284–1286.
15. M. G.: Turner, Ecology 91 (2010) 2833.
16. Oliver, T. H., Morecroft, M. D.: Interactions between climate change and land use change on biodiversity: Attribution problems, risks, and opportunities. *Wiley Interdisciplinary Reviews: Climate Change.* 5(3) (2014) 317–335. https://doi.org/10.1002/wcc.271.
17. Tomimatsu, H., Sasaki, T., Kurokawa, H., Bridle, J. R., Fontaine, C., Kitano, J., Yachi, S.: Sustaining ecosystem functions in a changing world: A call for an integrated approach. *Journal of Applied Ecology.* 50(5) (2013) 1124–1130. https://doi.org/10.1111/1365-2664.12116.
18. van Vuuren, D. P., Kok, M. T. J., Girod, B., Lucas, P. L., de Vries, B.: Scenarios in global environmental assessments: key characteristics and lessons for future use. *Global Environmental Change.* 22 (2012) 884–895.
19. O'Neill, B. C., Kriegler, E., Ebi, K. L.: The roads ahead: narratives for shared socioeconomic pathways describing world futures in the 21st century. Global Environmental Change, (2015). 1–10. https://doi.org/10.1016/j.gloenvcha.2015.01.004.
20. Bellard, C., Bertelsmeier, C., Leadley, P., Thuiller, W., Courchamp, F.: Impacts of climate change on the future of biodiversity. *Ecol. Lett.* 15 (2012) 365–377.
21. Jones, K. R., Watson, J. E. M., Possingham, H. P., Klein, C. J.: Incorporating climate change into spatial conservation prioritization: A review. *Biol. Conser.* 194(2016) 121–130.
22. Kerr, J. T., Pindar, A., Galpern, P., Packer, L., Potts, S. G., Roberts, S. M., Rasmont, P., Schweiger, O., Colla, S. R., Richardson, L. L.: Climate change impacts on bumblebees converge across continents. *Science.* 349(2015) 177–180.
23. Henne, P. D., Bigalke, M., Büntgen, U. et al.: An empirical perspective for understanding climate change impacts in Switzerland. *Reg Environ Change.* 18 (2018) 205–221 https://doi.org/10.1007/s10113-017-1182-9.
24. Potter, K. A., Arthur, W. H., Pincebourde, S.: Microclimatic challenges in global change biology. *Global Change Biol.* 19(2013) 2932–2939.
25. Dawson, T. P., Jackson, S. T., House, J. I., Prentice, I. C., Mace, G. M.: Beyond Predictions: Biodiversity Conservation in a Changing Climate. *Science.* 332 (2011) 53–58.
26. Angert, A. L., Crozier, L. G., Rissler, L. J., Gilman, S. E., Tewksbury, J. J., Chunco, A. J. Do species' traits predict recent shifts at expanding range edges? *Ecol. Lett.* 14 (2011) 677–689.
27. Chen, I., Hill, J. K., Ohlemüller, R., Roy, D. B., Thomas, C. D.: Rapid Range Shifts of Species Associated with High Levels of Climate Warming. *Science.* 333 (2011) 1024–1026.
28. Frenne, P. D., Rodríguezsánchez, F., Coomes, D. A., et al.: Microclimate moderates plant responses to macroclimate warming. *Proc. Natl. Acad. Sci.* USA. 110(2013) 18561–18565.
29. Bertrand, R., Lenoir, J., Piedallu, C., Riofrío-Dillon, G., De, R. P., Vidal, C., Pierrat, J. C., Gégout, J. C.: Changes in plant community composition lag behind climate warming in lowland forests. *Nature.* 479 (2011) 517–520.
30. Walther, G. R.: Community and ecosystem responses to recent climate change. *Philos. Trans. R. Soc. London* 365(2010) 2019–2024.
31. Bennett, E. M., Cramer, W., Begossi, A., Cundill, G., Díaz, S. et al.: Linking biodiversity, ecosystem services, and human well-being: three challenges for designing research for sustainability. *Curr. Opin. Environ. Sustainability.* 14(2015) 76–85.
32. Mace, G. M., Norris, K., Fitter, A. H.: Biodiversity and ecosystem services: a multilayered relationship. *Trends Ecol. Evol.* 27 (2012) 19–26.
33. Soliveres, S., van der Plas, F., Manning, P., Prati, D., Gossner, M. M., Renner, S. C., Alt, F., Arndt, H., Baumgartner, V., Binkenstein, J.: Biodiversity at multiple trophic levels is needed for ecosystem multifunctionality. *Nature.* 536 (2016) 456–462.
34. Newbold, T., Hudson, L. N., Hill, S. L., Contu, S., Lysenko, I., Senior, R. A., Borger, L., Bennett, D. J., Choimes, A., Collen, B.: Global effects of land use on local terrestrial biodiversity. *Nature.* 520 (2015) 45–52.

35. Garibaldi, L. A., Steffandewenter, I., Winfree, R., Aizen, M. A., Bommarco, R., Cunningham, S. A., Kremen, C., Carvalheiro, L. G., Harder, L. D., Afik, O.: Wild Pollinators Enhance Fruit Set of Crops Regardless of Honey Bee Abundance. *Science.* 339 (2013) 1608–1613.
36. Cardinale, B. J., Srivastava, D. S., Duffy, J. E., Wright, J. P., Downing, A. L., Sankaran, M., Jouseau, C.: Effects of biodiversity on the functioning of trophic groups and ecosystems. *Nature.* 443 (2006) 989–992.
37. Mori, A. S., Lertzman, K. P., Gustafsson, L. Biodiversity and ecosystem services in forest ecosystems: a research agenda for applied forest ecology. *J. Appl. Ecol.* 54 (2016) 12–27.
38. Bateman, I. J., Termansen, M. Bringing Ecosystem Services into Economic Decision-Making: Land Use in the United Kingdom. *Science.* 341 (2013) 45–50.
39. Moran, E. V., Alexander, J. M. Evolutionary responses to global change: lessons from invasive species. *Ecol. Lett.* 17 (2014) 637–649.
40. Gaston, K. J. Valuing Common Species. *Science.* 327 (2010) 154–155.
41. Adger, W. N., Brown, K., Nelson, D. R., Berkes, F., Eakin, H., Folke, C., Galvin, K., Gunderson, L., Goulden, M., O'Brien, K.: Resilience implications of policy responses to climate change. *Wiley Interdiscip. Rev. Clim. Change.* 2 (2011) 757–766.
42. Mantyka-Pringle, C. S., Martin, T. G., Rhodes, J. R.: Interactions between climate and habitat loss effects on biodiversity: a systematic review and meta-analysis. *Global Change Biol.* 19 (2013) 1642–1644.
43. Stott, P. A., Christidis, N., Otto, F. E. L., Sun, Y., Vanderlinden, J.P., van Oldenborgh, G. J., Vautard, R., von Storch, H., Walton, P., Yiou, P., Zwiers, F. W.: Attribution of extreme weather and climate-related events. *Wiley Interdiscip. Rev. Clim. Change.* 7 (2016) 23–41.
44. Leonard, M., Westra, S., Phatak, A., Lambert, M., Hurk, B. V. D., Mcinnes, K., Risbey, J., Schuster, S., Jakob, D., Smith, S. M. A.: compound event framework for understanding extreme impacts. *Wiley Interdiscip. Rev. Clim. Change.* 5 (2014) 113–128.
45. Gilchrist, G. W.: Specialists and Generalists in Changing Environments. Fitness Landscapes of Thermal Sensitivity. *Am. Nat.* 146 (1995) 252–270.
46. Reyer, C. P. O., Leuzinger, S., Rammig, A., Wolf, A., Bartholomeus, R. P., Bonfante, A., Lorenzi, F. D., Dury, M., Gloning, P., Jaoude, R. A. A.: Plant's perspective of extremes: terrestrial plant responses to changing climatic variability. *Global Change Biol.* 19 (2013) 75–89.
47. Houghton, R. A., House, J. I., Pongratz, J., van der Werf, G. R., DeFries, R. S., Hansen, M. C., Le Quere, C., Ramankutty, N.: Carbon emissions from land use and land-cover change. *Biogeosciences.* 9 (2012) 5125–5142.
48. Smith, P., House, J. I., Bustamante, M., Sobocka, J., Harper, R., Pan, G., West, P. C., Clark, J. M., Adhya, T., Rumpel, C.: Global change pressures on soils from land use and management. *Global Change Biol.* 22 (2016) 1008–1028.
49. Tang, L., Zeng, G. M., Shen, G. L., Li, Y. P., Zhang, Y., Huang, D. L.: Rapid detection of picloram in agricultural field samples using a disposable immune membrane-based electrochemical sensor. *Environ. Sci. Technol.* 42 (2008) 1207–1212.
50. Harley, C. D. G., Hughes, A. R., Hultgren, K. M., Miner, B. G., Sorte, C. J. B., Thornber, C. S., Rodriguez, L. F., Tomanek, L., Williams, S. L.: The impacts of climate change in coastal marine systems. *Ecol. Lett.* 9 (2006) 228–241.
51. Przeslawski, R., Byrne, M., Mellin, C.: A review and meta-analysis of the effects of multiple abiotic stressors on marine embryos and larvae. *Global Change Biol.* 21 (2015) 2122–2140.
52. Nash, K. L., Graham, N. A. J., Jennings, S., Wilson, S. K., Bellwood, D. R.: Herbivore cross-scale redundancy supports response diversity and promotes coral reef resilience. *J. Appl. Ecol.* 53 (2016) 646–655.
53. Ockinger, E., Schweiger, O., Crist, T. O., Debinski, D. M., Krauss, J., Kuussaari, M., Petersen, J. D., Poyry, J., Settele, J., Summerville, K. S. Life-history traits predict species responses to habitat area and isolation: a cross-continental synthesis. *Ecol. Lett.* 13 (2010) 969–979.
54. McCune, J. L.: Species distribution models predict rare species occurrences despite significant effects of landscape context. *J. Appl. Ecol.* 53 (2016) 1871–1879.

55. Simons Legaard, E. M., Harrison, D. J., Legaard, K. R.: Habitat monitoring and projections for Canada lynx: linking the Landsat archive with carnivore occurrence and prey density. *J. Appl. Ecol.* 53 (2016) 1260–1269.
56. Saura, S., Bodin, O., M. Fortin, J.: Stepping stones are crucial for species' long-distance dispersal and range expansion through habitat networks. *J. Appl. Ecol.* 51 (2014) 171–182.
57. Mitchell, M. G. E., Suarez Castro, A. F., Martinez Harms, M., Maron, M., McAlpine, C., Gaston, K. J., Johansen, K., Rhodes, J. R.: Reframing landscape fragmentations effects on ecosystem services. *Trends Ecol. Evol.* 30 (2015) 190–198.
58. Castello, L., Macedo, M. N.: Large-scale degradation of Amazonian freshwater ecosystems. *Global Change Biol.* 22 (2016) 990–1007.
59. Shoo, L. P., Olson, D. H., Mcmenamin, S. K., Murray, K. A., Van Sluys, M., Donnelly, M. A., Stratford, D., Terhivuo, J., Merino Viteri, A., Herbert, S. M.: Engineering a future for amphibians under climate change. *J. Appl. Ecol.* 48 (2011): 487–492.
60. Schwartz, M. W., Martin, T. G., Ann. N. Y. Translocation of imperiled species under changing climates. *Acad. Sci.* (2013): 1286, 15–28.
61. Kreiling, R. M., Schubauer Berigan, J. P., Richardson, W. B., Bartsch, L. A., Hughes, P. E., Cavanaugh, J. C., Strauss, E. A.: Wetland Management Reduces Sediment and Nutrient Loading to the Upper Mississippi River. *J. Environ. Qual.* 42 (2013) 573–583.
62. Villard, M. A., Metzger, J. P.: Beyond the fragmentation debate: a conceptual model to predict when habitat configuration really matters. *J. Appl. Ecol.* 51 (2014) 309–318.
63. Loarie, S. R. Duffy, P. B., Hamilton, H., Asner, G. P., Field, C. B., Ackerly, D. D.: The velocity of climate change. *Nature.* 462 (2009) 1052–1061.
64. Colwell, R. K., Brehm, G., Cardelus, C. L., Gilman, A. C., Longino, J. T. Global warming, elevational range shifts, and lowland biotic attrition in the wet tropics. *Science.* 322 (2008) 258–261.
65. He, X., Liang, J., Zeng, G., Yuan, Y., Li, X.: The Effects of Interaction between Climate Change and Land-Use/Cover Change on Biodiversity-Related Ecosystem Services. *Global Challenges.* 3 (2019) 1800095. https://doi.org/10.1002/gch2.201800095
66. Sharma, M., Thakur, R., Sharma, M.: Ethnobotanical survey of medicinal plants of unexplored hilly areas of district Ramban (J&K). *International Journal of Botany Studies.* 5(3) (2020) 55–63.
67. Kaushal, A., Sharma, M., Navneet, Sharma, M.: Ethnomedicinal, phytochemical, therapeutic and pharmacological review of the genus Erythrina. *International Journal of Botany Studies.* 5(6) (2020) 642–648.
68. Sharma, M., Thakur, R., Sharma, M. Ethnomedicinal, Phytochemical and Pharmacological Properties of *Crocus sativus* (Saffron). *Journal of Indian Botanical Society.* 99(3–4) (2020) 46–55.
69. Sharma, M., Thakur, R., Sharma, M., Sharma, A.K., Sharma, A.K. Changing Scenario of Medicinal Plants Diversity in Relation to Climate Changes: A Review. *Plant Archives.* 20(2) (2020) 4389–4400.
70. Sharma, A. K., Kamboj, V., Sharma, A. K., Thakur, R., Sharma, M.: Water quality and its impact on phytoplankton diversity: A case study of tehri reservoir, Garhwal Himalayas. *Science Archives.* 1(3) (2020) 166–173. https://doi.org/10.47587/SA.2020.1315.
71. Malik, D. S., Sharma, A. K., Sharma, A. K., Thakur, R., Sharma, M. A.: review on impact of water pollution on freshwater fish species and their aquatic environment. In: Advances in Environmental Pollution Management: Wastewater Impacts and Treatment Technologies, Volume 1, Eds. Kumar, V., Kamboj, N., Payum, T., Singh, J. and Kumar, P. (2020). pp. 10–28. https://doi.org/10.26832/aesa-2020-aepm-02.
72. Sharma, M., Navneet, Sharma, M.: Impact of Agricultural Pollutants on Aquatic Ecosystem and Its Management: A Critical Review. In: Current Research and Innovations in Life Sciences, Eds. Sharma, A.K., Thakur, R., Sharma, M. and Sharma, M. (2021) pp.15–35.
73. Sharma, A. K., Sharma, A. K., Sharma, M., Navneet, Sharma, M.: A Review on Impact of Water Pollution on Aquatic Ecosystem. In: Current Research and Innovations in Life Sciences, Eds. Sharma, A.K., Thakur, R., Sharma, M. and Sharma, M. (2021). pp.129–148.

Assessment of Pollution Load and Its Impact on Aquatic Biodiversity of River Ganga: A Review

D. S. Malik, Aditi Bisht, and Ruchika Chauhan

Abstract This review paper presents experimental results of some researchers who have observed Ganga River Water from Gangotri to West Bengal in the past 21 years. An attempt has been made to understand the gradual degradation of Gangajal and its present pollution level in relation to various uses. To assess water of Ganga, numerous measurements have been taken from time to time to study the physico-chemical condition of this sacred river by various researchers. These studies show that all is not well and the quality of Ganga is getting worse day by day. The reason could be the nostalgic mood of the industry and common man who continue to follow bad practices causing severe pollution of the Ganga. Though there are government laws to control river pollution, the implementation is not adequate as the laws in India are too volatile, too stringent and too inexpedient.

Keywords Pollution types · Water pollution · Aquatic biodiversity · Ganga river

1 Introduction

The Ganges River, also called in hindi as Ganga, is a prodigious river of the plains of the north Indian subcontinent. Since time immemorial it has been the holy river of Hindus. For maximum of its course it is a widespread and sluggish stream, flowing through the greatest fertile and thickly populated areas in the world [1]. In spite of its importance, its has an extent of 1560 miles (2510 km) which is moderately short compared to the other rivers of Asia or the world. The Ganges is a sacred river of India and Bangladesh. The Bhagirathi River flows from the base of Gangotri Glacier, which is known as Gaumukh. The river Bhagirathi is said to be the actual basis in Hindu culture [2]. It has over-all length 2525 km, it rises in the western Himalayas in Uttarakhand, and flows south and east through the Plain of Northern India and reaches Bay of Bengal through Bangladesh. It is India's longest river and second

D. S. Malik · A. Bisht · R. Chauhan (✉)
Department of Zoology and Environmental Science, Gurukula Kangri (Deemed to be University), Haridwar, Uttarakhand 249404, India
e-mail: ruchikachauhan69@gmail.com

K. D. Bahukhandi et al. (eds.), *Environmental Pollution and Natural Resource Management*, Springer Proceedings in Earth and Environmental Sciences,
https://doi.org/10.1007/978-3-031-05335-1_6

largest river in the world by water discharge. The basin of Ganga River is the most densely inhabited basin in the world, with over 400 million. In 2007 the river Ganga was graded amongst the five greatest polluted rivers of the world.

The mainstream of Ganga is formed when Alaknanda and Bhagirathi streams unite at Devaprayag, which penetrates southwestward through the Shivalik range at the north border of the Indo-Gangetic Plain to appear from the foothills at Rishikesh [3]. It then moves to the plain at Haridwar which is a sacred place for Hindus. The magnitude of Ganga water rises evidently as it procures more streams and arrives a area of heavier precipitation. Since, April to June the melting snow of Himalayas feed the river, and in the monsoon period i.e., from July to September the monsoon feed the river sometime causing flood situations also. During winter season the flow of river declines.

In Uttar Pradesh the river Ganga receives right bank tributaries i.e., Yamuna and Tons which intersects the Ganges just below Prayagraj. There are also three main left-bank tributaries, which are the Ramganga, the Gomati, and the Ghaghara.

The next state in which the river enters is the state of Bihar. Its chief tributaries are the Ghugri, Gandak, the Burhi Gandak, and the Kosi river. The important tributary is Son River. The river then borders the Rajmahal Hills and flows southeast to Farakka in central West Bengal state. West Bengal is the last state that the Ganga arrives, and, afterward it flows into Bangladesh [4]. The Ganga is called a Padma in West Bengal. The western distributaries of the delta are the Bhagirathi and Hoogly river. As the Ganges entitles from West Bengal into Bangladesh, many distributaries branch off to the south. The Ganges is merged to Brahmaputra in Bangladesh near Goalundo ghat.

The Ganges area was said to be densely forested. Ancient literatures specify that in the sixteenth and seventeenth centuries hunting of wild elephants, buffalo, bison, rhinoceroses, lions, and tigers was common in this area. The unique natural vegetation has vanished from the basin of Ganga, and the land is now cultivated to meet the requirements of food of the population. Few wild animals are now commonly cited like for deer, boars, and wildcats, wolves, jackals, and foxes. Bengal tigers, crocodiles, and marsh deer are seen in Sundarbans.

Fish flourish in all the rivers, specifically in the delta area, where they form an important share of the residents diet. In West Bengal common fishes comprise feather backs, walking catfish, milk fish, gourami, and barbs [5]. The Ganga Dolphin or *Platanista gangetica*, which is unsighted cetaceous can be seen all over the Ganga–Brahmaputra basin, but now it comes under endangered category due to its illegal poaching. Numerous diversities of birds are found, such as parrots, myna, crows, partridges, kites, ducks, snipes and fowls.

The use of the Ganga water for drinking and irrigation is common since the ancient times. The use of water in irrigation is defined in sacred scripture and mythological books inscribed more than 2000 years ago. The Irrigation facility and canal system was developed in the period of Muslim rule in the twelfth century onward. The channel system was further extended by the British [6]. The cash crop production of U.P. and Bihar in Ganges valley is benefited from the of irrigation canals. In olden times the Ganges and some of its tributaries, particularly in the east, remained significant transport routes. By the nineteenth century, irrigation-cum-navigation channels designed the chief routes of the transport system through water. The arrival of paddle steamships transformed internal transportation, motivating the evolution of indigo manufacture in Bengal and Bihar. Regular steamer facilities contended from Kolkata to Prayagraj and far beyond, along with to Agra on the Yamuna and up the Brahmaputra water transportation decline in mid-nineteenth century with the emergence of Railway transportation. Bangladesh and Bengal remain to depend on the waterways to transport jute, tea, grain, and other agricultural products. The Ganga's hydroelectric potential is massive, which is estimated to have fluctuated from some 51,700 to 128,700 MW. Some of that potential has been exploited in India, like in Uttarakhand, on Yamuna River and its tributaries in Himachal Pradesh, and in downstream, in Chambal and Rihand rivers.

The river ganga faces so much stress due to various human activities and development. Pollution in the Ganga River, possess noteworthy threats to health of human and the environment. Extremely contaminated with human waste and industrial pollutants, the river delivers water to approximately 40% of the population of the country in 11 states. Nowadays, the Ganga is well-thought-out to be the fifth-most polluted river in the world [7]. Many initiatives have been taken to clean the river, but all of them were unsuccessful to bring anticipated results.

2 Causes of Pollution in Ganga

All through festival periods, over 70 million individuals take dip in the Ganga to clean themselves from their previous sins. Few used and unused materials like food, plastic waste, flowers, leaves, dead bodies, clothes, human and animal waste are left in the Ganges and its shores by the toursist and piligrims which are also accountable for its contamination [8]. As per available data 12,000 MLD (million litres a day) of sewage from from municipal and industrial areas is drained into Ganga basin daily. Not only the sewage, nearly all human activities produce some type of environmental disorder that pollute the surrounding water [9]. For, convience the large majority of sources of water pollution are divided in broad categories of waste:

(A) Industrial, (B) Agricultural, (C) Domestic waste, (D) Religious waste.

Industrial waste: Wastes from industry serve as maincause of contaminantsbecause of (i) Manufacturing and drugs industry (ii) Power generation (iii) Mining and construction (iv) Food processing industries.

The Manufacturing industries includes the industries dealing with chemical, oil refining, organic chemicals and heavy metals. The example of such industry includes textile, leather tanning, plastics, pharmaceutical and paper pulp industries. The by product from such industry are extremely poisonous to many creatures, including humans. Many pharmaceuticals involved in making drugs like antibiotics and various other medicine [7]. The endproduct or the waste from such industries get mixed with water as the industries dispose them in the river and it increases the BOD and COD.

Power producing industries are chief contributors of heat and radioactivity. Radioactivity from nuclear power plants can contaminate surface as well as ground water in a variety of ways. The water gets contaminated by minning of uranium and enrichment plants. Minning industry contribute in sediments and acid drainage. As, main segment comprises vegetables, fruits, dairy, alcholic and non alcholic drinks and packed food [10]. The common worry in the industries are consumption and discharge of water, chemicals and compounds used in processing and cleaning, packing, removal and food leftovers.

Agricultural waste: These are produced by the agriculture and animal rearing. Agriculture is a chief source of organic chemicals, specially pesticides. The pesticides penetrate in surface and under ground water bodies. The usage of nitrogen manures upsurge the concentration of nitrates in ground water, leading to high nitrate levels in ground water sources [11]. The water may penetrate deep or run off may occur which may harm the flora and fauna of the marine biodiversity [10].

Domestic waste: These are the waste produced by households. Domestic waste Also include septic tank waste which ends up of in natural waters. These days numerous individuals dump their garbage into rivers, ponds, lakes and streams, thus creating the water body as the ultimate resting place of cans, bottles, plastics and other products from households [12]. Maximum of nowadays washing products are detergents and cleaners which comes from the petro chemicals industries. The chemicals substance contained in washing powders and other cleaners effect the well-being of all forms of life in the water.

Religious waste: All through the festive periods, more than 70 million individuals take holy dip in the Ganga to wash themselves of their former sins. Some materials like flowers, clothes, and wastes are left in the Ganga and its banks which are also accountable for its contamination [13]. Traditional beliefs of people say that being burnt on its banks and ashes floating down the Ganga will rinse the sins of those individuals who expire and this bring them straight to salvation. In Varanasi alone, an approximately forty thousand bodies are burned every year and are put into the Ganga [14]. As numerous families do not have enough money for cremation wood, countless bodies are placed into the Ganges are only half-burnt.

3 Pollution in the Major Cities from Which River Ganga Flows

3.1 Rishikesh

The town of Rishikesh is situated in the foothills of the Himalayas. It is recognized as the Gateway to the Garhwal Himalayas and is also famous as "Yoga Capital of the World". It is identified as the pilgrimage town and observed as one of the holiest places for Hindus. Sages and saints have stayed as well as visited Rishikesh since early periods to meditate in search of higher knowledge [15]. Ganga river subsequently flowing 249 km through its slender Himalayan valley, it arises from the foothills at Rishikesh, then move out onto the Gangetic Plain at the pilgrimage town of Haridwar [16]. The river Ganga is admired by millions of people in India but still unprocessed sewage is directly dumped into its water in Rishikesh. The intreated waste of hotels, ashrams, temples and shops is discharged in the river daily. The river Song flows through Dehradun meet Ganga River near Lakshman Jhula which is close to town of Rishikesh. The quality water beyond this point is finest. The water quality is degraded after this point as untreated sewage flowing in Song go in the Ganga and causes contamination. The quality of water was tested in 15 locations between Gangotri and Haridwar water in 7 sites above Lakshman Jhula was found to be of finest quality and appropriate for drinking. Whereas water quality of other sites tested between Rishikesh to Haridwar was not found of good quality and was not suggested for consumption and other domestic purposes without subjecting it to purification [3]. According to researchers, the main source of water pollution in Ganga is due to mass bathing, urban waste and effluent from sewage treatment, domestic sewage, discharge from hotels and ashrams reaches the Ganga, causing additional contamination.

3.2 Haridwar

In Sanskrit, the ceremonial meaning of Haridwar is Hari means "Lord Vishnu" and Dwar means "gateway". So, Haridwar is said to be "The Gateway to Lord Vishnu". Haridwar got this name because it is usually the place where devotees initiate their Char Dham Yatra for attainment of God. The city is positioned on the right bank of the river Ganga, at the bases of the Shivalik ranges. It is observed as a holy place for Hindus, holding chief spiritual events and serving as an entrance to numerous prominent places of worship [17]. One of the significant of the events is the Kumbh Mela, which is in every 12 years. Millions of tourists, devotees, and pilgrims assemble here to perform ceremonial bathing on the banks of the river to wash away their sins to attain moksha. The water quality in this area is degrading year by year even after lots of efforts of government to clean it. The degrading quality if water is due to the pilgrims and visitors who come to the city to wash their sins and leave the city with the garbage disposed in the river. Also, the untreated sewage, chemicals from industries and waste water is major cause of polluted water in Haridwar [18]. Many institutes like Gurukul Kangri, IIT, Roorkee, PSI, Dehradun, Doon University checked the water quality on various parameters and found the water quality of this region is not good and is unfit for drinking and do not fulfil the criteria od CPCB for bathing during festive seasons also.

3.3 Uttar Pradesh

Uttar Pradesh is a state of northern India, having over 200 million inhabitants, it is the most heavily settled state in India furthermore the densely populated province subdivision on the globe. The two important rivers of the state are Ganges and Yamuna which joins at the confluence of Triveni in Prayagraj (Allahabad) [19]. These two foremost tributaries, the Yamuna, which proceed by the Delhi (capital region) and afterwards abruptly aligned the south eastward flow of the river Ganges prior joining it near Prayagraj [20], including Tons, which go along north from the Vindhya range in Madhya Pradesh and unite the river Ganges just beneath Prayagraj. The major left bank tributaries in UP are the Ramganga, Ghagra and the Gomti river. The Ganga river arrives Uttar Pradesh at Bijnor and it flows through Bulandshahr, Hapur, Meerut, Muzaffarnagar, Aligarh, Kasganj, Sambhal, Amroha, Badaun, Shahjahanpur, Hardoi and Raebareli. It flows through 27 districts in U.P. [21]. The reports shows when river reaches Bijnor, water quality was B standard, which is suitable for outdoor bathing as the biological oxygen demand (BOD) or dissolved oxygen level was 1.4 mg per litre at the place. Quality of water kept on decreasing as it moves forward. The quality of water is C, but when the Ganga arrives at the territorial area of Kannauj, the quality of water changed from C to D, as it moves forwards. The quality of water remained in category D from Bithoor in Kanpur to Prayagraj [22]. In Noida, the Yamuna river water quality was testified to be E, which can be used for irrigation, industrial cooling and waste disposal in controlled way. The BOD level at Vrindavan and Mathura was testified to be of D water quality. Kanpur has the maximum pollution load i.e., 75% amongst all of the towns of Uttar Pradesh. The pollution load is due to the drains, sewage treatment plants, tanneries, industries, etc. [14]. Approximately, all classes of wastes are impending out of Varanasi, viz. all kind of biodegradable as well as non-biodegradable wastes, sewage, industrial waste, animal corpses, unclaimed human dead bodies. Water quality in Varanasi is intolerable for drinkable or for farming and it is getting worse [20].

3.4 Bihar

Bihar is a state in the eastern part of India. It is the 12th wide—ranging by the territory (94,163 km^2) and is ranked 3rd largest state by population. The river Ganges splits the Bihar plain, which flows west to east. Magadh, Mithila, and Bhojpur are the three main cultural provinces of the Bihar state. In Bihar quality of water is alkaline in nature with more total solids and also above saturated with dissolved oxygen. The chlorides are well in the permitted limits of drinking water quality standards. Whereas, some of the selected parameters are to some extent higher than the prescribed limit by WHO which makes it unfit for household and commercial purposes. The hardness of water is Slightly high in some sites which makes it unfit

Table 1 Physicochemical characteristics of ganga at various location

Parameters	Unit	Rishikesh	Haridwar	UP	Bihar	West Bengal
pH		7.5	7.5	7.40	7.75	6.06
EC	dSm^{-1}	140.4	0.29	0.44	0.007	
DO	$mg\ L^{-1}$	9.67	9.8	6.40	7.425	3.4
BOD	$mg\ L^{-1}$	2.25	1.3	3.3	2.475	4.51
COD	$mg\ L^{-1}$	5.32	4.00	19.20	15.125	26.15
TDS	$mg\ L^{-1}$	55.68	109	166.92	250.25	254.1
Turbidity	NTU	7.79	46.72	166	–	147.83
Chloride	$mg\ L^{-1}$	5.11	7.90	2.80	35.05	119.83
References		[25, 26]	[27]	[28]	[29]	[30, 31]

for drinking purpose also. Though high pH, TDS, Hardness, DO, BOD and MPN standards recommends cleansing is essential for domestic consumption [23, 24].

3.5 West Bengal

West Bengal is a province in the eastern part of India, in between the Himalayas and the Bay of Bengal. And its capital, Calcutta hold on architectural and cultural remnants of its formerly like an East India Company, the trading post and the capital of British Raj. Afterwards Bihar, the river evades the hills of Rajmahal to the south and proceed southeast to the Farakka in the West Bengal, at the crest of the delta [24]. The West Bengal is the last Indian state in which the Ganges enters, and then it flows to the Bangladesh, the Mahananda river and gets joined to the north. In India, West Bengal as well as in Bangladesh, the river Ganges is pronounced as the 'Padma' 0.1779 MLD wastewater is discharged into the River Ganga through 54 canals in West Bengal. The results show that the capacity utilization of West Bengal requires immediate attention. Grossly Polluting Industries West Bengal generates maximum effluent of 75.5% in terms of water consumed [14] (Table 1).

4 Impact of Pollution on Aquatic Biodiversity

Enormous quantity of river Ganges water is also used for irrigation purpose due to which the Ganga downstream of Narora (UP) is being assassinating. The establishment of non-coastal waterways leads to expand in pollution and homicide the aquatic life [3]. In West Bengal, the Farakka barrage access the saline water and cause flood in Bihar. The experts found that river Ganges is getting highly polluted as people are dumping a lot of wastes, toxic elements which causes the depletion of dissolved oxygen in the water and stimulates the circumstances of hypoxia, which is quite harmful for aquatic life. The most polluted part of the river Ganges is at Kanpur, it is the most populated city in Uttar Pradesh where the people in the locality washes their dirty clothes, bath their animals, urinate and discharge sewage in the water which makes the water impure [32].

The scientists have found a lot of quantity of mercury in the river Ganges, as they collected specimens contains the amount mercury, which results in the tended muscles of the fishes which assembles in the body of the aquatic organisms [5]. Approximately 50–80% of mercury is organic and it causes a connection between the food habit, mercury levels in muscles and the length of the fish. The 'Ganga River dolphin' is one of the fresh water dolphins in the world is listed as endangered species [33]. Irrigation dams and hydroelectric powers prevent the dolphins from proceeding up and down along the river Ganges, which is the most important reason for their depleting population. Another fresh water species of the river Ganges that is 'Ganges Softshell Turtle' (*Nilssonia gangetica*) which can also be seen in the river Indus, Mahanadi River, systems of Pakistan, Bangladesh and Southern Nepal [34]. These turtles are basically found in the deep rivers, lakes of freshwater. And as these turtles consists of prolonged agedness and elevated trophic levels in the food web of the aquatic life. But they are at risk due to the various mode of pollution.

5 Challenges and Remedial Measures

The Prime Minister of India Narendra Modi made it his own plan and set a aim to clean Ganga by 2019, which was then extended to 2020. Namami Gange is being executed by the National Mission for Clean Ganga (NMCG), and its state counterparts to achieve the twin purpose of effective reduction of pollution, management and rejuvenation of Ganga [35] there are well proportioned 221 projects under the scheme, which comprise STP, ghat development, cleaning of surface, afforestation, cleanliness, and public awareness, out of which only 58 have been completed.

River ecosystem have agonized from human interference resulting in habitat loss and degradation and as a consquence numerous fish species of freshwater have become vigorously jeopardized specific in Ganges basin [27]. In India 2246 Indigenous finfishes have been portrayed of which 765 has a place with fresh water fish. 143 species have a placed with 11 orders, 72 genera and 32 families were recorded across the Ganges which is about 20% of fresh water fishes proclaimed in India. Out of 143 species, 133 species were local to waterway Ganges and its tributaries and rest 10 species were exotics. Cleaning waterway ganga is a monstrous project which has gained critical progress after 2014 yet at the same time has far to go [36].

6 Challenges for Failure of Namani Gange Project

There are few challenges which Namani Gange project face which may lead to its failure. First challenge 2000 million litres capacity STPs had to be transformed of which only 328 MLD have been completed claimed by Namani Gange project. The status of project brings a doubt that whether the government would even accomplish its aim. Second Challenge is that the river self purifies it when there is constant flow of water. The river fails this elementary test except during rainy season. Therefore, it's not just about contaminated Ganga. It is about the survival of Ganga. Third challenge is the contamination from the cities which flows through a network of small and increasingly larger open drains, which ultimately flows into the Ganga. None of the three metropolises has a system for administration of solid waste, most of which is discarded in the roads, clogging exposed drains and adding to the contamination. Fourth challenge is scrubbing up the huge stretch of 2525 km that the Ganga crosses is a programme where regulating the funds becomes as giant issue. The UP SAAP in 2016 said that the towns in Ganga basin would require Rs. 5794 crore just for the formation of sewer networks in the state. Last challenge is the cleaning of the Ganga needs continuous synchronization between the agencies accountable for carrying out various tasks. This demands a visualization and a clear-cut strategy. The water resources department engaged MOUS with 10 ministries for better application of Namami Gange.

The cleaning of Ganga is not only the responsibility of government but every individual in the community. Basic remedial measures which can be followed to help clean Ganga can be done by setting up an approchable local authority to manage the responsibilty and budget for cleaning area near the ghats, records should be maintained of the tourist and devotees visiting the area and the traffic should be maintained and also local people as well as toursist should be made aware of the responsibilty of keeping the river clean.

7 Conclusion

The Ganga river agonizes from countless problems, most significant ones being the pollution of water. Release of untreated and partly treated sewage and industrial wastewater into the river is a main issue. Change of water flow of river through Upper and Lower Ganga canals, leaving nearly very little flow in the main river making the dilution difficult after treatment.

In Uttar Pradesh, there is requirement of handling of sewage water and accessibility of suitable conveyance system for sewage. It also requires minimum flow for its subsistence in the stretch of Uttar Pradesh. Subsequently a river is a existing eco-system and consequently final goal should be to protect the operation of the river eco-system.

Main industries like Tannery, Sugar, Pulp and Paper mills contributes noteworthy contamination load to river and its tributaries. There is instant requirement of strong environment investigation in order to check their compliance with the standards. It is also essential that minimum flow through the year should be maintained to support eco-system of river and aquatic life. It would be advisable to generate additional water storage capability for river system and release water in the thin period to efficiently sustain minimum flow in the river.

Acknowledgements The authors are highly grateful to the Department of Zoology and Environmental Science, Gurukula Kangri Vishwavidyalaya, Haridwar, India for providing infrastructural facilities.

References

1. Purkait S., Ganguly M., M. W. Aktar, D. Sengupta, A. Chowdhury A.: Impact assessment of various parameters polluting Ganga water in Kolkata region: A Study for quality evaluation and environmental implication. *Environment Monitoring Assessment*. Volume 155. (2009) 443e454
2. Bhattacharji, S; Bandyopadhyay, R.: Legends of Devi. *Orient Blackswan*.(1995) 54. ISBN 978-81-250-0781-4
3. Haritash, A. K., Gaur, S., Garg, S.: Assessment of water quality and suitability analysis of River Ganga in Rishikesh, India. *Applied Water Science*. Volume 6(4). (2016) 383–392
4. Dutta, S., Kole, R. K., Ghosh, S., Nath, D., Vass, K. K. Impact assessment of lead on water quality of River Ganga in West Bengal, India. *Bulletin of Environmental Contamination & Toxicology*. Volume 75(5). (2005)
5. Vass, K. K., Mondal, S. K., Samanta, S., Suresh, V. R., Katiha, P. K.: The environment and fishery status of the River Ganges. *Aquatic Ecosystem Health & Management*. Volume 13(4). (2010) 385–394
6. D'Souza, R. Water in British India: the making of a Colonial Hydrology. *History Compass*. Volume 4(4). (2006) 621–628
7. Roy, M., Shamim, F.: Research on the Impact of Industrial Pollution on River Ganga: A Review. *International Journal of Prevention and Control of Industrial Pollution*. Volume 6(1). (2020) 43–51
8. Dwivedi, S., Mishra, S., Tripathi, R. D.: Ganga water pollution: a potential health threat to inhabitants of Ganga basin. *Environment International*. Volume 117. (2018) 327–338
9. Rai, B.: Pollution and conservation of Ganga River in modern India. *International Journal of Scientific and Research Publications*. Volume*3*(4). (2013) 1–4
10. Priyadarshi, N.: Ganga River pollution in India — A Brief Report. American Chronicle. (2009) (http://www.americanchronicle.com/articles/view/109078)
11. Loehr, R.: Agricultural waste management: problems, processes, and approaches. *Elsevier*. (2012)
12. Trivedi, P., Bajpai, A., Thareja, S.: Evaluation of water quality: physico–chemical characteristics of Ganga River at Kanpur by using correlation study. *Nature and Science*, Volume1(6). (2009) 91–94
13. Alley, K. D.: Images of waste and purification on the banks of the Ganga. City Society. Volume 10(1). (1998) 167–182
14. Singh, A. N., Shrivastava, R., Mohan, D., & Kumar, P. Assessment of spatial and temporal variations in water quality dynamics of river Ganga in Varanasi. *Pollution*. Volume4(2). (2018) 239–250

15. Ghose, A.: Rishikesh. DK Eyewitness Travel Guide: India. DK.(2011).184. ISBN 978-0-75667-026-9. ISSN 1542-1554
16. Reiss, J., Abbott, C., Gold, S. F.: SACRED WATERS: A Pilgrimage Up the Ganges River to the Source of Hindu Culture (Book Review). *Publishers Weekly*. Volume 248(42). (2001) 68–69
17. Karar, A.: Impact of pilgrim tourism at Haridwar. *The Anthropologist*. Volume 12(2). (2010) 99–105
18. Joshi, D. M., Bhandari, N. S., Kumar, A., Agrawal, N.: Statistical analysis of physicochemical parameters of water of River Ganga in Haridwar district. *Rasayan Journal of Chemistry*.Volume 2(3). (2009) 579–587
19. Ahamed, S., Sengupta, M. K., Mukherjee, A., Hossain, M. A., Das, B., Nayak, B., ... and Chakraborti, D.: Arsenic groundwater contamination and its health effects in the state of Uttar Pradesh (UP) in upper and middle Ganga plain, India: a severe danger. *Science of the Total Environment*. Volume 370(2–3). (2006) 310–322
20. Rai, P. K., Mishra, A., Tripathi, B. D.: Heavy metal and microbial pollution of the River Ganga: A case study of water quality at Varanasi. *Aquatic Ecosystem Health & Management*. Volume 13(4). (2010) 352–361
21. Sinha, A., Isaac, R. K.: An analytical FIS model to check the quality of drinking water. In 2017 3rd International Conference on Computational Intelligence & Communication Technology (CICT). (2017) 1–6
22. Naseema, K., Masihur, R., Husain, K. A.: Study of seasonal variation in the water quality among different ghats of river Ganga, Kanpur, India. *Journal of environmental research and development*. Volume 8(1). (2013) 1
23. Tiwary, R. K., Rajak, G. P., Mondal, M. R.: Water quality assessment of Ganga River in Bihar Region, India. *Journal of environmental science & engineering*. Volume 47(4). (2005) 326–335
24. Rahman, M. M., Rahaman, M. M.: Impacts of Farakka barrage on hydrological flow of Ganges River and environment in Bangladesh. *Sustainable Water Resources Management*. Volume4(4). (2018) 767–780
25. Matta, G., Naik, P., Kumar, A. Gjyli, L., Tiwari, A. K. Machell J.: Comparative study on seasonal variation in hydro-chemical parameters of Ganga River water using comprehensive pollution index (CPI) at Rishikesh (Uttarakhand) India. *Desalination and Water Treatment* Volume118. (2018) 87–95
26. Kumar, A., Garg, V. Heavy metal and physico-chemical characteristics of river Ganga from Rishikesh to Brijghat, India. *Journal of Environment and Bio-Sciences*. Volume 33(2). (2019) 243–250
27. Kumar, V., Kumar, S., Srivastava, S., Singh, J., Kumar, P. :Water quality of River Ganga with reference to physico-chemical and microbiological characteristics during Kanwar Mela 2017, at Haridwar, India: A case study. *Archives of Agriculture and Environmental Science*. Volume *3*(1). (2018) 58–63
28. Abed, S. A., Jazie, A. A.: Assessment of some physico-chemical properties for water in Ganga River at Varanasi, India. *Al-Qadisiyah Journal for Engineering Sciences*. Volume *7*(2). (2014) 148–164
29. Satya, K., & Narayan, C. Study of physico-chemical and biological characteristics of the water of river Ganga at Patna, India. *Current World Environment*. Volume 13(3). (2018). 374
30. Basu, S., Banerjee, T., Manna, P., Bhattacharyya, B.,Guha, B.: Influence of physicochemical parameters on the abundance of Coliform bacteria in an Industrial Site of the Hooghly River, India. In Proceedings of the zoological society. *Springer-Verlag*. Volume 66 (1). (2013) 20–26
31. Ali, S. Y., Sunar, S., Saha, P., Mukherjee, P., Saha, S.,Dutta, S.: Drinking water quality assessment of river Ganga in West Bengal, India through integrated statistical and GIS techniques. Water Science and Technology. (2021)
32. Nirmal, K. J. E.: Impact of pollution on aquatic biodiversity. *Journal of Ecology, Environment and Conservation*. 3(3-4). (2007) 20–217
33. Singhania, N.: Pollution in River Ganga. Department of Civil Engineering, Indian Institute of Technology Kanpur. (2011)

34. Choudhury, B. C., Bhupathy, S.: Turtle trade in India: A Study of Tortoises and Freshwater Turtles (New Delhi: World Wildlife Fund for Nature-India/ TRAFFIC-India. (1993) 50
35. NMCG. Namami Gange. Projects along river Ganga available under private funding. National Mission for Clean Ganga (NMCG). (2017).CSR e-book. Available at https://nmcg.nic.in/csr/csrebook/csrebook.html#p=7
36. NMCG. Ganga river basin management plan. Interim report. Prepared by Consortium of 7 Indian Institute of Technologies (IITs). Ministry of Environment and Forests (MoEF), GOI, New Delhi. (2013)

Non-timber Forest Products: Current Status and Development

Faheem Ahamad, Rakesh Bhutiani, Mukesh Ruhela, and Nishant Rai

Abstract Non-Timber Forest Products (NTFP's) are biologically originated non-wood products derived from forests and constitute a vital source of livelihood. In India, approximately 275 million poor rural people inhabiting in harsh climatic conditions depend on NTFP's for their survival and cash livelihoods. There is a lack of accurate global level statistics about NTFPs from 1972 to 1995. After 1995, renewed interest in NFTP's may be due to increased interest in the value of biodiversity, Carbon sequestration, increased awareness about the use of NTFP's, and some environmental functions provided by NTFP's. NTFP's were called "Minor" forest products and undervalued in the past but nowadays the scenario is different. During recent decades much attention is paid on the exploitation of NTFP's without addressing the ecological factors and impact on the environment. In India, approx 40% of total official forest revenues and 55% of forest-based employment is provided by NTFP's. In the present era and coming future era NTFP's sometimes may be considered as misnomer due to severe degradation of forest areas. In the present investigation, an attempt has been made to elaborate on the NTFP's, their utilization, availability, and development along with ecological and environmental implications generated due to the exploitation of NTFP's.

F. Ahamad
Department of Environmental Sciences, Keral Verma Subharti College of Science (KVSCOS), Swami Vivekanand Subharti University, Meerut, (UP) 250005, India

R. Bhutiani (✉)
Limnology and Ecological Modelling Lab. Department of Zoology and Environmental Science, Gurukula Kangri Vishwavidyalaya, Haridwar, (UK) 249404, India
e-mail: rbhutiani@gmail.com

M. Ruhela
Department of Environmental Engineering, Subharti Institute of Technology and Engineering (SITE), Swami Vivekanand Subharti University, Meerut, (UP) 250005, India

N. Rai
Department of Biotechnology, Graphic Era Deemed to be University, Dehradun, (UK) 248002, India

K. D. Bahukhandi et al. (eds.), *Environmental Pollution and Natural Resource Management*, Springer Proceedings in Earth and Environmental Sciences,
https://doi.org/10.1007/978-3-031-05335-1_7

Keywords Economic valuation · Ecological factors · Misnomer · NTFP's · Sustainable development

1 Introduction

With 2.4% of the total world's geographic area and 17.5% contribution in the world population, India has only 1.8% contribution to the world's forestry sector [1]. In India, an estimated 67% of people are involved in the agriculture sector. Approximately 350–400 million people living in 0.152 Million rural villages adjacent to forest depend on the forest for their various needs [2]. According to a Forest Research Institute report (2017), India has 1,47,127 forest fringe villages in 275 districts of India [3]. The maximum number of forest fringe villages was in Madhya Pradesh (18,263).

In global context forest covers approximately 31% (4.06 billion) of the total geographical area. Approximately 80 million green jobs were provided by forests worldwide more than and support the livelihood of approximately 880 million peoples life by providing different products [4] while in India 8.20% of the total geographic area (about 25 Million Hectare) is under agroforestry [5]. Forests are an important source of non-timber forest products (NTFPs) for health care, domestic consumption, and cash or non-cash income generation (Approx 50%) for one-third of the population living in the rural communities [6–8]. Available literature shows that NTFP's are not defined according to the wide range of potential they possess and it also shows the lack of consistency in the available literature. NTFP's are biologically originated non-wood products derived from forests, and trees outside forests. NTFP's may be collected from the wild or produced after processing or products of agroforestry. They include Food, fodder, medicine, oils, resins, fibres, dyes, and raw materials for baskets, traditional paper, houses, brooms, mats, and numerous other items [9]. In the last two decades, more than 2100% increase was observed in NTFP's related research item [10] and most of them are from rural areas till 2012 but in recent times it has been observed that the market of NTFP's shifted from rural to urban centres [11–13].

2 Definitions of NTFP

The term NTFP consists of three essential terms and aspects:

1. **Non-Timber:** The part non-timber of the term NTFP indicates the exclusion of all woody materials as timber, chips, charcoal, and fuelwood.
2. **Forest:** The part forest of the term NTFP indicates that these products should be derived from forests and related land areas.
3. **Products:** The part products of the term NTFP corresponds to biologically originated goods such as plants, animals, and their products other than wood.

4. Certain other functions such as ecotourism, grazing, bio-prospecting (Forest services), and soil conservation, soil fertility, watershed protection (forest benefits) are not included in the term NTFP.

In different parts of the country and at different time NTFP's are known by different other names such as [9]:-

1. Minor forest products
2. Other forest products
3. Other economic forest products
4. Special forest products
5. Non-wood forest benefits
6. Non-wood goods and services
7. Non-wood forest products

A very old definition of NTFP's was provided by de Beer and McDermott [14]. They define NTFP as "all biological materials extracted or obtained from forests for human use other than timber used for commercial purposes". The definition was given by the authors when they were working in the forest, thus they focus on the biological resources (bushmeat, seeds, mushrooms, resins, bulbs, thatch grass, insects, and bark). The definition is wrong with respect to the present scenario as abiotic resources (such as water, carbon, sand, and stone, to mention a few) and social resources (aesthetic use) are not included in it. Although the identity of forest products as NTFP's is due to the harvesting and consumption of these products in human day to day life if human harvest products of NTFP's group excessively then the NTFP's produce is known as crop or livestock [7]. After reviewing the literature we can say that a lot of definitions of NTFP's are available but to date, there is no exact definition available for NTFP's.

3 Importance of NTFP's

NTFP's are of various types and play a different role in the livelihood of millions of rural and urban people across the globe [10, 15–17]. The approximate role of NTFP's in different spheres of life of local people is given in the value ladder given in Fig. 1. Nearly 80% of the population of the developing world depend on NTFP's for their nutritional and health requirement. The people of developing countries usually put their hopes on potential gains generated from the marketing of NTFP's for poverty alleviation and for more conservation of the natural resource base. Approximately 1 billion peoples worldwide depend on the different wild foods such as meat, insects, fruits, mushroom, medicine and fishes to some extent [4].

The importance of NTFP's is greater in daily activities while they get the least attention in cultural activities. NTFP's can be an important source of income that can supplement farming and other activities. Their economic, cultural, and ecological

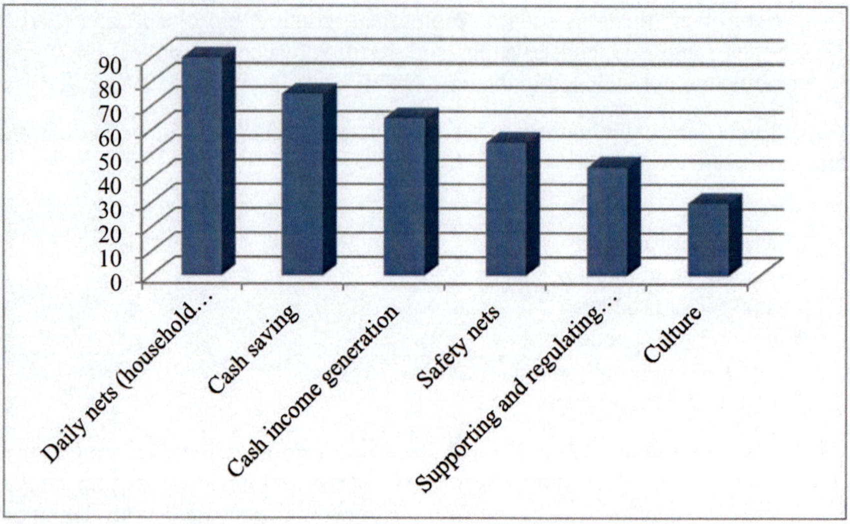

Fig. 1 Showing the role of NTFP's in different spheres of life of local people

value, when considered in aggregate, makes managing NTFP's an important component of sustainable forest management and the conservation of biological and cultural diversity. NTFP's also reduced incentives to convert forests for cattle ranching and farming. Therefore, NTFP's are a dependable source of income and food supply in rural areas.

Daily net refers to the basic necessities such as food, energy, shelter, and medicine [18] Female-headed households use more NTFP's in comparison to male-headed due to poor economic conditions [19, 20]. Similarly, rural areas collect more NTFP's than urban areas while consumption is highest in urban areas [21, 22]. NTFP's act as a safety net or self-insurance to the livelihood of peoples due to support in extreme climatic conditions such as drought and floods. During these calamities, these products help people in two ways. One is the increased use of already included products for their daily needs to reduce the expenditure and the second one is the use of non-included products in their daily nets [23]. Replacement of Kerosene oil or Electricity by the use of firewood is an example.

In certain areas or communities, NTFP's are inseparable parts of their daily essential activities. Some NTFP's are associated with cultures and are necessary for certain rituals and ceremonies. The association of NTFP's with the culture is both beneficial and harmful with respect to the conservation of these valuable products. Particular NTFP's associated with the culture or traditions of a particular community are conserved by that community and those areas become important from the biodiversity point of view [24, 25]. NTFP's also provide supporting and regularities services directly and indirectly. For example *Phragmites* reed and the fruit tree *Sclerocarya birrea* subsp. Caffra provides different services to other plants and animal and in turn support the communities [26, 27]. *Phragmites* and some other dense reeds help in water purification in streams and rivers by removing different nutrients from the water [28].

4 Challenges for NTFP's Development

1. Most of the NTFP's are not widely known by the peoples due to their traditional uses.
2. Use of NTFP's is not on the record as the production of most of the NTFP's is seasonal; therefore their potential is underestimated to date.
3. Trade of NTFP's falls under the category of the unorganized and informal sector due to their transaction mostly in households and small-scale units or outside the well-established marketing /channels systems and this is a major bottleneck in the development of NTFP's.
4. Policy planners and forestry professional mostly focuses on timber-orientation and promote timber forestry.
5. Lack of knowledge regarding the role of NTFP's in the livelihood of local people is a major constraint.

5 Current Status and Development of NFTP's

Worldwide 12 countries are blessed in terms of diversity of vegetation and animals and termed as mega biodiversity countries and India in one of them (ranked 12) due to its unique climatic and physiographic factors. During the recent decades, NTFP's become popular in developed as well as developing countries due to shifting of human beings from conventional to non-conventional sources. Vegetation is divided into tropical, subtropical, temperate, alpine and sub-alpine type. NTFP's can be derived from approx. 3000 species of plants. Less than half (about 40%) of all the NTFP's are accounted into revenue generation and the remaining large fraction is consumed by local people for their livelihood [29]. As per the National Forest Policy of 1988, NTFP's should be protected and improved as they sustained the local communities. In northern Brazil NTFP (*Himathantus drasticus* -janaguba -Apocynaceae) contribute 30–50% in the household income.

Based on their importance and marketing, NTFP's are classified as nationalized and non-nationalized. Nationalised NTFP's is controlled by the state and have high economic values while the state has no control over non-nationalized NTFP's and their market is controlled by local communities. Minor Forest Product Federation promotes in-situ conservation, ex-situ cultivation, propagation, value addition, processing, and marketing of NTFP's to provide benefit to the local communities. Tendu (*Diospyrus melanoxylon*) leaves also known as bidi patta, sal seed, bamboo, etc. are nationalized NTFP's while mahua, aonla, neem, mahul patta, chironji, tamarind, and honey are non-nationalized NTFP's [2] are extracted from NTFP's

[30]. More than 85% of herbal drugs used are derived from medicinal plants [31] and the production of these drugs provides livelihoods to millions of people in the Indian Himalayan region [32]. NTFP's play an important role in the economic upgradation of India because NTFP's helps in the livelihood of millions (nearly 400 million) of peoples living in forest areas such as tribal, landless people, women, and other rural poor [15, 33, 34]. About 70% of the total NTFP's is collected in the tribal belt.

Demand of basket increased from 4791 in 1992 to 15,000 in 2007 in Uganda [35]. Welford and le Breton [36] reported 15–20% increase per annum in global natural products industry. Nearly 1.6 billion US dollar was generated from a commercial collection of NTFP's in India, at the forest gate in 2010 [37, 38]. NTFP's contribute to about 75% of the total forest export revenue. NTFP's contribute to about half of the cash income for 30% of rural forest-dependent communities across India [39–41]. Nearly 17% of landless laborers get their wage labor from the collection of NTFP's directly and nearly 39% indirectly. During the last 15 years, world has shown tremendous progress in the contribution of forests in empowering people by providing ownership of NTFP's, which results in their poverty remediation.

Among all the medicinal plants of, some plants are used in the preparation of medicines and raw extracts for Ayurveda, Unani, modern medicines, Siddha, Homeopathy, and other herbal health products [42]. Threat along with employment also increased due to the demand for medicinal plants in national and international markets for the preparation of herbal medicines and nutritious food products [43]. The revenue generated from the global market of the Medicinal, Aromatic, and Dye Plants (MADP's) and other NTFP's beside the household consumption is about 59 billion US$ per year [29]. NTFP's used in the pharma sector developed to a large extent in comparison to other NTFP's.

5.1 Management of NTFP's

Due to logging and shifting, a large amount of forest area has degraded as a result NTFP's are continuously degrading, affecting the plant species distribution [44]. After recognizing the importance of indigenous plant species to improve the bread and butter and income of poor people, a program for the domestication of all these was initiated in the early 1990s, now enter in the fourth decade [45]. The new term used for the domestication of trees is Horticulture, in which the local people are actively engaged which ultimately empowers the rural peoples [46–49].

In India, it is the joint responsibility of national and state governments to manage the non-timber forest resources. Ownership of NTFP's was provided to Gram Sabhas/panchayats (village assemblies) by "Panchayats (Extension to Scheduled Areas) Act 1996" (PESA) while the Scheduled Tribes and other Traditional Forest Dwellers (Recognition and Forest Rights) Act 2006 popularly known as Forest Right Act 2006 was enacted by which a right was provided to forest dweller for inhabitation, cultivation on the forest land occupied before 13th December 2005 to manage, protect and regenerate the forest land. This also allows the collection and disposition of minor forest products (NTFP's) for their development. Here the word "ownership" means the right to net revenue after retaining all the additional charges (access and controlling charges) and administrative expenses of the department.

5.2 Contribution in the Economy

At global level, NTFP's generate 88 billion US dollar in 2011 [39]. No report or study is available about the exact contribution of the NTFP's to total employment or in the revenue generated in a particular state. A report suggests that NTFP's contribute to about 25% in rural household income in developing world [50]. More than 64% contribution in household income in Northeast Peru was reported by L'Roe and Naughton-Treves [51] while less than 5% was reported from Northeast and Eastern South Korea by Van Gevelt [52]. According to a report of NCFRA [53] around 100 million people in India derive their source of livelihood directly from the collection and marketing of NTFP's. Most of NTFP's collections take place in the six-state of India like Maharashtra, Madhya Pradesh, Chhattisgarh, Bihar, Orissa, and Andhra Pradesh [54] and in small amount from Uttarakhand, Arunachal Pradesh and Uttar Pradesh. Production and collection potential of Rs 3777 crore and Rs 1908 crores per annum is estimated from 14 different NTFP's. Tendu leaf, Bamboo, Mahuwa seed, and flower and Gum karaya are the major contributor NTFP's in the economy of different countries worldwide. According to an estimate, equivalent to 2.7 billion US dollars per year revenue is contributed by all these biological resources in India. NTFP's contributed to about 55% in the total forestry sector employment, 70% in the forest-based export income, and about 50% in forest revenues [55]. In the 60 billion US$ global herbal industry, 40 billion US$ is contributed by pharmaceuticals, 5.9 billion US$ by spices and herbs, 7 billion US$ by natural cosmetics, and 4 billion US$ by the essential oil. The industry is expected to reach 5 trillion US$ by the year 2050 with the present growth rate of % per year [56].

Sal seed possesses high commercial values and is a constituent of different products as soap, oil, animal or poultry feed, chocolates, rocket fuel, and tanning processes. Chhattisgarh, Madhya Pradesh, and West Bengal are considered as a major producer of sal seed among all these, the average yearly production of Chhattisgarh (14,000 Tonnes/year) is the highest. Collection of sal seed is a complex task and an individual is able to collect approx. 8–10 kg of seed per day worth INR 50, far less than daily minimum wages given in the National Rural Employment Guarantee Scheme of the Government of India. Thus the production of sal seed decreases in recent years [1]. Economically and socially weaker sections of the community of hilly regions mostly depend for their livelihood on Cane or Rattans. Rattans are mainly used for the production of ropes, furniture, walking sticks, umbrella handles, polo sticks, baskets, mat, sports goods, wickerwork, stuffing, and packaging. They are also used for medicinal purposes.

With 65% of the world's total lac production, India stands on the first rank. The annual production of India is estimated at 16,000 Tonnes per Year. Major lac producing states are Chhattisgarh, Jharkhand, West Bengal, Madhya Pradesh, Odisha, Uttar Pradesh, and Gujarat. Chhattisgarh and Jharkhand contribute to about 40% of India's Lac production. Lac is mainly used in pharmaceuticals, food, perfumes, varnishes, cosmetics, adhesives, polishes, paints, and textile industries. Commercial production of Gums is restricted to *Leguminosae, Sterculiaceae,* and *Combretaceae,* although it can be produced by a large number of families. Gums are also extracted from seeds, seaweeds, micro-organisms, and Aloe barbadensi, wood chips of Laris accidentals, seed coat of wheat, brans, barley, rice, and soybean. Maximum gum production in the country is contributed by Maharashtra (30%), Madhya Pradesh (21%), Jharkhand (16%), and Telangana (10%).

Resins are mostly extracted from branches and cones spontaneously and sometimes from wounded plants. Hard, oleo, and gum resin are three categories of resins. The highest amount of resin in India is collected from tapping from pine trees. Uttarakhand, Jammu & Kashmir, Himachal Pradesh, and Arunachal Pradesh are the major resin producing states of India, among all these states, the Major contributor in Uttarakhand. More than 70% of the total consumption is exported from China due to the policy to reduce the number of tapping trees. Tendu (*Diospyrus melanoxylon*), is a high commercial value nationalized NTFP used in the Bidi industry. Major producing states are Chhattisgarh, Jharkhand, West Bengal, Madhya Pradesh, Odisha, Uttar Pradesh, Karnataka, Arunachal Pradesh, and Gujarat. The estimation of the contribution of any sector in the economy of a country is listed in detail in the National Industrial Classification (NIC) abstracted from the International Standards of Industrial Classification (ISIC). Forestry is a subgroup of Section A (Agriculture, forestry, and fishing) and a part of Division 02 and named as Forestry and logging (Table 1).

5.3 Forest Certification

The public becomes more conscious of environmental issues both in developed and developing countries [57, 58]. Forest certification is a tool to link sustainable management practices with that of the environmentally-conscious person [59–61]. Forest certification is a major challenge for the sustainable management of NTFP's. Forest management unit certification (FMU) and chain-of-custody certification (CoC) are two separate processes in forest certification. FMU is the process in which it is verified that the area from where the NTFP's are extracted is being managed by a pre-defined standard while the process of certification that the product has originated from a certified forest is known as CoC [58]. Forest Stewardship Council (FSC) and the Program for Endorsement of Forest Certification Schemes (PEFC) are two international certification programs available for forest certification to date. Besides these two, there are other certification programs working on the national level such as American Tree Farm System (ATFS), Canadian Standards Association (CSA), Sustainable Forestry Initiative (SFI), Malaysian Timber Certification Council (MTCC), and China Forest

Table 1 Detailed industrial structure in respect of forestry and logging sector according to the NIC [1]

Section A		Agriculture, forestry and fishing
Division 02		**Forestry and Logging**
Group 021		**Silviculture and other forestry activities**
Class 2010		Silviculture and other forestry activities **Exclusion:** Growing of Christmas trees, Operation of tree nurseries, Production of wood chips and articles
	Sub-class 02,101	Growing of standing timber,
	Sub-class 02,102	Operation of forest tree nurseries
	Sub-class 02,109	Other forestry activities including growing of pulp and fire wood etc
Group 022		**Logging**
Class 0220		Logging **Exclusion:** Growing of Christmas trees, Growing of standing timber, gathering of wild growing NWFP's, Production of charcoal through distillation of wood
	Sub-class 02,201	Gathering and preparation of firewood
	Sub-class 02,202	Logging camps and loggers primarily engaged in felling timber and producing wood in the rough such as pitprops, split poles, pickets hewn
	Sub-class 02,203	Railway ties
	Sub-class 02,209	Production of charcoal in the forest and other logging activities
Group 023		**Gathering of NWFP's**
Class 0230		Gathering of NWFP's **Exclusion:** Managed production of any of these products, growing of mushroom or truffles, growing of berries and nuts, gathering of firewood, production of wood chips
	Sub-class 02,301	Gathering of Tendu leaves
	Sub-class 02,302	Gathering of Lac, Resins and Rubber like Gums
	Sub-class 02,303	Gathering of wild growing truffles, mushroom, nuts, berries, cork, balsams, vegetable hair, eelgrass, mosses, lichens
	Sub-class 02,309	Gathering of NWFP's
Group 024		**Support services to forestry**
Class 0240		Support services to forestry **Exclusion:** Operation of forest tree nurseries
	Sub-class 02,401	Forestry services activities
	Sub-class 02,402	Logging services activities such as transport of logs within the forest

Certification Council (CFCC). At the global level, two-third of the forest area is certified by PEFC while 28% is by FSC. The first forest certification (FSC CoC) in India was reported in 2001 (to a toy manufacturer in UP for Babul (*Acacia nilotica*) and Shisham (*Dalbergia sissoo*). There were 328 FSC CoC certifications and 11 PEFC CoC certifications issued till July 2012. However, to date, there is no certification reported for NTFP's. Shanley et al. [62] reported that standards for certification have been approved for approximately forty-six commercial NTFP's and the evaluations are under process. At the international level, the first certification of NTFP was reported in 1998 to SmartWood's Chicle operation in Mexico, in 1999 [63]. The first FSC certified NTFP- Chicle gum failed in the market place due to explosive global demand for chewing gum [64]. There are 17 criteria and 55 indicators in certification including the ecological, social, and economic perspective of NTFP's. The major challenges in the certification of NTFP's are the availability of secure forest land and insufficient scientific knowledge about the distribution, life cycle, population density, regeneration, and level of sustainable harvesting.

5.4 Marketing or Trade of NTFP's

Literature suggests that tribal involvement in NTFP's collection since time immemorial for their daily need but later they start collection and harvesting at the commercial level [65]. The use of NTFP's started with human existence. In the old era, the domestication of wild varieties of plants and animals starts to such an extent that the modern world forgot the natural origin of most of the modern-day staple foods [66]. Selected NTFP's such as medicinal plants (US$ 689.9 million), nuts (593.1), ginseng roots (389.3), cork and cork products (328.8), and essential oils (312.5) contributing to international trade [67]. Despite India's rich biodiversity (45,000 plant species in 16 Agro-climatic zones), only 3000 NTFP species yield is found. Among them, marketability of only 126 has developed [68–70]. NTFP and eco-tourism contribute to about 16% of the Forestry sector's gross value [71, 72]. At present, about 150 NTFP's are important in terms of national as well as international trade in India. The international market for medicinal use NTFP's is estimated to be 60 billion US dollars with a growth rate of 7% per annum [1, 73]. Vietnam exports NTFP's to about 90 countries [74]. The following two channels were identified for the marketing of NTFP's by Kumar and Meena [65].

Channel I: Tribe seller's → Village traders → Regulated Market (Laghu Van Upaj Mandi).

Channel II: Tribe seller's → Large Sized Aadivasi Multipurpose Co-operative Society (LAMPS).

Trade-in case of NTFP's is categorized as in the house, local level, regional level, national level, and international level. The income generated at each level of trade varies depends upon the engagement of a person in particular full-time or part-time and raw or processed NTFP's [73]. The rich people focus on the trade of low volume and high value while poor people trade in the opposite patterns [75]. The poor people focus on the collection of NTFP's of their own while rich people focus on buying instead of the collection [76, 77]. The increasing awareness in India regarding Ayurveda and PM Modi's Aatma Nirbhar Bharat Abhiyan will make many folds increase in the business of NTFP's.

6 Harvesting of NTFP's

Harvesting and extraction of NTFP's provide income to local communities without destroying the habitat [78, 79]. In olden times, the collection and utilization of NTFP's were on a low scale but as the awareness about their potential has increased among the peoples, their collection and utilization also increased which raised the issue of their sustainability. According to the definition of sustainability, sustainable harvesting of non-timber forest resources is defined as the harvesting of NTFP's in which the nuts, fruits, barks, latexes, roots, rhizomes, and different other products can be harvested in undefined quantity from a limited area of the forest without any impact on the structure and dynamics of the particular plant population. In other words, sustainable harvesting is defined as "the harvesting of particular species without any change in the species composition to maintain its population at natural or near-natural levels".

The increasing demand for the commercialized herbal industry at the national and international levels put pressure on the management and harvesting procedures for the maintenance of NTFP's. The first and most important step in sustainable harvesting is sustainable collection. Sustainable collection of NTFP's requires skilled labor. Due to low prices, the collectors use unskilled labor and use unfriendly harvesting processes to increase the income. Due to a lack of knowledge and skills, they destroy the whole plant in spite of the required parts of the plant.

Conservation of genetic diversity, habitat loss, quality, consistency, and certification of products is a major concern that arises due to increasing demand for NTFP's. Inadequate supply of quality raw material and processing infrastructural facilities, documentation, storage and transportation, standards operating procedures, marketing linkages, regulation of trade, and conservation of resources are the big challenges in front of NTFP's sector [29]. The tendency to collect unripe fruits, damaging whole tree, repeated forest fires, uncontrolled grazing, technical issues, institutional issues, lack of market transparency, and illicit removal are some of the problems in the sustainable harvesting of NTFP's. All these problems occur due to lack of knowledge, competition between the collectors, and unhealthy forest policies. On the basis of the parts used, the NTFP's species have been classified as follows [29, 80]:-

1. Fruit/seeds
2. Flowers
3. Leaves
4. Root / rhizome leaves and flower
5. Bark

6. Gum/ resin
7. Entire plant (All plant parts being useful)

Collection of used parts, stage of collection, time of collection, method of collection, and quantity required are five important factors for the sustainable harvesting of NTFP's. Checklist to assess harvesting sustainability [29]:-

1. Knowledge of the natural distribution of the species
2. Frequency of occurrence or abundance
3. Population structure (age/size/class distribution)
4. Dynamics of the species (growth and reproduction rates)
5. Variation among habitats
6. Role played within the ecosystem

7 Impact of Harvesting on the Biodiversity and the Environment

Harvesting of NTFP's has short and long-term effects on the ecosystem and overall biodiversity. Short term effects of harvesting are the growth rate or reproduction capacity of the plant while long term effects can be observed on ecosystem levels. Harvesting of NTFP's can be sustainable when sufficient time is given between the two harvestings so that there will be no negative impact on the ecosystem functions, abundance species, or community [81]. The two basic problems in the study of the impact of harvesting on the ecosystems are the duration of the study {assessment needs long study period and a separate study of impact (as in nature it is very difficult to study the impact of harvesting of particular NTFP's on other species)}. In practical cumulative impacts are observed. The other factors affecting sustainability is the harvesting of plant parts, time of harvesting, and care taken during harvesting. The trees with a large number of fruits are least affected by the harvesting while the impact is opposite with less number of fruits [81, 82].

Although justified harvesting of NTFP's is useful for the livelihood of rural peoples and tribal. Overexploitation of these natural commodities exerts a negative impact on the environment. A study conducted in Rajasthan reported the degradation of the environment due to the exploitation of NTFP's [83, 84]. The persons trapped in the confusion of socio-economic dilemma often work innovatively and cooperatively as reported by Ostrom [85, 86] and other common property [87] and collective action theorists [88]. A study conducted by Murli et al. [89] also reported the degradation of species yielding NTFP's resulting in environmental degradation. Ticktin [90] also reported the impact on different ecological processes from individuals to the ecosystem level.

8 Conclusion

The present paper aimed to find out the exact definition, importance, harvesting, certification, current status, and development of NTFP's in India. To solve the objectives of the present study, a survey of the available literature on NTFP's at national and international was performed. NTFP's plays an important role in poverty reduction by the sustainable and judicious uses of the product other than wood. Among all the sectors, pharma is the largest sector of NTFP's contributing to the economy of the country which results in the economic up-gradation of the country as well as rural and tribal people. Among all the products, the production of tendu leaves is highest in recent years. In India certification of forest products was started long back but certification of NTFP's is not started yet. Certification of NTFP's will help in the development of NTFP's. There is a need for sustainable harvesting of NTFP's as the exploitation of NTFP's effect on the environment and biodiversity to a large extent. Training and awareness programs in the rural and adjacent areas of the forest region will help in sustainable harvesting with more economic benefits and fewer environmental impacts.

References

1. ICFRE (Indian Council of Forestry Research and Education): Forest sector Report India 2019. (2019c) 1–207
2. MoEF&CC: State of environment report New Delhi. Ministry of Environment, Forest and Cimate Change, Government of India (2009)
3. FRI (Forest Research Institute): Forest Resource Dependence and Ecological Assessment of Forest Fringes in Rainfed Districts of India, National Report, Forest Research Institute, Dehradun (2017)
4. FAO (Food and Agriculture Organization of the United Nations): The state of the world's forests. (2020) 1–30
5. Dhyani, S.K., Handa, A.K.,Uma.: Area under agroforestry in India: An assessment for presenty status and further perspective. Indian Journal of Agroforestry, (2013) 15(1): 1–11
6. Shackleton, C.M., Shackleton, S.E., Buiten, E., Bird, N.: The importance of dry forests and woodlands in rural livelihoods and poverty alleviation in South Africa. Forest Policy and Economics, (2007a) 9: 558–577
7. Shackleton, C.M., Ticktin, T., Pandey, A.K.: The need to understand the ecological sustainability of non-timber forest products harvesting systems. Ecological Sustainability for Non-Timber Forest Products: Dynamics and Case Studies of Harvesting. Routledge, London, UK, ISBN-13: 978–0415728591, Ed by Charlie M Shackleton, Ashok K. Pandey, Tamara Ticktin : (2015a) 1–295
8. Angelsen, A., Jagger, P., Babigumira, R., Belcher, B., Hogarth, N.J., Bauch, S., Borner, J., Smith-Hall, C., Wunder, S.: Environmental income and rural livelihoods: a global comparative analysis. World Development. https://doi.org/10.1016/j.worlddev.2014.03.006 (2014)
9. ICFRE (Indian Council of Forestry Research and Education): Cultivation of Selected NWFP's-A manual: (2019a) 1–218
10. Shackleton, C.M., Pandey, A.K., Ticktin, T.: Non-timber forest products in Livelihoods. Ecological Sustainability for Non-Timber Forest Products: Dynamics and Case Studies of Harvesting. Routledge, London, UK, ISBN-13: 978–0415728591, (2015b) Pages: 294

11. Lewis, J.A.: The power of knowledge: information transfer and acai intensification in the peri-urban interface of Belem, Brazil. Agroforestry Systems (2008) 74: 293-302
12. Padoch, C.E., Brondizio, S., Costa, M., Pinedo-Vasquez, R., Sears, R., Siqueira, A.: Urban forest and rural cities: multi-sited households, consumption patterns, and forest resources in Amazonia. Ecology and Society (2008)13(2): 2 (online)
13. McMullin, S., Phelan, J., Jamnadass, R., Iiyama, M., Franzel, S., Nieuwenhuis, M.: Trade in medicinal tree and shrub products in three urban centres in Kenya. Forests, Trees and Livelihoods (2012) 21: 188–206
14. de Beer, J.H., McDermott, M.J.: The Economic Value of Non-Timber Forest Products in Southeast Asia (2nd ed.) IUCN, Amsterdam (1996)
15. Pandey, A.K., Tripathi, Y.C., Kumar, A.: Non Timber Forest Products (NTFPs) for Sustained Livelihood: Challenges and Strategies. Research Journal of Forestry, 1–7. DOI: https://doi.org/10.3923/rjf.2016 (2016)
16. Pandey, A.K., Bhargava, P., Negi, M.S.: Sustainable management of non-timber forest produce through joint forest management. Indian Forester (2011) 137: 105-113
17. Malhotra, K.C., Bhattacharya, P.: Forest and Livelihood. CESS Publisher, Hyderabad, India, (2010) Pages: 246
18. Shackleton, C.M., Shackleton, S.E.: The importance of non-timber forest products in rural livelihood security and as safety-nets: evidence from South Africa. South African Journal of Science (2004) 100: 658–664
19. Shackleton, S.E., Campbell, B., Lotz-Sisitka, H., Shackleton, C.M.: Links between the local trade in natural products, livelihoods and poverty alleviation in a semi-arid region of South Africa. World Development (2008) 36: 505–526
20. Paumgarten, F., Shackleton, C.M.: The role of non-timber forest products in household coping strategies in South Africa: the influence of household wealth and gender. Population and Environment (2011) 33: 108–131
21. Ward, C.: Livelihood changes along the rural-urban continuum with an emphasis on natural resources. MSc thesis, Rhodes University, Grahamstown, 151 pages (2012)
22. Kaoma, H., Shackleton, C.M.: Collection and use of urban tree products by households in poorer residential areas of three South African towns. Urban Forestry and Urban Greening (2014) 13: 244–252
23. Takasaki, Y.: Do the commons help augment mutual insurance among the poor? World Development (2011) 39: 429–438
24. Cocks, M.L., Dold, T., Vetter, S.: God is my forest–Xhosa cultural values provide untapped opportunities for conservation. South African Journal of Science (2012) 108: 880
25. Byers, B.A., Cunliffe, R.N., Hudak, A.T.: Linking the conservation of culture and nature: a case study of sacred forests in Zimbabwe. Human Ecology (2001) 29:187–218
26. Hall, J.B.O., Brien, E.M., Sinclair, F.L.: *Sclerocarya birrea*: a monograph. School of Agricultural and Forest Science 19, University of Wales, Bangor, 157pages (2002)
27. Weis, J.S., Weis, P.: Is the invasion of the common reed, *Phragmites australis*, into tidal marshes of the eastern US an ecological disaster? Marine Pollution Bulletin (2003) 46: 816–820
28. Findlay, S., Groffman, P., Dye, S.: Effects of *Phragmites australis* removal on marsh nutrient cycling. Wetlands Ecology and Management (2003) 11: 157–165
29. ICFRE (Indian Council of Forestry Research and Education): Sustainable Harvesting of Non Timber Forest Produce (NTFPs)-An Operational Manual: (2019b) 1–102
30. Chen, Shi-Lin., Hua, Y., Hong-Mei, L., Qiong, W., Chun-Fang, L.A., Steinmetz.: Conservation and sustainable use of medicinal plants: Problems, progress and prospects. Chinese Medicine (2016) 11(1): 37
31. Prasad, R., Bhattacharya, P.: Sustainable harvesting of medicinal plant resources. In S.B. Roy, ed. Contemporary studies in natural resource management in India New Delhi, India, Inter-India Publications (2003) 168–198
32. Phondani, P.C., Negi, V.S., Bhatt, I.D.: Promotion of medicinal and aromatic plants cultivation for improving livelihood security: a case study from West Himalaya, India. International Journal of Medicinal and Aromatic Plants (2011) 1: 245- 252

33. Stanley, D., Voeks, R., Short, L.: Is Non-Timber Forest Product Harvest Sustainable in the Less Developed World? A Systematic Review of the Recent Economic and Ecological Literature. Ethnobiol. Conserv. (2012) 1: 2012–2018
34. Shackleton, C.M., Pandey, A.K.: Positioning non-timber forest products on the development agenda. For. Policy Econ, (2014) 38: 1–7
35. Muhwezi, O., Cunningham, A.B., Bukenya-Ziraba, R.: Lianas and livelihoods: the role of fibrous forest plants in food security and society around Bwindi Impenetrable National Park, Uganda. Econ. Bot. (2009) 63: 340–352
36. Welford, L., le Breton, G.: Bridging the gap: Phytotrade Africa's experience of the certification of natural products. For Trees Livelihoods (2008) 18: 69–79
37. Kumar, A., Kumar, R., Lakhchaura, P.C.: Global forest resources assessment 2015: country report India. Rome (2015)
38. Joshi, S.: Super market, secretive. Exploitative, is the market in the minor forest produce unmanageable? Down to Earth. (2003) 28
39. FAO (Food and Agriculture Organization of the United Nations): The state of the world's forests, 2018: forest pathways to sustainable development. Rome: Food and Agriculture Organization of the United Nations, (2018)
40. Meinhold, K., Darr, D.: The Processing of Non-Timber Forest Products through Small and Medium Enterprises—A Review of Enabling and Constraining Factors. Forests, 10, 1026; doi:https://doi.org/10.3390/f10111026 (2019)
41. Bhattacharya, P.: Non-timber forest products (NTFPs) related policy and legal issues in India: promoting NTFP-based livelihoods for forest dwellers: results of a practitioner workshop. Hyderabad, Telangana (2015)
42. Rastogi, R.P., Mehrotra, B.N.: A Compendium of Medicinal Plants Vol. 1. Central Drugs Research Institute, Lucknow and Publicationand Information Directorate, New Delhi-497 (1993)
43. Mukherjee, P.K.: Evaluation of Indian traditional medicine. Drug Information Journal (2001) 35: 623-632
44. Leakey, R.R.B.: Non-Timber Forest Products—A Misnomer? Journal of Tropical Forest Science (2012) 24(2): 145-146
45. Leakey, R.R.B., Tchoundjeu, Z., Schreckenberg, K., Shackleton, S., Shackleton, C.: Agroforestry Tree Products (AFTPs): targeting poverty reduction and enhanced livelihoods. International Journal of Agricultutral Sustainability (2005) 3: 1–23
46. Tchoundjeu, Z., Asaah, E., Anegbeh, P.O., Degrande, A., Mbile, P., Facheux, C., Tsobeng, A., Atangana, A.R., Ngompeck, M.L.: Putting participatory domestication into practice in west and central Africa. Forests, Trees and Livelihoods (2006) 16: 53–70
47. Tchoundjeu, Z., Degrande, A., Leakey, R.R.B., Simons, A.J., Niminog, K.E., Asaah, E., Facheux, C., Mbile, P., Mbosso, C., Sado, T., Tsobeng, A.: Impact of participatory tree domestication on farmer livelihoods in west and central Africa. Forests, Trees and Livelihoods (2010) 19: 219–234
48. Asaah, E.K., Tchoundjeu, Z., Leakey, R.R.B., Takousting, B., Njong, J., Edan, I.: Trees, agroforestry and multifunctional agriculture in Cameroon. International Journal of Agricultutral Sustainability (2011) 9: 110-119
49. Lombard., C., Leakeyy, R.R.B.: Protecting the rights of farmers and communities while securing long term market access for producers of non-timber forest products: experience in southern Africa. Forests, Trees and Livelihoods (2010) 19: 235–249
50. Wunder, S., Borner, J., Shively, G., Wyman, M.: Safety nets, gap filling and forests: a global comparative perspective. World Development, www.sciencedirect.com/science/article/pii/S0305750X14000710, DOI: https://doi.org/10.1016/j.worlddev.2014.03.005 (2014)
51. L'Roe, J., Naughton-Treves, L.: Effects of a policy-induced income shock on forest-dependent households in the Peruvian Amazon. Ecol Econ (2014) 97: 1-9
52. Van Gevelt, T.: The economic contribution of non-timber forest products to South Korean mountain villager livelihoods. For Trees Livelihoods (2013) 22: 156-169

53. NCFRA: Report of the National Committee on Forest Rights Act, 2011 (2011)
54. Mamidi, P.: A reprt on the role of Non-Timber Forest Products (NTFP) in Tribal Economy of India – An Exploratory Study. (2016) 1–29
55. Chauhan, K.V.S., Sharma, A.K., Kumar, R.: Non-timber forest products subsistence and commercial uses: trends and future demands. International Forestry Review (2008) 10: 201–216
56. Ministry of Commerce and Industry: Export of Herbs and Herbal Products http://pib.nic.in/newsite/PrintRelease.aspx?relid=187-278 (2019)
57. Perera, P., Vlosky, R.: A history of forest certification. Retrieved April 9, 2009, from http://www.lfpdc.lsu.edu/unece/certificateEnvironment/documents/2003-2006/ce03_001.pdf (2006)
58. Yadav, M., Dugaya, D.: Non-timber forest products certification in India: opportunities and challenges. Environ Dev Sustain, 15:567–586, DOI https://doi.org/10.1007/s10668-012-9393-1 (2013)
59. Mallet, P., Karmann, M.: Certification of non-timber forest products: An emerging field. www.fallsbrookcentre.ca. (2000)
60. Rametsteiner, E.: Sustainable forest management certification. Frame conditions, system design and impact assessment. Vienna: Ministerial Conference on the Protection of Forests in the Europe, European Commission. (2000)
61. Rametsteiner, E., Simula, M.: Forest certification—an instrument to promote sustainable forest management? Journal of Environmental Management (2003) 67: 87–98
62. Shanley, P., Pierce, A., Laird, S., Robinson, D. et al.: Beyond timber: Certification of non-timber forest products.Ajoint publication of Forest Trends, CIFOR and International People and Plants (pp. 1–68). Retrieved January 5, 2012, from http://www.cifor.org/online-library/browse/view-publication/publication/2543.html (2005)
63. SmartWood: Non timber forest product certification standards addendum. SmartWood, Practical Conservation through Certified Forestry, November, 1999. Retrieved March 1, 2011, from http://www.rainforest-alliance.org/forestry/documents/ntfp-addendum-generic-guidelines.pdf (1999)
64. Shanley, P., Pierce, A.R., Laird, S.A., Robinson, D. et al.: Beyond timber: Certification and management of non-timber forest products. Bogor, Indonesia, Center for International Forestry Research (CIFOR). 144p. ISBN: 978–979–1412–44–5 (2008)
65. Kumar, S., Meena, G.L.: Procurement and marketing of non-timber forest products. Indian Forester (2018) 144 (3): 252-259
66. Laws, B.: Fifty Plants that Changed the Course of History, Struik Nature, Cape Town (2011)
67. Walter, S.: Certification and benefit-sharing mechanisms in the field of non-wood forest products – an overview. Medicinal Plant Conservation, Volume 8, Newsletter of the IUCN Species Survival Commission, Medicinal Plant Specialist Group Bonn (2003)
68. Maithani, G.P.: Management perspectives of Minor Forest Produce. MFP News, October-December,1994 Dehradun (1994)
69. FAO (Food and Agriculture Organization of the United Nations): Non-wood Forest Products in 15 countries of Tropical Asia: An overview, P. Vantomme, A. Markkula & R. N. Leslie, eds. Bangkok. (also available at www.fao.org) (2002)
70. Yadav, M., Basera, K.: Status of Forest Products Production and Trade. Report of Centre for Sustainable Forest Management and Forest Certification, Indian Institute of Forest Management, Bhopal (2013)
71. MoEF: Report of the National Forest Commission. MoEF. GoI, available at http://envfor.nic.in/nfc/nfc.htm (2006)
72. Choudhury, P.R.: Forest-route to poverty alleviation- Myths and Realities: Analysis of NWFP-livelihood linkages in some Indian States (2007)
73. Cunningham, A.B.: Non-timber products and markets: Lessons for exportoriented enterprise development from Africa', in Shackleton, S.E., Shackleton, C.M. and Shanley, P. (eds) *Non-Timber Forest Products in the Global Context*. Springer, Heidelberg (2011) 83–106
74. Nguyen, T.V., Hua, L.V.J., Huyen, V.T.T., Zhang, B.: Determinants of Non-Timber Forest Product Planting, Development, and Trading: Case Study in Central Vietnam. Forests 11: 116; doi:https://doi.org/10.3390/f11010116 (2020)

75. Mulenga, B.P., Richardson, R.B., Tembo, G., Mapemba, L.: Rural household participation in markets for non-timber forest products in Zambia. Environment and Development Economics (2013) 19: 487–504
76. Shackleton, C.M., Shackleton, S.E.: Household wealth status and natural resource use in the Kat River Valley, South Africa. Ecological Economics (2006) 57: 306–317
77. Shackleton, S.E., Shanley, P., Ndoye, O.: Invisible but viable: recognising local markets for non-timber forest products. International Forestry Review (2007b) 9: 697–712
78. Leisher, C., Sanjayan, M., Blockhus, J., Kontoleon, A., Larsen, S.N.: Does Conserving Biodiversity Work to Reduce Poverty? A State of Knowledge Review, Nature, Conservancy, Washington, DC (2010)
79. Ticktin, T.: The ecological sustainability of non-timber forest product harvest: Principles and methods. Ecological Sustainability for Non-Timber Forest Products: Dynamics and Case Studies of Harvesting. Routledge, London, UK., ISBN-13: 978–0415728591, Pages: 294 (2015)
80. IIFM: Bhopal-India process as the basis for developing standards and integrated approach of forest certification. C&I India Update, 6(1): 9. Bhopal: Indian Institute of Forest Management. www.iifm.ac.in/sfmindia (2007)
81. Ticktin, T., Johns, T.: Chinanteco management of Aechmea magdalenae (Bromeliaceae): implications for the use of TEK and TRM in management plans. Economic Botany (2002) 56: 43–57
82. Rist, L., Kaiser-Bunbury, C.N., Fleischer-Dogleyd, F., Edwards, P., Bunbury, N., Ghazoul, J.: Sustainable harvesting of coco demer, Lodoicea maldivica, in the Vallee de Mai, Seychelles. Forest Ecology and Management (2010) 260: 2224–2231
83. Ray, S.: Management of natural resources – institutions for sustainable livelihoods: the case of Rajasthan. New Delhi: Academic Foundations (2008)
84. Natcher, D., Kalagnanam, V., Rawal, R., Johnston, M., Mamun, A.A.: Non-timber forest products and village livelihoods in Rajasthan, India: adaptation in a changing environment, International Journal of Sustainable Development & World Ecology, DOI: https://doi.org/10.1080/13504509.2017.1304462 (2017)
85. Ostrom, E.: Community and the endogenous solution of commons problems. Journal of Theoretical Politics (1992) 4: 343–351
86. Ostrom, E.: A behavioral approach to the rational choice theory of collective action presidential address, American Political Science Association. American Political Science Review. 92:1–21. Platteau JP. 2004. Monitoring elite capture (1998)
87. Borrini-Feyerabend, G., Pimbert, M., Favar, M.T., Kothari, A., Renard, Y.: Sharing power: learning-by-doing in co management of natural resources throughout the world. London: Earthscan; (2007) 473
88. DeCremer, D., van Vugt, M.: Collective identity and cooperation in a public goods dilemma: a matter of trust or self-efficacy? Current Research in Social Psychology (1998) 3: 1–9
89. Murali, K.S., Shankar, U., Shaanker, R.U., Ganeshaiah, K.N., Bawa, K.S.: Impact of NTFP extraction on regeneration, population structure, and species composition1 Economic Botany (1996) 50(3): 252–269
90. Ticktin, T.: The ecological implications of harvesting non-timber forest products. Journal of Applied Ecology (2004) 41: 11–21

Assessment of Water Quality and Biodiversity Status of Alaknanda River at Garhwal, Uttarakhand: A Case Study

Garima Tomar, D. S. Malik, Amit Kumar Sharma, Vishal Kamboj, and Vikas Kumar

Abstract Biotic and abiotic factors of a river compel an aquatic ecosystem, these factors are interrelated with each other. A healthy freshwater ecosystem denotes the assimilative capacity of a water body in terms of diverseness. Aquatic biodiversity of a freshwater ecosystem are very important for ecosystem services. Some environmental changes like climate change, anthropogenic factors, competition for survival, aquatic habitat loss etc. are responsible for the loss of biotic components in a water stream. The present study was aimed to assessed the interrelation between abiotic and biotic parameters of River Alaknanda. Water samples were collected from a 172 km long stretch of river during 2017–2018. The physico-chemical parameters viz. air temperature, water temperature, relative humidity, rainfall, light intensity, depth, pH, total dissolved solids, dissolved oxygen, biochemical oxygen demand and chemical oxygen demand were analysed by following the standards method. All the physico-chemical parameters were noted below the prescribed permissible limits of WHO/BIS. Biological parameters as phytoplankton, zooplankton, macroinvertebrates and fishes were identified with the help of identification keys. A total of 32 species of phytoplankton, 19 species of zooplankton, 33 species of macro benthos and 21 fish species were found in the sampling zones of river. All the biotic and abiotic parameters evaluated by performing canonical correspondence analysis on observed data. According to this study findings, CCA is suitable tool for aquatic biodiversity and ecological habitat parameters to scrutinize the river health. This tool is very useful to evaluate the significant relation between biotic components and physico-chemical parameters of an aquatic ecosystem.

G. Tomar (✉) · D. S. Malik
Environmental Hydrology Division, National Institute of Hydrology, Roorkee, India
e-mail: rs.garima@gkv.ac.in

G. Tomar · A. K. Sharma · V. Kamboj
Department of Zoology and Environmental Sciences, Gurukula Kangri Vishwavidyalaya, Haridwar 249404, UK

V. Kumar
School of Applied and Life Sciences, Uttaranchal University, Dehradun, India

K. D. Bahukhandi et al. (eds.), *Environmental Pollution and Natural Resource Management*, Springer Proceedings in Earth and Environmental Sciences,
https://doi.org/10.1007/978-3-031-05335-1_8

Keywords Planktonic community · Macro-benthos · Fishes · Canonical correspondence analysis · River Alaknanda

Abbreviations

CCA	Canonical Correspondence Analysis
DO	Dissolved Oxygen
BOD	Biochemical Dissolved Oxygen
COD	Chemical Oxygen Demand
RH	Relative Humidity
TDS	Total Dissolved Solids
AT	Air Temperature
WT	Water Temperature
WHO	World Health Organization
BIS	Beareau of Indian Standards

1 Introduction

Water is a crucial component of life on earth, and freshwater systems are the most valuable source for living beings here [1]. Freshwater ecosystems are intensively pressurized at this moment [2]. The water quality affects the species composition, abundance, and productivity of aquatic organisms in any water body [3]. Although, several anthropogenic activities affect the freshwater system's characteristics on a global and local level [4]. These factors are responsible to threat freshwater biodiversity by declining the aquatic biota [5]. The biodiversity of freshwater ecosystems are more susceptible to extinct then other ecosystem's biotic community [2]. The freshwater ecosystem provides natural and essential components to aquatic biota, maintains the ecosystem functions of the biota affected by the shrinking of biodiversity [6]. Generally, a stable freshwater ecosystem can encourage nutrient retention, productivity, and functioning [7].

Some naturally occurring environmental factors are also exploit the aquatic biodiversity as global climate change and disturbance in water body [8]. There are a lot of anthropogenic factors as municipal waste, industrial effluent, agricultural runoff, land use pattern and non-point source pollution disturbed the biodiversity on a worldwide scale. A few of the pollution indicators species show the municipal waste inlet and non-point sources addition in the river [9]. Uttarakhand is a small state, supports various specific features that's why this is different from other states of a country regarding development potential. It has natural resources. There is considerable development has done by the government for hydropower generation [10]. There are various natural water sources (rivers, lakes, streams, estuaries, etc.) in this northern region of India that supports a good variety of life itself. In hydrology, Alaknanda is considered the source stream of the Ganges on account of its greater length and discharge [10]. All the perineal rivers contain plentiful density in biotic species and aquatic fauna because these rivers carry water throughout the year. In terms of status of the water system, this river holds good biodiversity in terms of density and diversity. The present healthiness of any freshwater ecosystem is necessary to know for the restoration the aquatic biodiversity [11]. This study aims to evaluate the relationship status between abiotic and biotic components of the river Alaknanda.

2 Study Area

Alaknanda river occurs from the convergence of two major glaciers Bhagirath Kharak and Satopanth at Uttarakhand [12]. It is located at 30°29'28" N 79°05'08" E of Northern India and has an elevation of 161.27 km. Alaknanda river meets with Bhagirathi River at Devprayag by a distance traveled of 190 km, and basin area is 10,882 km^2. Along 172 km study stretch was divided into four sampling zones (two sampling points on each zone) for one year (2017–2018). Details of sampling locations on the river are given in the study map (shown the selected sites on river stretch) in Fig. 1. Water samples were collected by grab sampling technique in morning hours and stored in sterilized polyethylene sampling bottles. The Physico-chemical parameters as rainfall, air temperature, water temperature, relative humidity, light intensity,

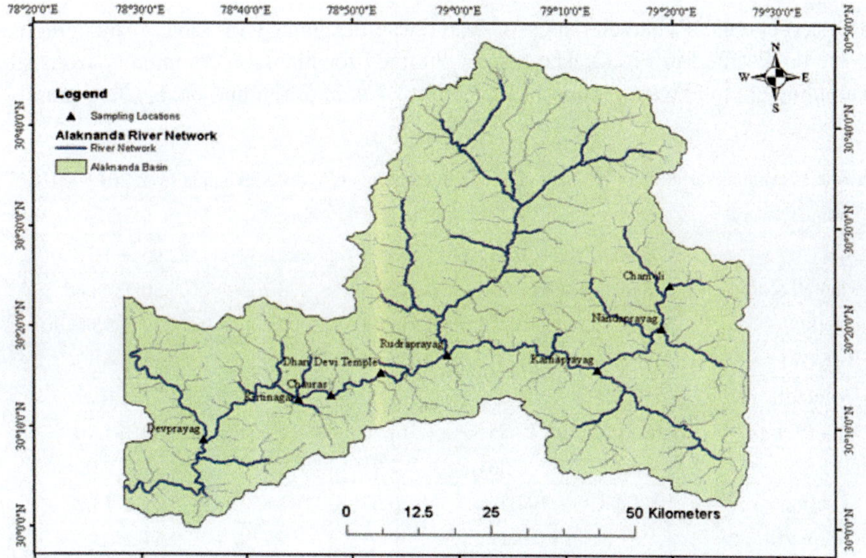

Fig. 1 Map of the sampling sites

depth, pH, Total Dissolved Solids, Dissolved Oxygen, Biochemical Oxygen Demand, and Chemical oxygen demand were carried out by standard methods as per [13]. pH, air temperature, and water temperature were measured by pH meter (HANNA, model: H196107). DO, BOD, and COD were analyzed by using the Winkler method. Double distilled water and merk analytical grade chemicals were used for the water samples analysis. A sampling of phytoplankton was done by using a timed scrapping technique proposed by Wetzel and Likens [14]. Water samples for phytoplankton and zooplankton were collected by filtering 100 L of sample through bolting silk net of size 20 μm and immediately preserved in 10% ethanol solution [15, 16]. Centrifuged sample was taken for the analysis of planktons in river. Remi-24 is used for centrifuging the sample by taking 30 mL concentrated sample. The counting and identification of the biotic species was done with the help of Sedgwick rafter counting cell and digital microscope Optika OPTIKAM pro 8LTCCD. Macro-benthos was collected by surber sampler with size 0.50 mm to a depth of 10 cm in a quadrate [17, 18]. Piscine samples were collected by cast net with the help of local fisher man, identified as per protocol given by [19, 20]. Piscine-fauna was collected with drag nets (mesh size 2.5 × 2.5 cm; 3 × 3 cm; 7 × 7 cm; length × breadth = 75 × 1.3 m; 50 × 1 m) and cast nets (mesh size 0.6 × 0.6 cm). Canonical correspondence analysis has been performed for this study to evaluate the relationships between biological assemblages of species and their environment.

3 Result and Discussion

Physico-chemical characteristics of water was designated in Table 1 and graphs were represented in Fig. 2. The results obtained for plankton community from all sampling stations were plotted in Figs. 3a, b, 4 a, b, 5 a, b and 6a, b. CCA graphs

Table 1 Annual variation of physico-chemical parameters of river Alaknanda from 2017–2018

Parameters/unit	Zone-A	Zone-B	Zone-C	Zone-D	Mean ± SD
A.T. (°C)	20.63	20.5	23.94	23.15	22.05 ± 1.76
Rainfall (cm)	128.99	152.7	72.75	78.73	108.30 ± 38.89
L.I. (lux)	1557.50	22,185.3	12,711.62	3202.43	9914.20 ± 9544.40
R.H. (%)	51.33	38.9	38.5	39.92	42.17 ± 6.13
Depth (m)	1.8	1.8	1.4	2.1	1.77 ± 0.27
W.T. (°C)	12.41	11.83	14.78	14.19	13.30 ± 1.40
pH	8.05	8.06	7.9	8.02	8.01 ± 0.07
TDS (mg/L)	106.03	102.59	76.69	98.85	96.04 ± 13.22
DO (mg/L)	9.7	9.29	9.39	9.77	9.54 ± 0.23
BOD (mg/L)	1.24	1.16	1.13	1.15	1.17 ± 0.04
COD (mg/L)	5.96	6.06	9.2	6.43	6.91 ± 1.53

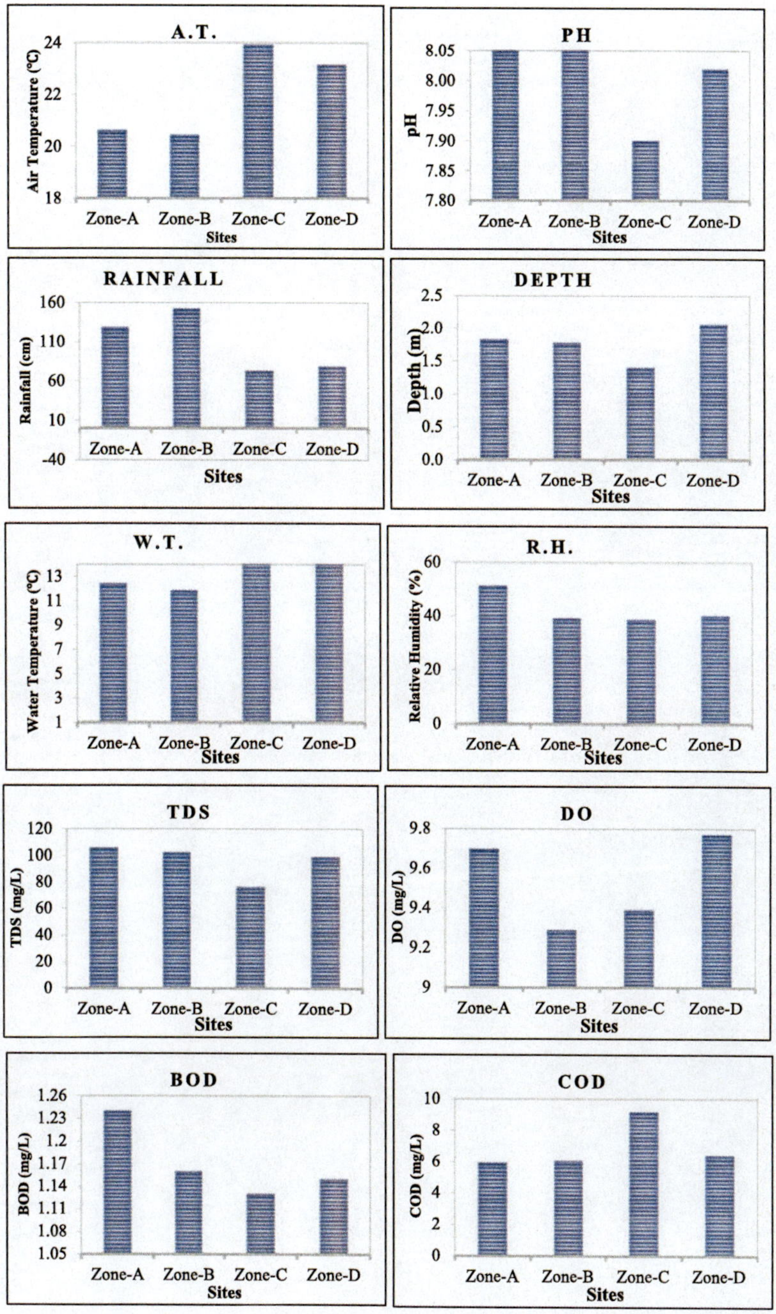

Fig. 2 Phyico-chemical parameters of river Alaknanda during 2017–2018

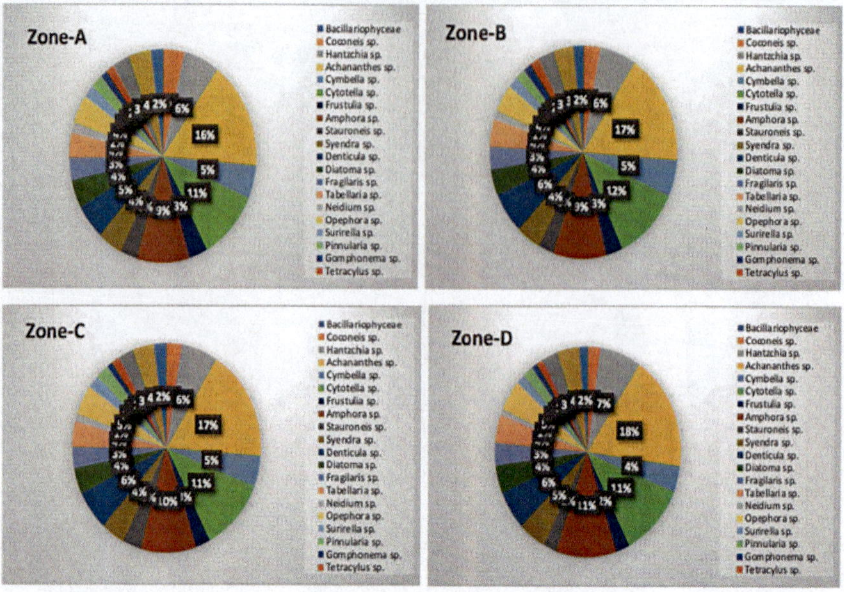

(a)

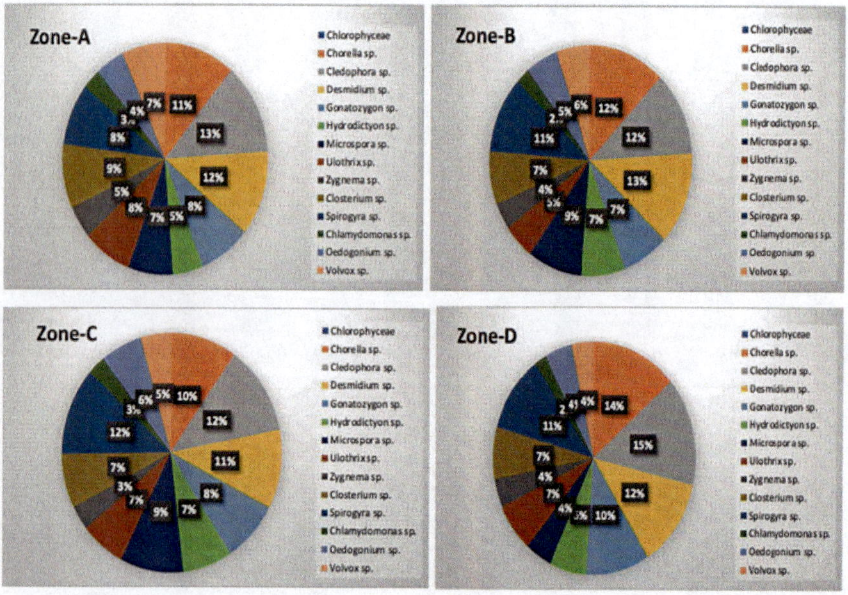

(b)

Fig. 3 **a, b** Phytoplankton at river Alaknanda during 2017–2018

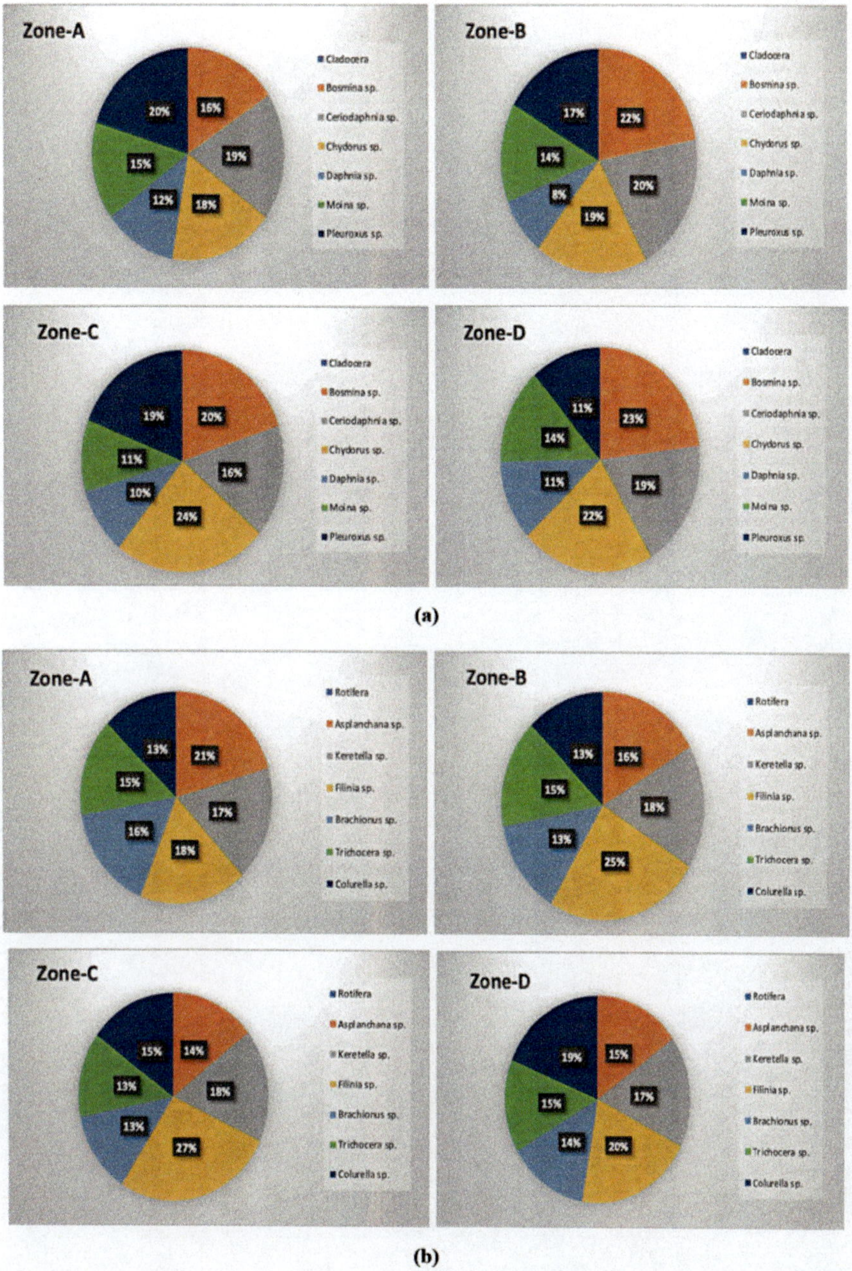

Fig. 4 **a, b** Zooplankton at river Alaknanda during 2017–2018

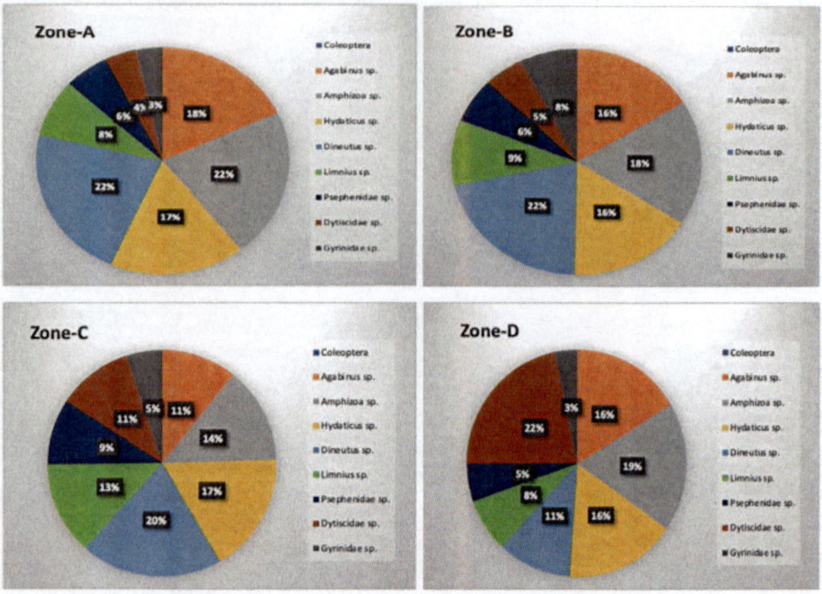

(a)

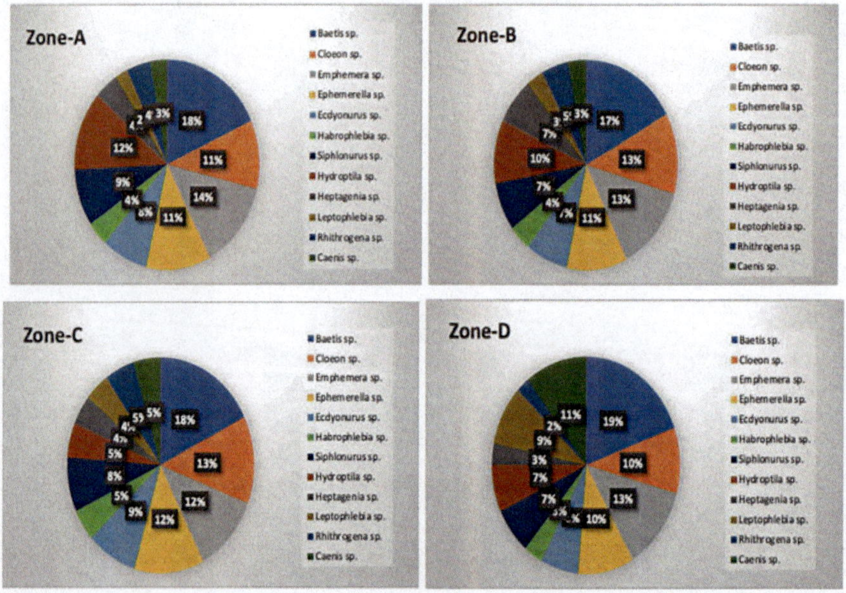

(b)

Fig. 5 **a, b** Macro benthos at river Alaknanda during 2017–2018

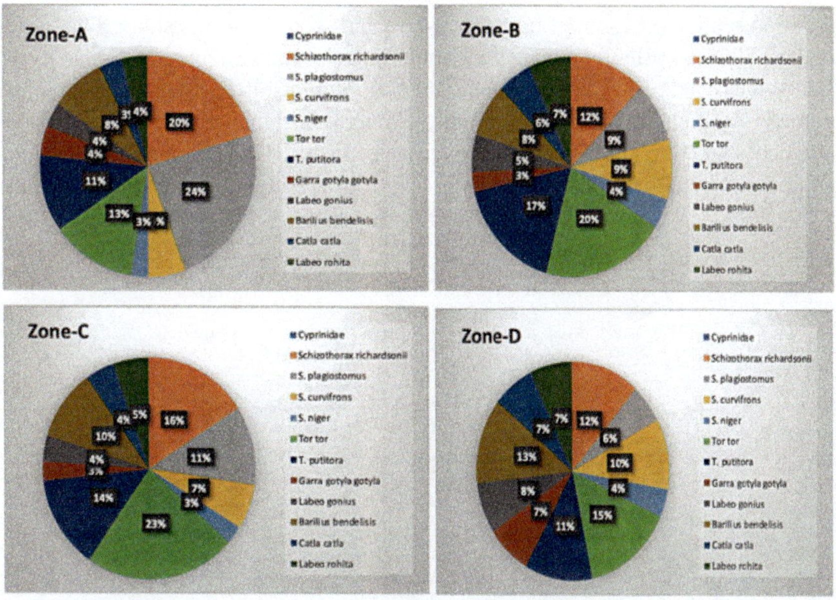

(a)

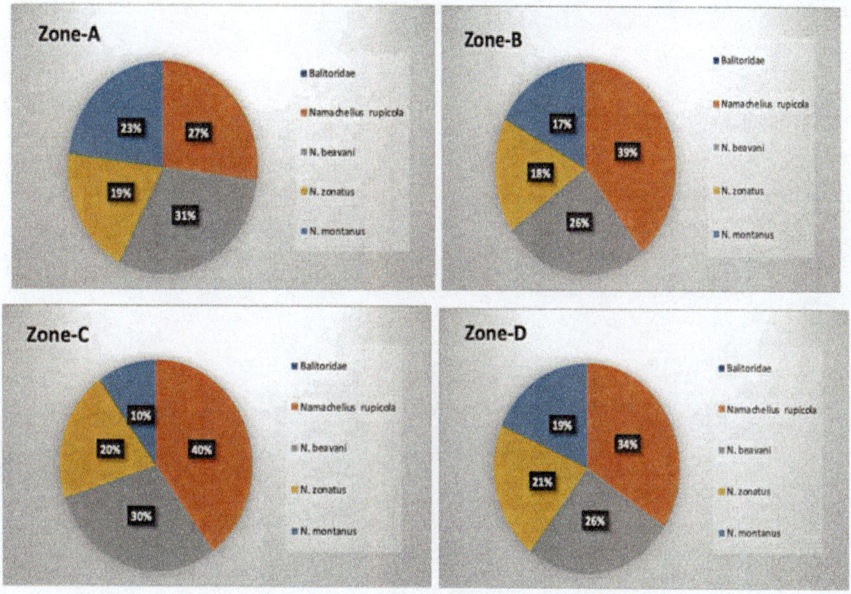

(b)

Fig. 6 **a, b** Fishes at river Alaknanda during 2017–2018 were represented in Figs. 7a, b, 8a, b, 9a, b and 10a, b for this study. The vector length of a given variable indicates the importance of the variable in CCA analysis. Eigen value of axis 1 (0.008) explained 46.44% relationship between environmental parameters and phytoplankton community. Whereas, Eigen value of axis 2 (0.004) explained 28.32% relationship. The longest vector of RH, TDS and DO showed a significant correlation with zone-B. High values of water temperature are associated

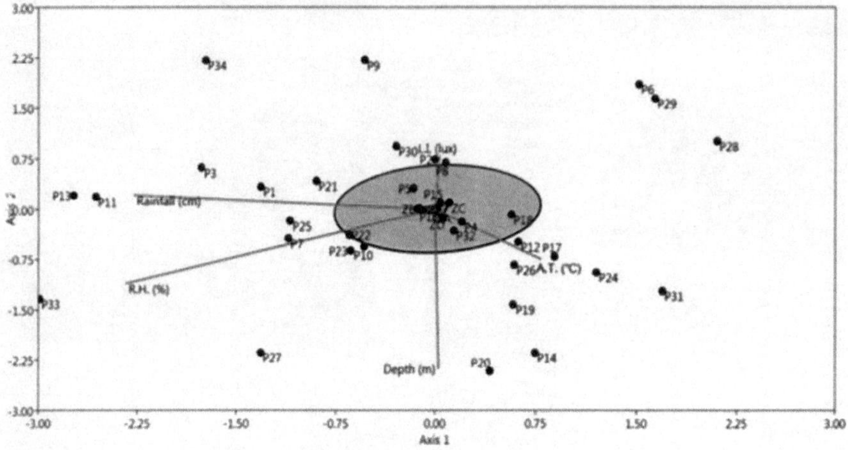

(a)

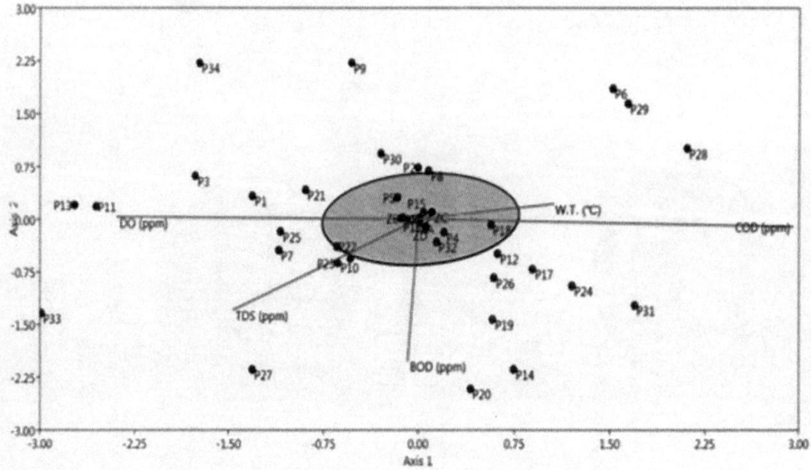

(b)

Fig. 7 a, b CCA biplot between habitat parameters and phytoplankton species at selected zones

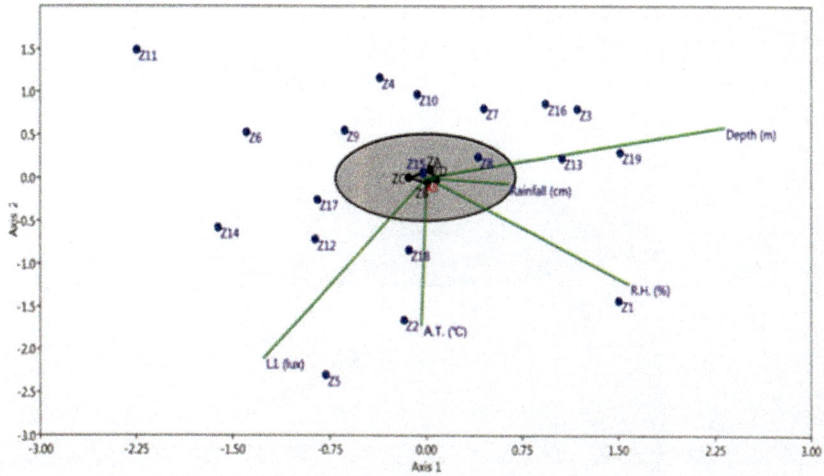

(a)

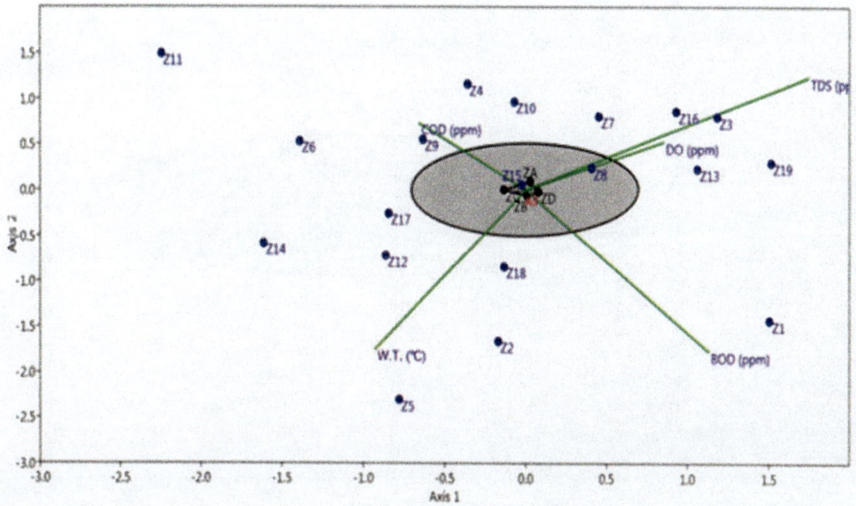

(b)

Fig. 8 **a, b** CCA biplot between habitat parameters and zooplankton species at selected zones

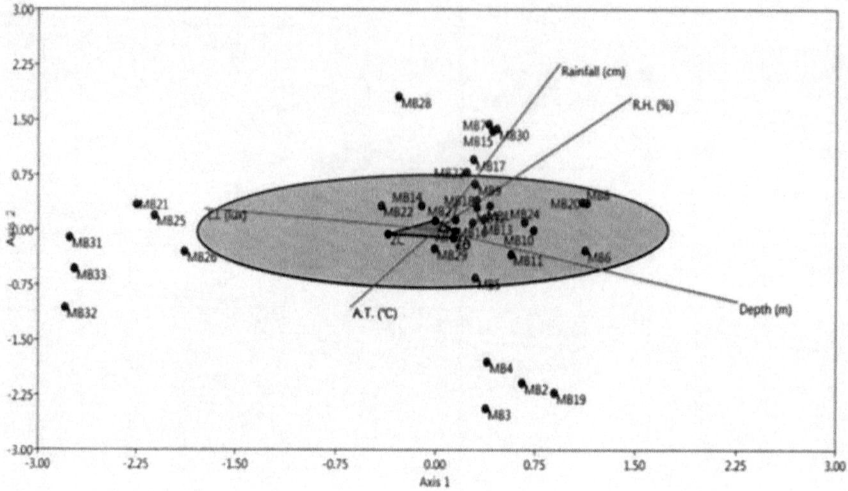

(a)

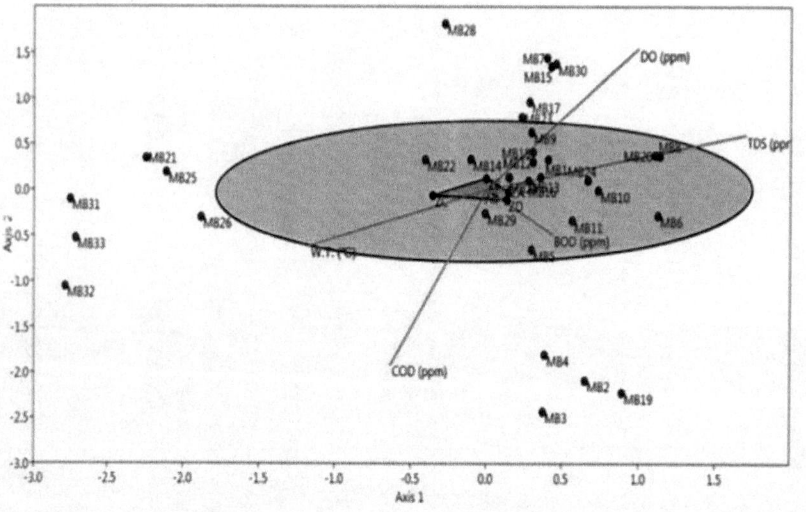

(b)

Fig. 9 **a, b** CCA biplot between habitat parameters and macro-benthic species at selected zones

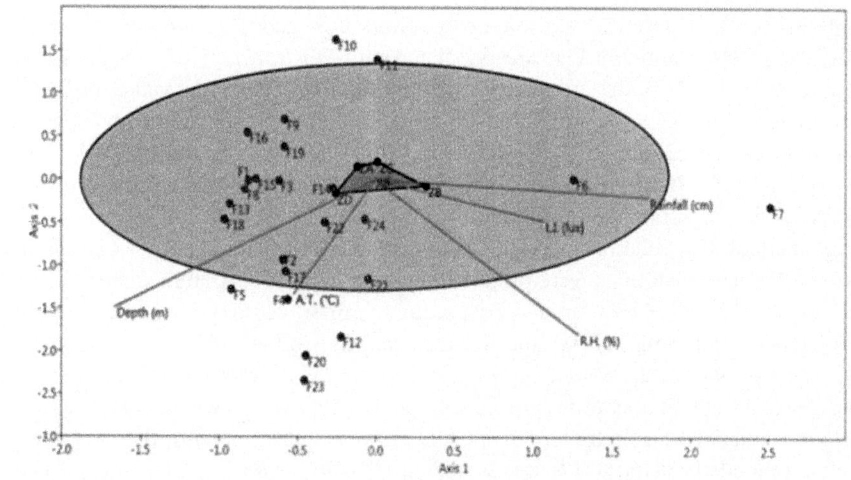

(a)

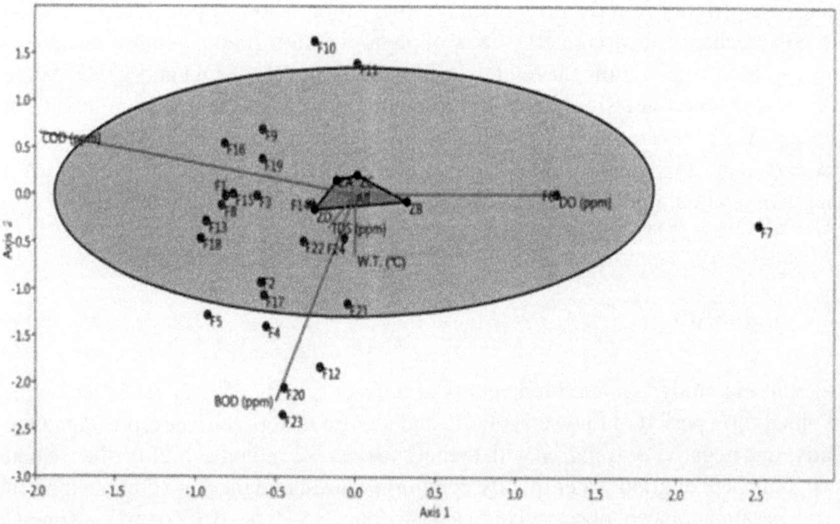

(b)

Fig. 10 **a, b** CCA biplot between habitat parameters and fish species at selected zones with *Navicula sp.*, DO are associated with *Neidium sp.* and BOD are associated with *Nitzchia sp.* High values of TDS are associated with *Tabellaria sp., Closterium sp. and Zygnema sp.,* COD values are associated with *Cledophora sp.* High values of DO are associated with *Mesocyclops sp.*, total dissolved solids are associated with *Chydorus sp.* and COD are associated with *Diaptous sp.* and *Keretella sp.* during 2017–2018. Eigen value of axis 1 (0.005) explained 48.86% relationship between environmental parameters and zooplankton community. Whereas, Eigen value of axis 2 (0.003) explained 28.75% relationship. The results obtained for macro benthos community from all sampling stations were plotted in Figs. 4a, b and 5a, b. High values of BOD are associated with *Culex sp.* and *Hydaticus sp.*, dissolved oxygen associated with *Siphlonurus sp., Simulium sp.* and *Antocha sp.* High values of TDS are associated with *Hesperocorixa sp., Emphemera sp.,* and *Gyrinidae sp.* Vector length of dissolved oxygen and water temperature showed significant with zone-B, whereas TDS showed relation with zone-D. Vector length of COD showed significant with zone-A, C. High values of total dissolved solids are associated with *Glyptothorax cavia* and dissolved oxygen associated with *Schizothorax rihardsonii.* High values of BOD are associated with *Glyptothorax pectinopeterus, Bagarius bagarius* and *Pseudochenies sulcatus.* Canonical correspondence analysis showed a significant relation in between aquatic habitat parameters and the aquatic community of a water body. Different biotic species were found in river during study observation, showed positive correlation with phsico-chemical parameters. A few of phytoplankton, macro benthos and fishes were recorded in a significant relation with pH, TDS, DO, BOD and COD. Some species were noted in a significant relation with depth, water temperature and light intensity. A very few researchers have been given the overview on CCA on the biotic flora and fauna throughout the world. Several studies on significant relation of a water ecosystem are published by different researchers viz: [21–25].

4 Conclusion

In the present study, aquatic biodiversity and water quality of river Alaknanda was examined for a period of one year. Biotic and abiotic factors of river represented the positive and negative correlation with them. The river was noted as highly diverse that is also indicator of good water quality condition. The abundance and composition of aquatic parameters were observed high to low Zone-A > Zone-B > Zone-D > Zone-C respectively on all sampling stations of river.

Acknowledgements The author is gratefully acknowledged to the Department of Science and Technology, New Delhi (IN), National Institute of Hydrology, Roorkee and Department of Zoology and Environmental Sciences, G.K. (Deemed to be University), Haridwar for their financial support to this research work.

References

1. Hilt, S., Brothers, S., Jeppesen, E., Veraart, A. J., & Kosten, S. (2017) Translating regime shifts in shallow lakes into changes in ecosystem functions and services. *BioScience*, 67(10): 928–936. https://doi.org/10.1093/biosci/bix106
2. WWF (2020) Living Planet Report, Bending the curve of biodiversity loss. Almond, R.E.A., Grooten M., Petersen, T. (Eds). WWF, Gland, Switzerland.
3. Mishra, B.K., Badola, R. & Bhardwaj, A.K. (2009) Social issues and concerns in biodiversity conservation: experiences from wildlife protected areas in India, *Tropical Ecology*, 50(1): 147-161.
4. Jenny, J.P., Anneville, O., Arnaud, F., Baulaz, Y., Bouffard, D., Domaizon, I., Bocaniov, S.A., Chèvre, N., Dittrich, M., Dorioz, J.-M., Dunlop, E.S., Dur, G., Guillard, J., Guinaldo, T., Jacquet, S., Jamoneau, A., Jawed, Z., Jeppesen, E., Krantzberg, G., Lenters, J., Leoni, B., Meybeck, M., Nava, V., Nõges, T., Nõges, P., Patelli, M., Pebbles, V., Perga, M.-E., Rasconi, S., Ruetz, C.R., Rudstam, L., Salmaso, N., Sapna, S., Straile, D., Tammeorg, O., Twiss, M.R., Uzarski, D.G., Ventelä, A.-M., Vincent, W.F., Wilhelm, S.W., Wängberg, S.A. & Weyhenmeyer, G.A., (2020) Scientists' Warning to Humanity: Rapid degradation of the world's large lakes. *J. Great Lakes Res.* 5: 1-15.
5. Reid, A.J., Carlson, A.K., Creed, I.F., Eliason, E.J., Gell, P.A., Johnson, P.T.J., Kidd, K.A., MacCormack, T.J., Olden, J.D., Ormerod, S.J., Smol, J.P., Taylor, W.W., Tockner, K., Vermaire, J.C., Dudgeon, D. & Cooke, S.J. (2019) Emerging threats and persistent conservation challenges for freshwater biodiversity. Biol. Rev. 94, 849–873.
6. Díaz, S., Settele, J., Brondízio, E.S., Ngo, H.T., Agard, J., Arneth, A., Balvanera, P., Brauman, K.A., Butchart, S.H.M., Chan, K.M.A., Garibaldi, L.A., Ichii, K., Liu, J., Subramanian, S.M., Midgley, G.F., Miloslavich, P., Molnár, Z., Obura, D., Pfaff, A., Polasky, S., Purvis, A., Razzaque, J., Reyers, B., Chowdhury, R.R., Shin, Y.-J., Visseren-Hamakers, I., Willis, K.J., Zayas, C.N., (2019) Pervasive human-driven decline of life on Earth points to the need for transformative change. Science 366.
7. Allan, E., Weisser, W., Weigelt, A., Roscher, C., Fischer, M. & Hillebrand, H. (2011) More diverse plant communities have higher functioning over time due to turnover in complementary dominant species. Proc. Natl. Acad. Sci. U. S. A. Pp-218.
8. Tian, W., Zhang, H., Zhang, J., Zhao, L., Miao, M., Huang, H., 2017. Biodiversity effects on resource use efficiency and community turnover of plankton in Lake Nansihu, China, *Environ. Sci. Pollut. Res,* 24: 11279-11288.
9. Kotnala, G., Dobhal, S. & Chauhan, J.S. (2016) Monitoring of self -purification capacity of river Alaknanda stretch at Srinagar, Uttarakhand, India, *International Journal of River Basin Management*, 5: 491–498.
10. Mittal, S., Tripathi, G. & Sethi, D. (2008) Development strategy for the hill districts of Uttarakhand. *Indian Council for Research on International Economic Relations*, 80–95.
11. Dhapola, M. Rao, S.N. & Bhadula, S. (2015) Hydro-biological characterization of Haripur Reservoir (Gularboj) in Uttarakhand with special reference to pollution status. *Report and Opinion*, 7(8): 12-19.
12. AHEC (2011) Study on assessment of cumulative impact of hydropower projects in Alaknanda and Bhagirathi basins up to Devprayag. Ministry of Environment and Forest, New Delhi, India, Chapter 7: 1-13.
13. APHA, American Public Health Association (2012) Standard methods for the examination of water and waste water. 22ndedition, Washington D.C. New York. Pp 1256.
14. Wetzel, R.G. & Likens, G.E. (1991) Limnological analyses. second edition Springer-verlag New York. Pp-175.
15. Palmer, C.M. (1972) A composite rating of Algae tolerating organic pollution load. Journal of Phycology 5:78-82.
16. Needham, J.G. & Needham, P.R. (1978) A guide to the study of freshwater biology,
17. Halden, Day. Inc. Pul. San. Francisco. Pp-225.

18. Pennak, R.W. (1953) Comparative limnology of eight Colorado mountain lakes, Univ. Colo. *Studies. Ser. Biol.* 2:1–75.
19. Edmondson, W.T. (1974) Trophic relation of the zooplankton. Trans. Am. Microse. Soc. 76: 225-246.
20. Jhingran, A. (1991) Inland fishes of India and adjacent countries. 2volumes. Oxford and IBH Publishing Co. Pvt. Ltd., New Delhi, 29: 1158.
21. Jayaram, K.C. (1999) The freshwater fishes of the Indian Region. Narendra Publishing House, Delhi. Pp-551.
22. Kamboj, V. & Kamboj, N. (2020) Spatial and temporal variation of zooplankton assemblage in the mining-impacted stretch of Ganga River, Uttarakhand, India. *Environmental Science and Pollution Research* 1–12.
23. Wang. J., Jiancheng, H., Zhengfei, L. & Zhicai, X. (2019) Seasonal shifts in the assembly dynamics of benthic macroinvertebrate and diatom communities in a subtropical river. *Ecology and Evolution* 1: 1-13.
24. Medupin, C. (2020) Spatial and temporal variation of benthic macroinvertebrate communities along an urban river in Greater Manchester, UK. *Environmental Monitoring Assessment* 192–84.
25. Huttunen, K.L., Muotka, T., Karjalainen, S.M., Laamanen, T. & Aroviita, J. (2020) Excess of nitrogen reduces temporal variability of stream diatom assemblages. *Science of the total environment* 1–32.

Efficient Management of Rice Straw Using Vermicomposting Technology: A Synergetic Approach of Agricultural Waste Management

Amrit Kumar, Nitin Kamboj, Vishal Kamboj, Aditi Bisht, Neeraj Pandey, and Manisha Bharti

Abstract India is a developing country and most of the population live in the rural area. In India, most of the population depends on agriculture practices for their survival. About 43.86 million hectares of land is used for paddy cultivation and it produces 104.80 metric tons of rice. By the excessive use of agriculture, huge organic waste (rice straw) is generated. Most of this organic waste is used as cattle feed but the remaining waste is burnt by local farmers. This harmed the environment by releasing various toxic elements into the atmosphere. Due to the burning of rice straw numerous primary and secondary air pollutants such as CO, CO_2, NO_X, SO_X, and NH_3 etc. release in the environment. The burning of rice straw in the open field also decreases soil fertility by killing microorganisms present in the soil. To overcome this burning problem of the rice straw farmer need new techniques to compost rice straw waste. For the degradation of the organic waste vermicomposting is an ecofriendly technique. The purpose of this chapter is to focus on the vermi-technology to decompose rice straw. Just because of vermicomposting technology is an easier, cheaper (cost-effective), ecofriendly approach. The future of vermi-technology is bright cause nowadays people move towards organic farming and avoid the chemicals used in agriculture for high yield. Vermicomposting decomposes the rice waste as well as provide a job for rural area and the urban area also.

Keywords Agricultural waste · Composting · *Eisenia fetida* · Rice straw · Vermicomposting

1 Introduction

A rapid increase in population leads to huge waste production in different sectors of society. An alarming increase in urbanization and industrialization in India is the primary cause of enormous waste production (biodegradable solid waste like

A. Kumar · N. Kamboj (✉) · V. Kamboj · A. Bisht · N. Pandey · M. Bharti
Department of Zoology and Environmental Science, Gurukula Kangri (Deemed to be University), Haridwar, Uttarakhand, India
e-mail: nkamboj@gkv.ac.in

K. D. Bahukhandi et al. (eds.), *Environmental Pollution and Natural Resource Management*, Springer Proceedings in Earth and Environmental Sciences,
https://doi.org/10.1007/978-3-031-05335-1_9

kitchen waste, modern waste, natural waste, municipal solid waste, and so on) which potentially harms the mankind. Organic waste act as a niche (habitat and trophic) for the pathogens. The consumptions of a large number of fruits and vegetables also lead to the enormous production of organic waste which is estimated 50% of the total biodegradable waste [1]. Agricultural waste is mainly produced by conventional farming methods. The poor techniques of farming ~~does~~ not properly decompose the agricultural waste which leads to generate a huge amount of biodegradable waste like harvest straw, husk, cobs, etc. In India, approximately 43.86 million hectares of land area ~~are~~ is used for paddy cultivation and approximately 104.80 metric tons of rice is produced [2]. Around the globe, rice straw is the largest harvested waste. In Asian nations, including India, the production of rice is high which leads into generation of rice straw. In North-West India, Punjab and Haryana states are the enormous producer of rice. In 2012–13 India generates 105 metric tons of rice (approximately 2390 kg/hectare) and secured the second position all over the world in rice production [2]. After obtaining the rice grains from the crop, the rice burned by the local farmers which increases the air pollution by creating smog (smoke + fog) and also destroys various micro and macronutrients of soil which are essential for crop growth [3]. Both Haryana and Punjab contribute 48% of total straw production in India and this straw is not decomposed naturally but openly burnt by the farmers [4]. Haryana and Punjab alone burn 30 million paddy straws each year which causes air pollution not only in Haryana and Punjab but also in nearby states like Delhi and Uttar Pradesh. Due to burning of paddy straws, emission of Greenhouse gases occur, which is potentially dangerous to mankind [5]. The burning of agricultural residues like paddy straw, husk, bran, stover, sugarcane tops, and bagasse generates numerous greenhouse gases that harm our environment by increasing global warming. The burning of rice straw is one of the main causes of the biodiversity loss in agricultural land and also declines the soil fertility. Smog generated by the straw burning cause health-related problems like eyes itching in humans [6]. The burning of paddy straw sharply increases the number of air pollutants (primary and secondary air pollutants) such as CO, CO_2, NO_X, SO_X, NH_3, volatile organic compounds, non-methane hydrocarbons, etc. This also results in the loss of organic carbon, nitrogen, potassium, sodium, many more organic and inorganic nutrients from the soil [7]. Ministry of Agriculture attributes that the agricultural waste is burning due to shortage of labor [2]. Jitendra [8] concluded that about 80% of the crop residues are procured via burning during the post harvested period in April–May, and November–December. The annual emission of gases CO, CO_2, N_2O, and NO_X gases from the paddy and wheat straw burning are 0.11, 2.306, 0.002, and 0.084 metric tons respectively [9, 10]. In India, on an average 500 metric tons of crop residues are produced annually [11]. The recycling of rice straw is a major challenge for farmers, however much of the rice straw is used by the farmers for his their cattle's but the rest are burned. Agricultural wastes especially rice bran and rice straw are rich in nutritive values. Containing much organic matter. Table 1: showing the chemical composition of the rice straw, in rice straw numerous inorganic and organic compounds are present. In developing countries, it is not properly decomposed due to the presence of high organic matter thus pollutes the environment (air, water, and soil) and has creates some human

Table 1 Chemical constituents of rice straw

Chemical composition of rice straw (% on the dry matter)			References
1	Organic matter	82	[12]
2	Crude protein	4	
3	Crude fibres	37	
4	Non-fatty esters	43	
5	Total ash	18	
6	Calcium	0.14	
7	Phosphorous	0.05	
8	Neutral detergent fibre	75	
9	Acid detergent fiber	54	
10	Cellulose	37	
11	Lignin	8	
12	Silica	8	[13]
13	Carbon	38.24	
14	Hydrogen	5.2	
15	Oxygen	36.26	
16	Nitrogen	0.87	
17	Sulfur	0.18	
18	Chloride	0.58	

health-related problems harmful effects. Table 2: describe the inorganic compound in the rice straw ash. In the rice straw ash SiO_2 is present in the highest percentage

Table 2 Properties of ash from rice straw

Properties of ash from rice straw compounds	Quantity	Source
SiO_2	74.67	[14]
Al_2O_3	1.04	
TiO_2	0.09	
Fe_3O_3	0.85	
CaO	3.01	
MgO	1.75	
Na_2O	0.96	
K_2O_3	12.3	
SO_3	1.24	
P_3O_5	1.41	

Gas emissions from rice straw and air pollutants releases after the rice straw burning in the field in Egypt followed by K_2O_3, CaO, MgO, P_3O_5, and SO_3, and so on with 74.67, 12.3, 3.01, 1.75, 1.41, and 1.24 and so on respectively.

2 Objectives of Chapter

As previously described, to overcome the agro-waste generated pollution problem like rice straw burning pollution vermicomposting is the best and eco-friendly method to decompose it without harming the environment (already mentioned eco-friendly). Rice Straw is organic biodegradable agro-waste that is generated after crop yield as a waste product. When rice straw decomposes in an open field due to high moisture content of paddy field, moisture it is digested anaerobically and produces biogas out of which 60% is methane and 40% is carbon dioxide [15]. Agro waste like rice straw open burning is a potential threat to the environment, open burning of rice straw releases a huge amount of greenhouse gases into the environment. No need the open burning of rice straw burning releases N_2O and CH_4 into the air and pollutes the environment. 13.1% of emission out of the total toxic gases emission release from the open burning of rice straw [16]. For this reason, vermicomposting technology is the best method to decompose rice straw very efficiently.

To minimize this problem environmental agencies and governments are looking for cheap, more efficient and eco-friendly techniques to manage these types of wastes very efficiently, in an eco-friendly, and at a very cheaper cost [17]. Table 3: represent the list of results of gas emission from the burning of rice straw in Egypt. Rice straw burning is the physical method of decomposing the rice straw. Burning f straw releases various primary and secondary air pollutants which harmed the environment and harmed human beings also no need to write again and again. Agricultural waste is the crop residues that remain after obtaining the crop yield, most of these waste

Table 3 Gas emissions rate of rice straw burning

SI. No	Pollutants	Emission factor	Emissions	Emissions in CO_2 Eq.	Reference
		g/kg straw, dry weight	*kg pollutant/ha*	*Ton CO_2 eq/ha*	[20]
1	CO_2	1460.00	9344.0	9.34	
2	CH_4	0.74	4.7	0.10	
3	N_2O	0.79	5.1	1.57	
4	CO	72.40	463.4		
5	NO_X	3.52	22.5		
6	SO_2	0.15	0.9		
7	$PM_2.5$ (fine particulate matter	12.95	82.9		
8	Total		9354	11	

are useless, it includes crops stalk, leaves, stems, fruits covers, etc. Rice straw and husks are the common example of crop/agricultural waste [18, 19].

Farm fire images of Haryana and Punjab were recorded by the NASA satellite in the last seven days 20–27 November 2018. Red dots indicate the stubble burning activity over Amritsar and Ludhiana located in Punjab, Karnal, Kurukshetra, and Ambala in Haryana and someplace in Western Uttar Pradesh.

Vermicomposting is a cheaper and eco-friendly composting process to degrade waste. In this process crop residues such as rice straw, bran, sugarcane bagasse, and wheat straw (biodegradable waste) are decomposed with the help of worms (*Eesina fetida, Perionyx excavatus*), which convert this waste into nutrients rich manure vermicast within 45–50 days [21, 22]. The vermicomposting technology is very useful and cost-effective to recycle biodegradable solid waste into humus human useful products. Vermicomposting increases soil fertility, soil texture, high and crop production [23]. It also improves the microbial activity and diversity in the soil that promote the crop yield [24]. Table 4: showing the nutritive value of garden and vermicompost. The inorganic elements that are essential for soil fertility are significantly higher in vermicompost and compared than garden compost. The biological activity in the soil is increased by the vermicompost which helps in the germination of seeds [25, 26]. Vermicompost increases the aeration in soil, water holding capacity, soil texture, and stabilizes the interaction between earthworm and microorganisms in by a non-thermophilic process [27]. Vermicompost is nutrients rich as it contains nitrates, phosphates, exchangeable calcium, and soluble potassium that help in plant growth [28, 29].

Table 4 Nutritive value of garden compost and vermicompost [30]

SI. No	Nutrients elements	Garden compost (%)	Vermicompost (%)
1	Organic carbon	12.2	9.8–13.4
2	Nitrogen	0.8	0.51–1.61
3	Phosphorous	0.35	0.19–1.02
4	Potassium	0.48	0.15–0.73
5	calcium	2.27	1.18–7.61
6	Magnesium	0.57	0.093–0.568
7	Sodium	<0.01	0.058–0.158
8	Zinc	0.0012	0.0042–0.110
9	Copper	0.0017	0.0026–0.0048
10	Iron	1.1690	0.2050–1.3313
11	Manganese	0.0414	0.0105–0.2038

3 Techniques Used to Decompose Rice Straw

Rice straw is decomposed normally using three methods; Physical decomposition, Chemical digestion, and Biological decomposition.

3.1 *Physical Decomposition*

Stubble burning and Pyrolysis are mainly two processes to degrade rice straw. In stubble burning, rice straw is burnt in the field this leads to the emission of a large number of air pollutants which negatively affects the environment most of these are greenhouse gases [31]. Pyrolysis is another procedure used to decompose rice straw. The irreversible thermal conversion process done above 300 ℃ in the absence of Oxygen is known as pyrolysis. Pyrolysis degrades the biomass into three products gases, bio-oil, and char [32, 33]. Products of pyrolysis can be used as fuel. Combustion is the process of burning organic and inorganic or anything substances in the presence of oxygen, which is another specific and important technique to decompose the agro-waste [34].

3.2 *Chemical Digestion*

Involves using 3% urea, 1% single super phosphate (SSP), and 0.1% CaO (w/w) on the chopped rice straw (1–2 cm length) and achieve the moisture contents of 70% of the treated material. The nutritive value does not much differ from the untreated rice straw in chemically treated rice straw so chemical treatment to degrade the rice straw is not a useful process as compared to fungal digestion of rice straw [35].

3.3 *Biological Decomposition*

3.3.1 Bacterial Digestion

The anaerobic digestion of rice straw generate 60% methane and 40% carbon dioxide both are greenhouse gases that harm the environment [15].

3.3.2 Ruminant Feeding

After the yield, the remaining rice straw is used for cattle as food by farmers.

3.3.3 Fungal Digestion

The fungal application on the rice straw reduced the solid substrate fermentation (SSF) from 4 to 2 weeks [35].

3.3.4 Vermicomposting

Vermicomposting is a very efficient and easy procedure to decompose organic waste like rice straw. Through vermicomposting with the help of worm rice straw converts into odorless and nutrient-rich manure called vermicompost [30].

Feeding material for the worms in the worm bins (worms bin is the rearing pots for vermiculture) are rice straw with various types of biodegradable organic wastes kitchen waste, paper wastes, agro wastes, and nitrogen-rich material like cattle dung, goat manure, ass manure, etc. [36]. Vermicomposting is an eco-friendly, cheap, and cost-effective way to recycle biodegradable organic wastes (agriculture waste, paper wastes, and municipal solid wastes) by the use of earthworms in vermicompost. We will describe the method of vermicomposting in more detail.

4 Earthworm

Earthworms are the soil engineers, they improve the soil structure, aeration, and increase soil porosity, water holding capacity, increase micro and macronutrients in the soil. Worms are grouped under the phylum Annelida, class Oligochaeta. These are segmented, bisexual creatures (bisexual means both the sex organ found in the single organisms), nightcrawlers and mainly reside under the soil [37, 38]. The whole body of earthworms is metamerically segmented with pointed anterior and posterior ends [39]. Earthworms live in the soil and eat natural vegetation, agro waste, living protozoan, rotifers, nematodes, and other microorganisms [40]. Over the globe, more than 1800 species of terrestrial worms under the class Oligochaeta are found [41–46]. There are three types of earthworms in the world; epigeic, endogeic, and anecic. Epigeic are surface feeders and do not build a burrow. Some of the Epigeic species of earthworms are *Eisenia. fetida, Eudrilus eugeniae, Perionyx excavatus,* and *Eisenia andrei.* All these four species are suitable for vermicomposting but *E. fetida* is the most widely used species in vermicomposting because of its wide temperature tolerance limit. *E. fetida* is also known as red worm, tiger worms, and red wiggler. It can survive at maximum 40 °C and lowest 10 °C environmental temperature. In optimum condition i.e. 28–30 °C it attain sexual maturity [47].

5 Types of Earthworms Used in Vermicomposting

Epigeic earthworms are the best species for vermicomposting. Epigeic species *E. fetida* is well-known species for vermicomposting because they reproduce very fast and tolerate adverse environmental conditions. *E. fetida* fecundity rate is very high and consumes daily own body weight agro-waste so that's why it is suitable species for vermicomposting [48, 49].

5.1 Selection of Site

Vermicomposting is done at any place with concealing, sufficient temperature, high humidity, appropriate air circulation, and so on. If composting is done in an open area, a shed is required to protect the Vermiculture tank/vermibed from direct sunlight and rain (Figs. 1, 2 and 3).

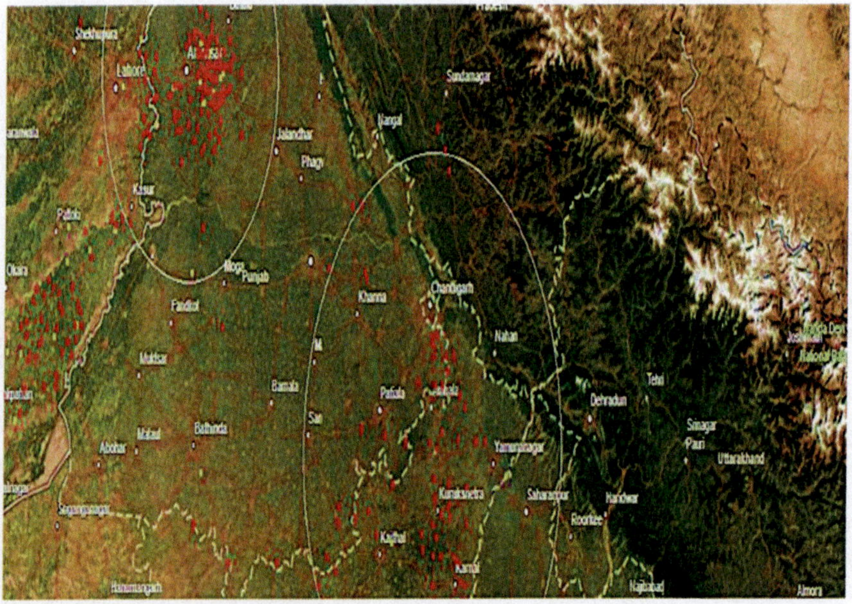

Fig. 1 The stubble-burning season starts in Punjab and Haryana [50]

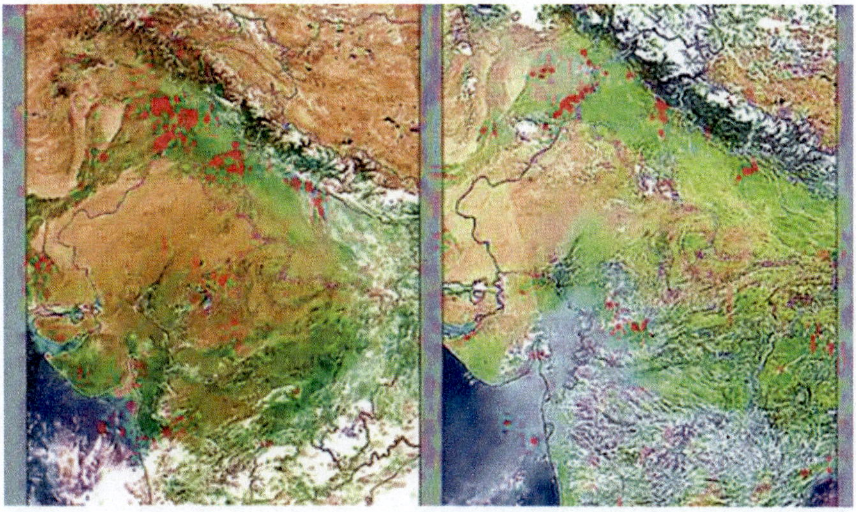

Fig. 2 The stubble-burning heat map diagram of Punjab and Haryana [50]

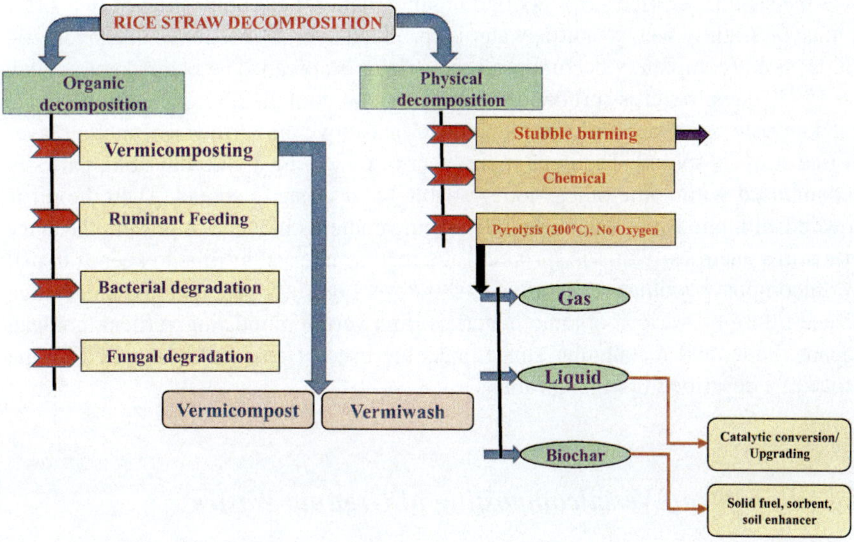

Fig. 3 Flow diagram for decomposing the rise straw

5.2 *Process of Vermicomposting*

Figure 4: showing the basic procedure of organic waste vermicomposting. Initially, for the composting of rice straw built a 6 inches height cemented wall to prevent the escape of earthworms. Sand soil is spread at the height of 3 inches. Rice straw spread

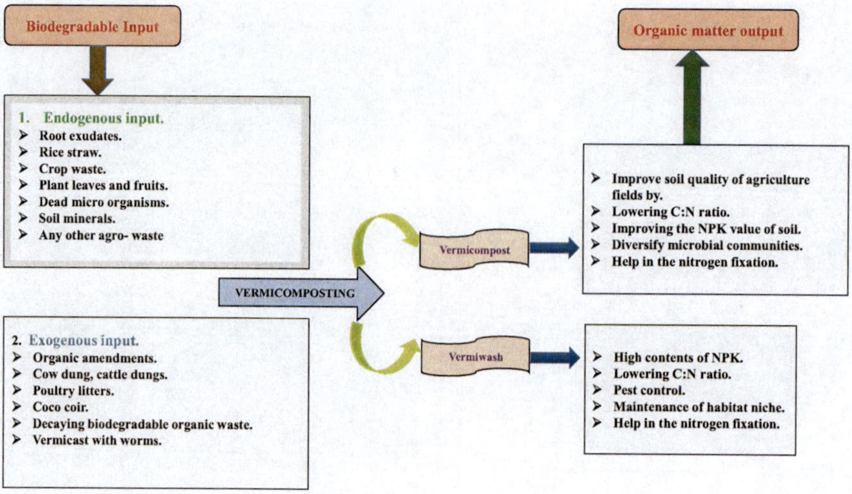

Fig. 4 Flow diagram of vermicomposting of agro-waste and rice straw

over the sand layer, rice straw is dried organic matter. For composting of rice straw it must be slightly wet, so a little water is sprinkled over it. Just above this layer, 20–25 days old (completely decomposed) cow dung is spread. The bed is kept as such for 15–20 days, water is sprinkled regularly to let it cool. 2–2.5 kg of earthworm for 135 kg waste, is release with vermicompost. Just above the earthworm, another layer of rice straw is spread. Then one more layer of cow dung 15–20 days old, partially decomposed with some other biodegradable solid waste is spread. Then the pit is covered with jute for maintaining the moisture contents in it and to prevent the entry of worm's enemies. Water is sprinkled at a regular interval of time to keep it moist. Vermicompost becomes ready after 70–80 days. Figure 6: give detailed knowledge of bed filling by various organic materials. In a vermicomposting vertical, conical, square shape, and rectangular shape tanks are used. These tanks are filled by the procedure described in Fig. 5 (Table 5).

5.3 *Benefits of Vermicomposting of Organic Wastes*

Vermicomposting provides jobs not only in an urban area even in rural areas also. By the use of vermicomposting, we can decompose bad odor organic matter, rice straw, paper waste, industrial sludge and convert them into vermicast. Vermicompost contains essential nutrients for plant growth. Enhances germination, plant growth, and crop yield. Improve root growth and structure. Vermicompost improves soil fertility by enhancing the soil texture [3]. Vermicompost is odorless, easy to apply, and easy to handling and store. Vermicompost enriches the soil with microorganisms. Microbial activity in worms' castings is increased 10–20 times higher

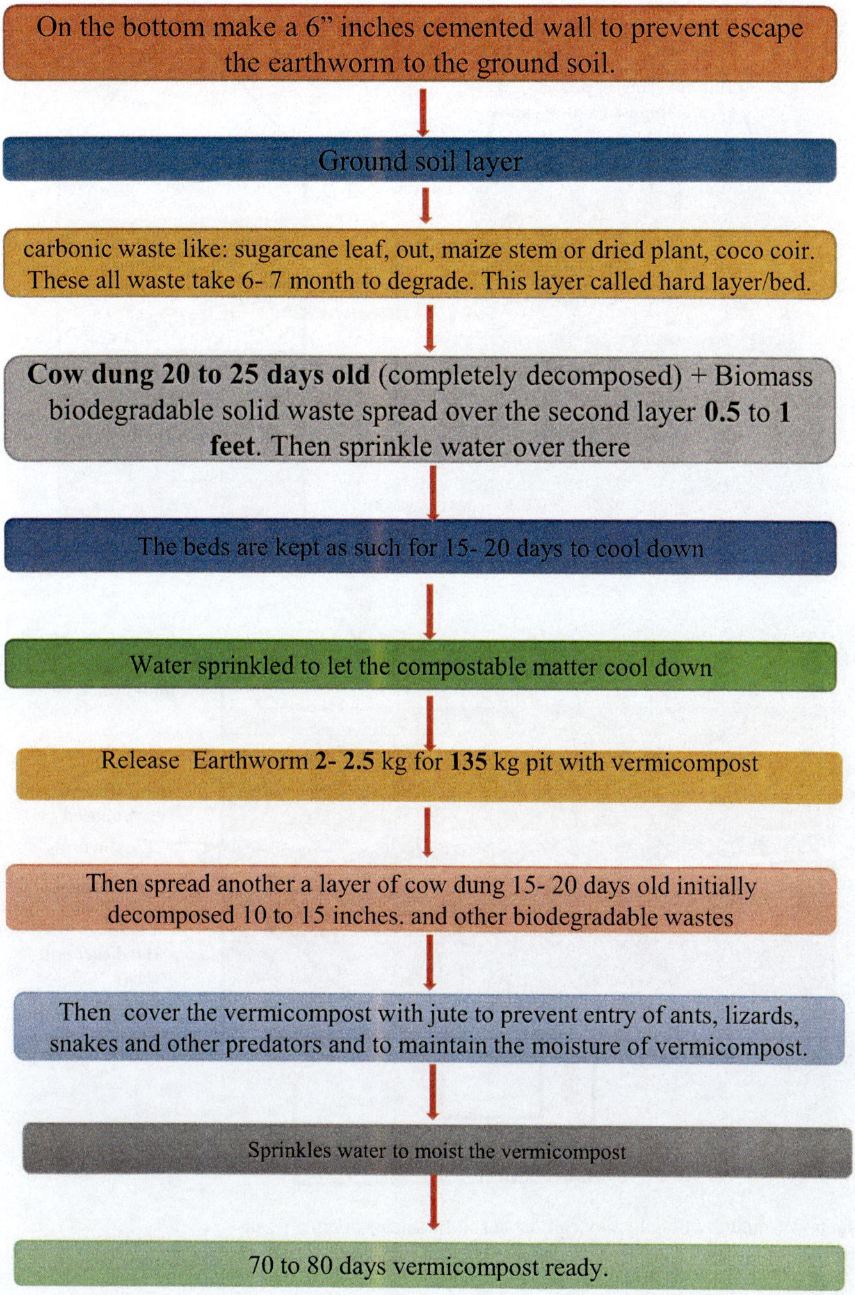

Fig. 5 Flow diagram of vermicomposting processing

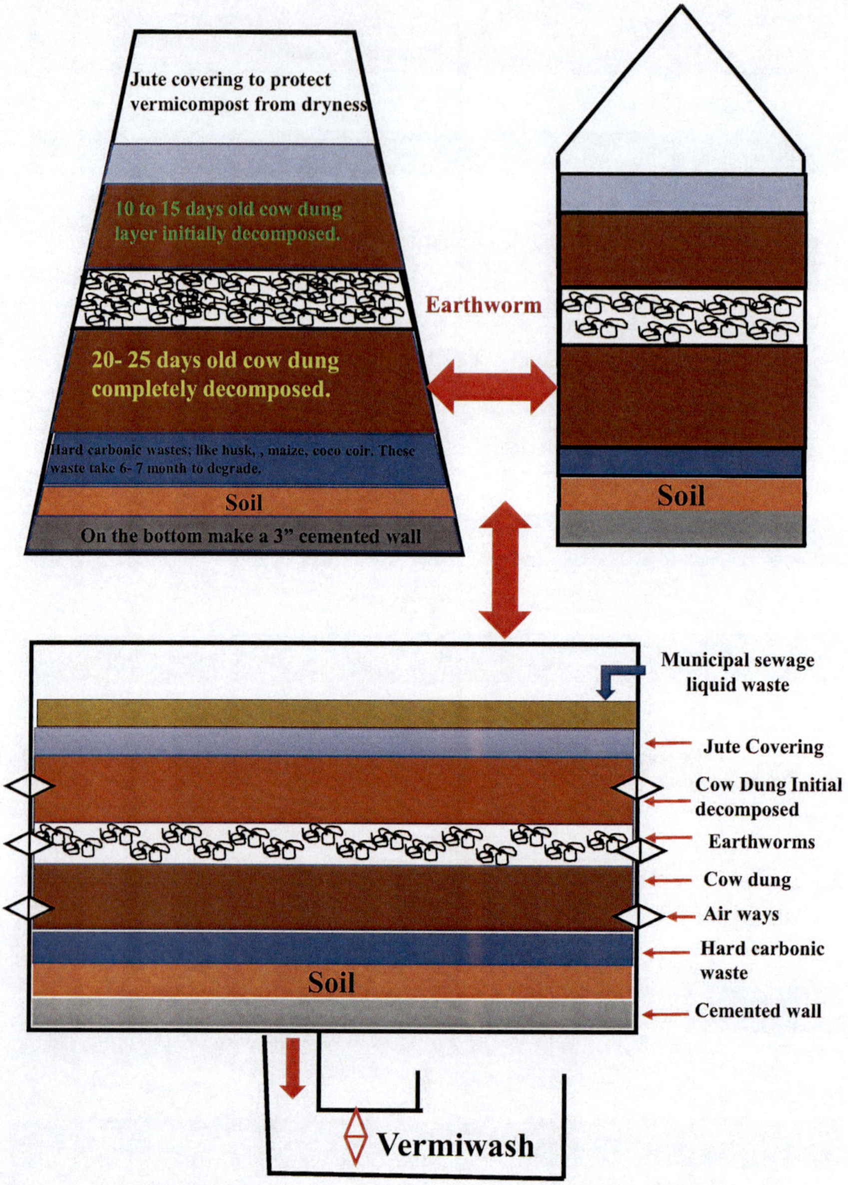

Fig. 6 Vermitank filling by biodegradable solid waste or organic waste

Table 5 Various Organic biodegradable wastes and their decomposition via vermicomposting technology

SI. No	Substrate	Waste	Time	Worms	References
1	Cow Dung	Distillery sludge, Tealeaf	45	*Eisenia fetida*	[51, 52]
2	Cow Dung and sawdust	Guar gum	60	*Perionyx sansibaricus*	[51, 53]
3	Cow Dung	Waste of citronella plant	50	*Eudrilus eugeniae*	[54]
4	Cow Dung	Preconsumed processing vegetable waste	50	*Eisenia fetida*	[55, 56]
5	Cow Dung	Sewage sludge	45–50	*Eisenia fetida*	[53, 57]
6	Cow Dung	The filter cake, trash, and bagasse	50	*Eudrilus eugeniae*	[58]
7	Cow Dung	Food industry sludge		*Eisenia fetida*	[59]
8	Primary sewage sludge	Paper pulp mill sludge	55	*Eisenia andrei*	[60]
9	Sugarcane trash	Sewage sludge	50	*Eisenia fetida*	[61]
10	Sewage sludge	Polycyclic aromatic hydrocarbons	60	*Eisenia fetida*	[62]
11	Cow Dung	Nonrecyclable post-consumer paper waste	60	*Eisenia fetida*	[63]
12	Poultry droppings	Textile mill sludge	50	*Eisenia fetida*	[64]
13	Anaerobically digested biogas plant slurry	Textile mill wastewater sludge	60	*Eisenia fetida*	[65]
14	Cow dung/agricultural residue	Solid textile mill sludge	55	*Eisenia fetida*	[66]
15	Sugar mill filter cake	Horse dung	45	*Eisenia fetida*	[67]
16	Cow dung	Solid textile mill sludge	55	*Eisenia fetida*	[64]
17	Cattle dung	Crop residue	45	*Eisenia fetida*	[68]
18	Wheat straw, cow dung, and biogas slurry	Vegetable solid waste	50	*Eisenia fetida*	[69]
19	Cow dung	Water hyacinth	50	*Eisenia fetida*	[70]
20	Cattle manure	Dairy industry sludge	55	*Eisenia andrei*	[60]
21	Cattle dung	Beverage industry sludge	55	*Eisenia fetida*	[71]

(continued)

Table 5 (continued)

SI. No	Substrate	Waste	Time	Worms	References
22	Cattle dung	Sugar industry waste	50	*Eisenia fetida , Perionyx excavatus, and eudrilus eugeniae*	[72]
23	Cow dung and agricultural residue	Leather processing sludge	60	*Eisenia fetida*	[73]
24	Sheep manure	Cotton industrial waste	50	*Eisenia fetida*	[74]
25	Cow manure	Oat straw	45	*Eisenia fetida*	[75]

than in the soil and organic matter that the worm ingests. Improve the Water Retention/Holding Capacity (WHO). Pest Suppression minimizes the incidence of pests and diseases. Plant growth regulation and higher yields [27]. Enhances the physical, chemical, and biological properties of soil. Vermicompost is free from pathogens, toxic elements, weed seeds, etc. Vermicompost contains earthworm cocoons and increases the population and activity of earthworm in the soil.

6 Discussion

Composting agro-waste rice straw via vermicomposting technology is an easier, efficient, cost-effective, and eco-friendly method. Throughout civilization huge amount of biodegradable and non-biodegradable wastes are generated from the various sectors of society like kitchen wastes, agriculture wastes. The biodegradable waste has a stinky smell that irritates and causes eyes and nasal-related problems in humans and other mammals. The decaying organic waste is the habitat of various disease-causing pathogenic microorganisms. India uses 43.86 million hectares of land for paddy cultivation and produces approximately 104.80 metric tons of rice (NFSE, 2016). With the production of rice tremendous amount of rice straw also generates. Most of these agro-wastes farmers feed their cattle and the rest are burned. The burning of rice straw pollutes the atmosphere by creating smog (fog + smoke) [27]. Rice straw burning generates smog which causes various health-related problems like eye itching, rainy nose, etc. in humans [6]. The burning of rice straw is also a major threat to the biodiversity loss of the agricultural land and also decline in the fecundity of the soil by destroying nutrients of the soil. The annual emission of the CO, CO_2, N_2O, and NO_X by burning rice straw in open land is 0.1, 2.306, 0.002, and 0.084 metric tons respectively [9, 10]. Proper decomposition of paddy straw is the major challenge nowadays because if it decomposed anaerobically it generates 60% methane and 40% carbon dioxide and releases it into the environment. Methane and carbon dioxide both are greenhouse gases and potentially increase global warming [15]. Previous research described that if it burned in an open field it also releases the N_2O, and CH_4 into the environment. About 13.1% emission out of the total is the release from the open burning of rice straw [16]. To minimize the release of greenhouse gases into the atmosphere generated through rice straw burning, vermicomposting is the cheaper, eco-friendly, and efficient composting technology. Through vermicomposting, biodegradable organic waste is converted into nutrients rich manure. Vermicompost is odorless and rich in micro and macronutrients, that help in improving soil fertility via changing the texture of soil and improving the NPK value of the soil [76]. The vermicomposting of biodegradable natural solid waste is a modest and effective strategy that deteriorates all the biodegradable waste into humus (vermicompost). The biological activity in the soil is increased by the vermicompost, this helps in the germination of seeds [26]. Vermicompost increment the air circulation in soil, water holding capacity soil texture and stabilize the interaction between earthworm and microorganisms in a non-thermophilic process [27]. Vermicompost is supplemented rich and contains nitrates, phosphates, replaceable calcium, and dissolvable potassium that help in plant growth [28].

7 Current Trend and Future of Vermicomposting

In 2019 the worldwide vermicompost market esteem was 63.55 million USD and it is projected to arrive at 222. 42 USD million by 2027. Developing at a CAGR of 16. 74% from 2020 to 2027 (Verified Research Market, August 2020). As compared to 2014 the share of the vermicompost market increase by 24.89–38.09% million USD worldwide in 2015 (Absolute Reports Pvt Ltd, August 2019). So the overall scenario is the vermicomposting is a rapidly growing market. All over the world the negative impact of fertilizer used in agriculture everyone has chosen organic farming for health concerns.

8 Conclusion

Today solid waste management is the most serious issue confronting the world. India is the mass maker of natural waste and we don't have effective innovation to defeat this issue. Vermicomposting is the modest and most effective strategy for delaying natural waste, it breaks down natural waste into humus products (Vermicompost). In vermicomposting by the use of night crawler we can easily change the natural waste into the vermicompost. Kitchen waste likewise a significant issue yet the vast majority of kitchen waste is effectively broken down by vermicomposting. The paddy field straw decomposition anaerobically or by burning generates a huge amount of greenhouse gases that affect the environment negatively. Rice straw burning increases the N_2O and CH_4 in the environment and also increases the carbon dioxide amount in the atmosphere, all of these are greenhouse gases and these are the main causes of global warming. By the use of vermiculture technology minimize the greenhouse gas emission into the atmosphere and convert the biodegradable waste into the nutrients rich manure that helps in the agriculture field.

References

1. FAO. Global food losses and food waste. www.fao.org/docrep/014/mb060e/mb060e00. (2011)
2. National Food Security Mission, A status Note on Rice in India, (2016)
3. Yan, Y.W., Azwady, A.N., Shamsuddin, Z.H., Muskhazli, M., Aziz, S.A., Teng, S. K.: Comparison of plant nutrient contents in vermicompost from selected plant residues. *African Journal of Biotechnology*. Volume 12(17). (2013)
4. Singh, J.: Paddy and wheat stubble blazing in Haryana and Punjab states of India: A menace for environmental health. *Environmental Quality Management*. Volume 28(2). (2018) 47-53
5. NPMCR. http://agricoop.nic.in/sites/default/files/NPMCR_1.pdf (2019)
6. Lohan, S. K., Jat, H. S., Yadav, A. K., Sidhu, H. S., Jat, M. L., Choudhary, M., Sharma, P. C.: Burning issues of paddy residue management in northwest states of India. *Renewable and Sustainable Energy Reviews*. Volume 81. (2018) 693-706
7. Jain, N., Bhatia, A., Pathak, H.: Emission of air pollutants from crop residue burning in India. *Aerosol and Air Quality Research*. Volume 14(1). (2014) 422–430. [CrossRef]
8. Jitendra and Others. India's Burning Issues of Crop Burning Takes a New Turn, Down to Earth. https://www.downtoearth.org.in/coverage/river-of-fire-57924 (2018)
9. Hayashi, K., Ono, K., Kajiura, M., Sudo, S., Yonemura, S., Fushimi, A., Tanabe, K.: Trace gas and particle emissions from open burning of three cereal crop residues: Increase in residue moistness enhances emissions of carbon monoxide, methane, and particulate organic carbon. *Atmospheric Environment*. Volume 95. (2014) 36–44. [CrossRef]
10. Gupta, P. K., Sahai, S., Singh, N., Dixit, C. K., Singh, D. P., Sharma, C., Garg, S. C.: Residue burning in rice-wheat cropping system: Causes and implications. *Current Science*. (2004) 1713–1717
11. Bhuvaneshwari, S., Hettiarachchi, H., Meegoda, J. N.: Crop residue burning in India: Policy challenges and potential solutions. *International journal of environmental research and public health*. Volume 16(5). (2019) 832
12. Singh, R. B., Saha, R. C., Singh, M., Chandra, D., Shukla, S. G., Walli, T. K., Kessels, H. P. P.: Rice straw, its production, and utilization in India. In: Handbook for straw feeding systems in livestock production (1995) 325–339. ICAR
13. Jenkins, B. M., Bakker, R. R., Wei, J. B.: On the properties of washed straw. *Biomass and bioenergy*. Volume 10(4). (1996) 177-200
14. Lim, J. S., Manan, Z. A., Alwi, S. R. W., Hashim, H.: A review on the utilization of biomass from the rice industry as a source of renewable energy. *Renewable and sustainable energy reviews*. Volume 16(5). (2012) 3084-3094
15. Allen, J., Pascual, K. S., Romasanta, R. R., Van Trinh, M., Van Thach, T., Van Hung, N.: Chivenge, P.: Rice Straw Management Effects on Greenhouse Gas Emissions and Mitigation Options. *Sustainable Rice Straw Management*. Volume 145. (2020)
16. Romasanta, R. R., Sander, B. O., Gaihre, Y. K., Alberto, M. C., Gummert, M., Quilty, J., Wassmann, R.: How does rice straw burning compare with other straw management practices in terms of on-field CH4 and N2O emissions? A comparative field experiment. *Agriculture, Ecosystems and Environmen*. Volume 239. (2017) 143-153
17. Shak, K.P., Wu, T.Y., Lim, S.L, Lee, C.A.: Sustainable reuse of rice residues as feedstock in vermicomposting for organic fertilizer production. *Environmental Science Pollution Research*. Volume 21. (2014) 1349-59
18. Agamuthu, P.: Challenges and opportunities in agro-waste management: An Asian perspective. In: Inaugural meeting of first regional 3R forum in Asia. (2009) 11–12

19. Kyoji, S.: "Report of the 2008 first world landslide forum on 18–21 November 2008 at UNU, Tokyo". (2009). 167–179
20. Bakker, R. R. C., Elbersen, H. W., Poppens, R. P., Lesschen, J. P.: Rice straw and wheat straw-potential feedstocks for the biobased economy. (2013) NL Agency
21. K, Shagun.: Stubble burning season starts in Punjab and Haryana; https://www.downtoearth.org.in/news/air/stubble-burning-season-starts-in-punjab-and-haryana-66979. (2019)
22. By Jayashree Nandi Hindustan Times, New Delhi. https://www.hindustantimes.com/india-news/fewer-farm-fires-this-year-till-now-show-nasa-images/storyL4RMmRCMWO5XGXjOOKT4dO.html. (2018)
23. Pigatin, L. B. F., Atoloye, I. O., Obikoya, O. A., Borsato, A. V., Rezende, M. O. O.: Chemical study of vermicomposted agroindustrial wastes. *Int. J. Recycl. Org. Waste Agric.*Volume 5. (2016) 55–63
24. Amouei, A.I., Yousefi, Z., Khosravi, T.: Comparison of vermicompost characteristics produced from sewage sludge of wood and paper industry and household solid wastes. *J Environ. Health Sci. Eng.* Volume 15:5. (2017) https://doi.org/10.1186/s40201-017-0269
25. Follet, R.H., Murphy, L.S., Donahue, R.L.: Fertilizers and soil amendments. Prentice-Hall. Engelwood Cliffs. (1981) 393–422
26. Barakan, F.N., Salem, S.H., Heggo, A.M., Bin-Shiha, M.A.: Activities of rhizosphere microorganisms as affected by application of organic amendments in a calcareous loamy soil 2. Nitrogen transformation. *Arid Soil Research and Rehabilitation.* Volume 9 (4). (1995). 467– 480
27. Arancon, N. Q., Lee, S., Edwards, C. A., Atiyeh, R.: Effects of humic acids derived from cattle, food, and paper-waste vermicomposts on the growth of greenhouse plants: the 7th international symposium on earthworm ecology· Cardiff· Wales· 2002. *Pedobiologia.* Volume 47(5-6). (2003) 741-744
28. Atiyeh, R.M., Edwards, C.A., Subler, S., Metzger, J.D., Earthworm processed organic wastes as components of horticultural potting media for growing marigolds and vegetable seedlings. *Compost Science and Utilization.* Volume 8 (3). (2000) 215–223
29. Edwards, C.A., Burrows, I.: The potential of earthworm composts as plant growth media. In: Edwards, C.A., Neuhauser, E. (Eds.), *Earthworms in Waste and Environmental Management.* SPB Academic Press. The Hague, Netherlands. (1988) 21–32
30. Adhikary, S.: Vermicompost, the story of organic gold: A review. *Agricultural Sciences.* Volume 03(07). (2012) 905-917. DOI: https://doi.org/10.4236/as.2012.37110
31. Edwards, C.A.: The use of earthworms in the breakdown and management of organic wastes. In: Edwards, C.A. (Ed.). Earthworm Ecology. CRC Press, Boca Raton. (1998) 327–354
32. Bridgwater, T.: Biomass for energy. *Journal of the Science of Food and Agriculture.* Volume 86(12). (2006) 1755–1768
33. Capareda, S.: Introduction to biomass energy conversions. (2013)
34. Munder, S., Karaj, S., Gummert, M., Haefele, S.M. and Müller, J.: Improving thermal conversion properties of rice straw by briquetting (Doctoral dissertation, Master Thesis. University of Hohenheim). 2013
35. Rai, S. N., Walli, T. K., Gupta, B. N.: The chemical composition and nutritive value of rice straw after treatment with urea or Coprinus fimetarius in a solid-state fermentation system. *Animal Feed Science and Technology.* Volume 26(1-2). (1989) 81-92
36. Kiyasudeen, K., Ibrahim, M. H., Quaik, S., Ismail, S. A.: Vermicomposting: An earthworm mediated waste treatment technique. In Prospects of Organic Waste Management and the Significance of Earthworms. (2016) 167-199. Springer, Cham
37. Lavelle, P.: The structure of earthworm communities. In Earthworm Ecology. (1983) (449-466). Springer, Dordrecht.
38. Lee, K. E.: Earthworms: their ecology and relationships with soils and land use. *Academic Press Inc.* (1985)
39. Edwards, Clive A., Lofty, J. R.: Biology of Earthworms. London: Chapman & Hall. (1977) ISBN 0–412–14950–8
40. Lofty, Clive A., Lofty, J. R.: Biology of Earthworms. London: Chapman & Hall. Volume 80. (1977) ISBN 0–412–14950–8

41. Sims, R.W., Gerard, B.M.: Earthworms: Keys and Notes for the Identification and Study of the Species. 1st Edn, *Bill Archive Publ.* Leiden. (1985) 36–80
42. Julka, J.M.: The Fauna of India and Adjacent Countries. Megadrile: Oligochaeta (earthworms) Haplotaxida: Lumbricina: Megascolecoidea: Octochaetdae. Zoological Survey of India, India. Volume 406. (1988)
43. Martin, J.P., Black J.H., Hawthorne, R.M.: Earthworm biology. http://edis.ifas.ufl.edu/in047. (2000)
44. Malek, M.: Preliminary study of earthworm taxonomy from Tehran province. In: *Proceedings of the 3rd International Oligochaete Taxonomy Meeting*. Platres. Cyprus. (2007) 22–22
45. Kooch, Y., Jalilvand, H., Bahmanyar, M.A., Pormajidian, M.R.: Abundance, biomass, and vertical distribution of earthworms in ecosystem units of hornbeam forest. *J. Boil. Sci.* Volume 8. (2008) 1033-1038
46. Tripathi, G., Bhardwaj, P.: Biodiversity of Vermiresources. In: Vermiresource Technology, Tripathi, G. (Ed.). Discovery Publishing House. New Delhi. India. (2003) 1–62
47. Domínguez, J., Edwards, C.A.: Biology and Ecology of Earthworm Species Used for Vermicomposting, Vermiculture Technology: Earthworms, Organic *Waste, and Environmental Management*. (2011)
48. Haimi, J., Huhta, V.: Capacity of various organic residues to support adequate earthworm biomass for vermicomposting. *Biol Fertil Soils.* Volume 2. (1986) 23-27
49. Kale, R. D., Bano, K., Krishnamoorthy R. V.: Potential of *Perionyx excavatus* for utilizing organic wastes. *Pedobiologia.* Volume 23. (1982) 419-425
50. Obi, F. O., Ugwuishiwu, B. O., Nwakaire, J. N.: Agricultural waste concept, generation, utilization, and management. *Nigerian Journal of Technology.* Volume 35(4). (2016) 957–964. [CrossRef]
51. Suthar, S.: Bioremediation of aerobically treated distillery sludge mixed with cow dung by using an epigeic earthworm *Eisenia fetida. The Environmentalist*, Volume 28(2). (2008) 76-84
52. Mahaly, M., Senthilkumar, A. K., Arumugam, S., Kaliyaperumal, C., Karupannan, N.: Vermicomposting of distillery sludge waste with tea leaf residues. *Sustainable Environment Research*, Volume 28(5). (2018) 223-227
53. Yadav, A., Garg, V. K.: Industrial wastes and sludges management by vermicomposting. *Reviews in Environmental Science and Bio/Technology.* Volume 10(3). (2011) 243–276
54. Deka, H., Deka, S., Baruah, C. K., Das, J., Hoque, S., Sarma, N. S.: Vermicomposting of distillation waste of citronella plant (Cymbopogon winterianus Jowitt.) employing *Eudrilus eugeniae. Bioresource Technology.* Volume 102(13). (2011) 6944–6950
55. Gupta, R., Garg, V. K.: Stabilization of primary sewage sludge during vermicomposting. *Journal of hazardous materials.* Volume 153(3). (2008) 1023-1030
56. Awasthi, M. K., Wong, J. W., Kumar, S., Awasthi, S. K., Wang, Q., Wang, M., Zhang, Z.: Biodegradation of food waste using microbial cultures producing thermostable α-amylase and cellulase under different pH and temperature. *Bioresource technology.* Volume 248. (2018) 160-170
57. Li, X., Xing, M., Yang, J., Huang, Z.: Compositional and functional features of humic acid-like fractions from vermicomposting of sewage sludge and cow dung. *Journal of hazardous materials.* Volume 185(2-3). (2011) 740-748
58. Sen, B., Chandra, T.: Chemolytic and solid-state spectroscopic evaluation of organic matter transformation during vermicomposting of sugar industry wastes, Bioresource Technology. Volume 98. (2007) 1680–1683
59. Yadav, A., Garg, V.K.: Feasibility of nutrient recovery from industrial sludge by vermicomposting technology, *Journal of Hazardous Materials.* Volume 168. (2009) 262–268
60. Elvira, C., Sampedro, L., Dominguez, J., Mato, S.: Vermicomposting of wastewater sludge from paper pulp industry with nitrogen-rich materials, *Soil Biology and Biochemistry.* Volume 29. (1997) 759–762
61. Suthar, S.: Vermistabilization of municipal sewage sludge amended with sugarcane trash using epigeic Eisenia fetida (Oligochaeta), *Journal of Hazardous Materials.* Volume 163. (2009) 199–206

62. Contreras-Ramos, S.M., Alvarez-Bernal, D., Dendooven, L.: Characteristics of earthworms (Eisenia fetida) in PAHs contaminated soil amended with sewage sludge or vermicompost, *Applied Soil Ecology*. Volume 41. (2009) 269–276
63. Gupta, R., Garg, V.K.: Vermiremediation and nutrient recovery of non-recyclable paper waste employing Eisenia fetida. *Journal of Hazardous Materials*. Volume 162. (2009) 430–439
64. Kaushik, P., Garg, V.K.: Vermicomposting of mixed solid textile mill sludge and cow dung with the epigeic earthworm *Eisenia foetida*. *Bioresource Technology*. Volume 90. (2003) 311–316
65. Garg, V., Kaushik, P., Dilbaghi, N.: Vermiconversion of wastewater sludge from textile mill mixed with anaerobically digested biogas plant slurry employing *Eisenia foetida, Ecotoxicology, and Environmental Safety*. Volume 65. (2006) 412–419
66. Mata-Alvarez, J., Dosta, J., Romero-G€uiza, M., Fonoll, X., Peces, M., Astals, S.: A critical review on anaerobic co-digestion achievements between 2010 and 2013, *Renewable and Sustainable Energy Reviews*. Volume 36. (2014) 412–427
67. Chen, G., Zheng, Z., Yang, S., Fang, C., Zou, X., Luo, Y.: Experimental co-digestion of corn stalk and vermicompost to improve biogas production. *Waste Management*. Volume 30. (2010) 1834–1840
68. Yadav, A., Garg, V.: Bioconversion of food industry sludge into a value-added product (vermicompost) using epigeic earthworm *Eisenia fetida*. *World Review of Science Technology and Sustainable Development*. Volume 7. (2010) 225–238
69. Suthar, S.: Vermicomposting of vegetable-market solid waste using *Eisenia fetida*: Impact of bulking material on earthworm growth and decomposition rate. *Ecological Engineering*. Volume 35. (2009) 914–920
70. Yadav, A., Garg, V.: Industrial wastes and sludges management by vermicomposting. *Reviews in Environmental Science and Bio/Technology*. Volume 10. (2011) 243–276
71. Rupani, P.F., Singh, R.P., Ibrahim, M.H., Esa, N.: Review of current palm oil mill effluent (POME) treatment methods: Vermicomposting as a sustainable practice. *World Applied Sciences Journal*. Volume 11. (2010) 70–81
72. Khwairakpam, M., Bhargava, R.: Bioconversion of filter mud using vermicomposting employing two exotic and one local earthworm species. *Bioresource Technology*. Volume 100(23). (2010) 5846-5852
73. Nogales, R., Cifuentes, C., Benitez, E.: Vermicomposting of winery wastes: A laboratory study. *Journal of Environmental Science and Health Part B*. 40 Volume (2005) 659–673
74. Albanell, E., Plaixats, J., Cabrero, T.: Chemical changes during vermicomposting (*Eisenia fetida*) of sheep manure mixed with cotton industrial wastes. *Biology, and Fertility of Soils*. Volume 6. (1988) 266–269
75. Singh, J., Kaur, A., Vig, A., Rup, P.: Role of *Eisenia fetida* in rapid recycling of nutrients from bio sludge of beverage industry. *Ecotoxicology and Environmental Safety*. Volume 73. (2010) 430–435
76. Sinha, R. K., Valani, D., Chandran, V., Soni, B. K.: Earthworms - The Soil Managers: Their Role in Restoration and Improvement of Soil Fertility. Volume (54). (2011 b). *Agricultural Issues & Policies*. NOVA Science Publishers, N.Y., USA

Effect of Ultraviolet (UV)-B Radiation on Morphology and Photosynthetic Activity of *Spirogyra* sp.

Davendra Singh Malik and Prachi Rathi

Abstract *Spirogyra* sp. is phytoplankton species belonging to *chlorophyceae* group and also use as a bioindicator. They are very sensitive to overexposure from solar ultraviolet (UV) radiation. In this study, *Spirogyra* sp. was collected from freshwater and then cultured in the Pringsheims Medium. The aim of the present study was to find out the impact of ultraviolet (UV-B) radiation on morphology and photosynthetic activity of *Spirogyra* sp. For this different experimental setup were divided on the basis of time duration. Exposure duration was 2 h/day for six groups except control. Reading of artificial UV-B was recorded at 5th, 10th and 15th day respectively and one is kept as control (covered with acrylic sheet to avoid any light penetration). After 15th days of artificial UV-B exposure it is observed that chlorophyll content decreased and carotenoid increased. Maximum effect and fragmentation were observed after 15th day of artificial UV-B exposure. The study provides basic scientific data about the impacts of UV-B radiations on aquatic ecosystem.

Keywords *Spirogyra* species · UV-B radiation · Chlorophyll · Carotenoid

1 Introduction

Phytoplankton is an autotrophic individual which found in upper most layer (up to 200 m) of the surface water body. In freshwater, various phytoplankton species found which are belonging to different phytoplankton groups. The most common is green algae which belong to *Chlorophyceae* group and are among the oldest group living on the earth's surface [5]. Algae are spread in soils, swamps, wetlands, and freshwater resources such as river and lakes. In green algae, *Spirogyra* sp. well known freshwater bioindicator species and also known for its pharmacological activities as an anti-hypertension resulted from its isolated gallic acid [8]. In modern era, one of the serious environmental problems on earth has been experiencing due to increase

D. S. Malik · P. Rathi (✉)
Department of Zoology and Environmental Science, Gurukul Kangri (Deemed to be University), Haridwar, Uttarakhand 249404, India
e-mail: rathi.pr94@gmail.com

K. D. Bahukhandi et al. (eds.), *Environmental Pollution and Natural Resource Management*, Springer Proceedings in Earth and Environmental Sciences,
https://doi.org/10.1007/978-3-031-05335-1_10

in ultraviolet (UV) radiation. The ozone layer protects us from UV radiation, but due to the higher pollution the amount of green house gases such as CFCs, CCS and organic bromides is increase and they are responsible for damage the ozone layer [6]. The solar radiation penetrates in it up to some depth and provides them an optical energy for the process of photosynthesis and production of food [10]. Phytoplankton organisms are preferred to move into the photic zone due to their energetic requirements for solar radiation and on the other hand they are affected by excessive solar radiation. The increased amount of UV radiation which penetrates the euphotic zone has affected the phytoplankton productivity. These autotrophic organisms live in an environment where factors such as light availability, uptake of nutrients, sinking and grazing pressure affects the growth and distribution. Since algae are simple, small sized organisms do not contain protective layers as the skin, so the effect of these rays will be fast and direct [9]. The effects of short ultraviolet rays in these organisms are shown by their diminished ability to adjust their position in the water column. So, keeping this view, we investigate the impacts of UV-B radiation on photosynthetic pigments in *Spirogyra* sp. on different time exposure.

2 Materials and Methods

2.1 *Identification and Culturing of* Spirogyra *sp.*

Methodology and culturing of *Spirogyra* sp. were shown in Fig. 1. One drop of collected plankton sample was putted on Sedgwick rafter scale (because Sedgewick rafter holds a much higher volume, and is good for large species and useful for the abundance study of mixed samples). Identifying under a light microscope (Optika B-150 series) with camera attached in to it. Identified *spirogyra* sp. was isolated and then grown in a Pringsheims medium, for multiple growths of selected species. The equal number of tested organisms was then transmitted into seven different sterilised petri dish by using magnifying glass (Carson triview 5x, 10x, 15 × folding loupe with built in case tv-15).

2.2 *Experiment Set up for Laboratory Experiment*

Seven wooden piece frames were taken of same dimension (45 cm long and 22 cm wide). Each has one circular open space (diameter = 18 cm) and fitted them on seven different and the test organisms may not pass through the net and remains at the surface as shown in a Fig. 2. Seven aquariums were taken for the experimental set up and divided them into seven groups. Group A (control), group B (expose to natural solar radiation for 5 days), Group C (expose to natural solar radiation for 10 days), Group D (expose to natural solar radiation for 15 days), Group E (UV-B exposure for

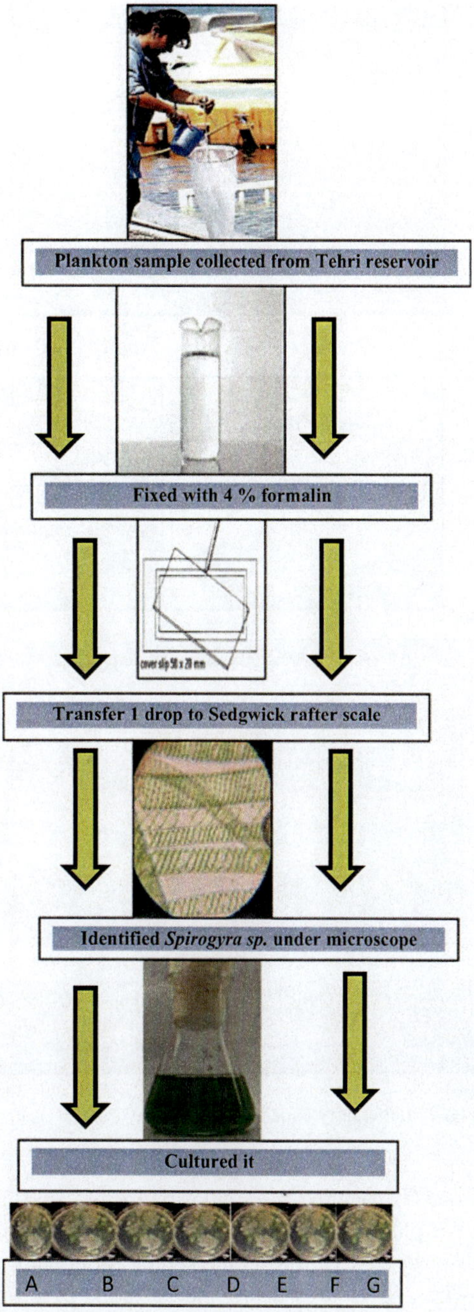

Fig. 1 Methodology of collection and culture of *Spirogyra* sp.

Fig. 2 Laboratory experiment set up

5 days) Group F (UV-B exposure for 10 days) Group G (UV B exposure for 15 days). Petri plates containing the material in 10 ml of culture solution were exposed to a distance of 15 cm from the source for equal irradiation of tested organisms. Group A was covered by acrylic sheet and kept in a laboratory. Group B, C and D were kept outside the lab for natural solar exposure and group E, F, G were kept in a lab to provide artificial UV-B through UV-B Philips lamp of intensity 0.638 at 312 nm. The exposure was given for two hour/day. Natural solar radiation was measured by coleparmer radiometer (WW-09811-56) daily.

2.3 Determination of Total Amount of Chlorophyll and Carotene Pigments

The effect of natural solar and artificial UV-B radiation on chlorophyll content, carotenoid and fragmentation were investigated in seven different groups. After exposure, samples were treated with acetone for the pigment extraction i.e., for chlorophyll a, b and total carotenoid. Samples were filtered and crushed with 90% acetone in mortar so that the pigment comes out of the cell. The obtained solution vortexed and then centrifuged at around 2000 rpm for 5 min after centrifugation, supernatant was separately collected and pigments (chlorophyll ‘a’, Chlorophyll ‘b’ and total carotenoid) was estimated spectrophotometrically using UV-VIS spectrophotometer. The chlorophyll concentration and the concentration of carotenoid were determined using equations of [12, 13].

3 Result and Discussion

Solar UV-B radiation reaching the earth surface exerted significant effect on microalgae [4]. Photosynthesis is potentially the main target of UV radiation due to multiplicity of possible effects [7]. Daily natural solar radiation variation during the study period was shown in Fig. 3. Effect of natural solar radiation and artificial UV B on photosynthetic pigments of *Spirogyra* sp. was shown in Table 1.

Natural solar radiation in Haridwar was measured in range from 0.512 to 0.955 mw/cm^2 (Fig. 3). Minimum natural solar radiation was measured on 9th day during the experimental study it was due to the cloudy weather on that day.

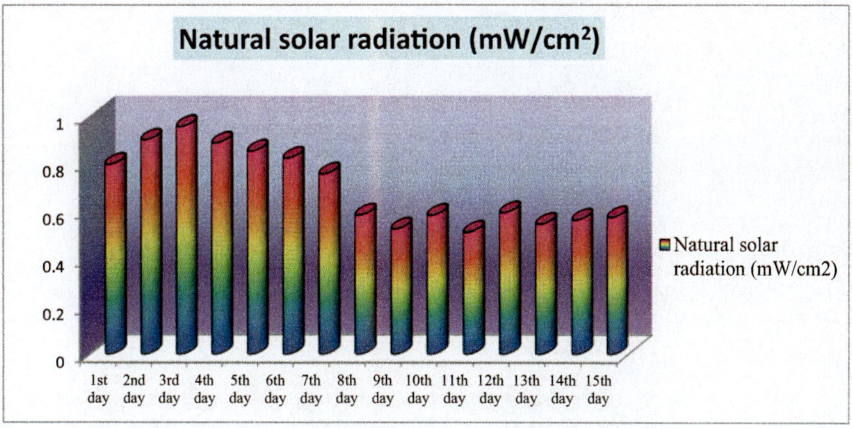

Fig. 3 Natural solar UV radiation variation of fifteen days in Haridwar

Table 1 Effect of solar and artificial ultraviolet radiation on *spirogyra* sp. in laboratory

Parameter/experiment setup for 15 days	Control (Group A)	Natural solar UV			Artificial UV-B exposure		
		After 5 days of exposure (Group B)	After 10 days of exposure ((Group C)	After 15 days of exposure (Group D)	After 5 days of exposure Group E)	After 10 days of exposure (Group F)	After 15 days of exposure (Group G)
Chlorophyll 'a' (mg/l)	10.46 ± 0.08	9.51 ± 0.35	9.45 ± 0.38	9.38 ± 0.32	6.42 ± 0.31	5.79 ± 0.28	4.63 ± 0.53
Chlorophyll 'b' (mg/l)	3.95 ± 0.79	3.82 ± 0.88	3.76 ± 0.28	3.65 ± 0.23	3.05 ± 0.65	2.66 ± 0.49	2.28 ± 0.45
Carotenoid (mg/g)	3.99 ± 0.17	4.31 ± 0.25	4.35 ± 0.18	4.42 ± 0.21	5.26 ± 0.21	6.33 ± 0.17	7.19 ± 0.09
Fragmentation	–	–	–		+	+	++

The effects of UV-B photosynthetic pigments chlorophyll content showed decreasing trend with increasing duration/time period of natural solar radiation and artificial UV-B radiation. The chlorophyll content decreased in all exposure either it is natural solar radiation or artificial UV-B exposure. It was found that highest chlorophyll content (10.46 $\pm$ 0.08) was found in Group A (control) and lowest (4.63 $\pm$ 0.53) in Group G (15 days of UV-B exposure). Chlorophyll 'b' pigment was found maximum (3.95 $\pm$ 0.79) in Group A (control) and minimum (2.28 $\pm$ 0.45) in Group G (15 days of UV-B exposure).

Carotenoids are the accessory pigments in plants that absorb light energy for use in photosynthesis. Highest carotenoid content (7.19 $\pm$ 0.09) was found in a group G (after 15 days of UV-B exposure) while lowest carotenoid content (3.99 $\pm$ 0.17) was found in group A (control). They also protect chlorophyll from photo damage [1]. The increase in the amount carotenoids when exposing algae to UV-B is due to its important role in protecting chlorophyll. Carotenoids cover the chlorophyll and protect it from the negative impact of these rays. Carotenoid content increases as the stress or duration of solar radiation on phytoplankton increases [14].

Fragmentation of filaments, shrinkage, deformation and degeneration of chloroplasts were frequently noticed in all treatments of UV-B radiation but not in an exposure of natural solar radiation and in control. The degree of the above effects was found to be linearly increased with increase in the time period to which the algae was exposed. The responses of *Spirogyra* sp. are depending on duration/time period of UV-B radiation exposure and protective mechanism of the organisms being exposed. *Spirogyra* sp. was exposed to UV radiation (artificial and solar) for 5, 10 and 15 days. There is no fragmentation in *spirogyra* sp. cell wall was found in control and natural solar but on UV-B exposure low fragmentation was found on 5th day and the fragmentation increases as the number of days increases. Highest fragmentation was found after 15 days. Morphological changes upon UVR exposure may be due to the release and accumulation of reactive oxygen species that oxidizes the lipids of sheath or cell membrane in the presence of UV radiation [11]. Deformation in *Spirogyra* sp. first observed in chloroplast and then breakage of filaments (after 15 days treatment). It showed deformed chloroplast and lack of usual spiral arrangement in the UV exposed *Spirogyra* sp. filaments. Besides, UVR also induced photo orientation in *Spirogyra* sp. it shows migration to its lit area. This behaviour may be considered as a tolerant strategy against UV-B radiation [2, 3, 15].

4 Conclusion

In the present study it can concluded that artificial UV-B possess potential impact on morphology and physiology of *Spirogyra* sp. when exposed for 5th, 10th and 15th days. Morphological changes include deformation in *Spirogyra* sp. was observed in chloroplast along with the breakage of filaments (after 15 days treatment). It showed deformed chloroplast and lack of usual spiral arrangement in the UV exposed *Spirogyra* filaments. Besides, UV radiation also induced photo orientation in *Spirogyra* sp. it shows migration to its little area. This behaviour may be considered as a tolerant strategy against UV-B radiation. The fragmentation in the selected filamentous algae was corresponding to its exposure to UV radiation. No fragmentation was observed in Group A, Group B, Group C and Group D but fragmentation increased as we move from Group E to Group G due to the harmful UV-B radiation effect on *Spirogyra* sp. cell wall it may be due to the release and accumulation of reactive oxygen species that oxidizes the lipids of sheath or cell membrane.

Acknowledgements The Authors are highly gratefully to department of Zoology and Environmental Science, Gurukul Kangri (deemed to be university), Haridwar for provide the laboratory facility for the research.

References

1. Armstrong, G.A., Hearst, J.E.: Carotenoids 2: Genetics and molecular biology of carotenoid pigment biosynthesis. *FASEB J.* 10(2). (1996) 228–237.
2. Bebout, B.M., Garcia-Pichel, F.: UV-B induced vertical migrations of cyanobacteria in microbial mat. *Appl. Environ. Microbiol.* 61 (12). (1995) 4215–4222.
3. Castenholz, R.W. Multiple strategies for UV tolerance in cyanobacteria. Spectrum, 10, (1997). 10–16.
4. Hadar, D.P., Kumar, H.D., Smith, R.C., Worrest, R.C. Effects of solar UV radiation on aquatic ecosystems and interactions with climate change. *Photochem. Photobiol. Sci.* 6(3). (2007) 267–285.
5. Halfar, J., Hetzinger, S., Adey, W., Zack, T., Gamboa, G., Kunz, B., Williams, B., Jacob, D.E. Coralline algal growth-increment widths archive North Atlantic climate variability. *Palaeogeography, Palaeoclimatology, Palaeoecology.* 302 (1-2). (2011) 71–80.
6. HE, Y.Y., Haeder D.P. UV-B induced formation of reactive oxygen species and oxidative damage of the cyanobacterium *Anaebaena* sp.: Protective effectsof ascorbic acid and N-acetyl L-cysteine. *Journal of Photochemistry and Photobiology B: Biology.* 66(2). (2002) 115–124.
7. Holzinger, A., Karsten, U., Lutz, C., Wiencke, C. Ultrastructure and photosynthesis in the supralittoral green macroalga Prasiola crispa (Lightfoot) Kutzing from Spitsbergen (Norway) under UV exposure. *Phycologia.* 45(2). (2006) 168–177.
8. Kang, J.W., Lee, N.Y., Cho, K.C., Lee, M.Y., Choi, D.Y., Park, S.H., Kim, K.P. Analysis of nitrated proteins in Saccharomyces cerevisiae involved in mating signal transduction. *Proteomics.* 15(2–3). (2015) 580–90.
9. Karsten, U., Dummermuth,A., Hoyer, K. and Wiencke, C. Interactive effects of ultraviolet radiation and salinity on the ecophysiology of two Arctic red algae from shallow waters. *Polar biology*. 26(4), (2003) 249–258.
10. Klisch M., Haeder D.P. Mycosporine like amino acids (MAAs) profile of a rice field cyanobacterium Anabaena doliolum as influenced by PAR and UVR. *Planta.* 229(1). (2008) 225–233.
11. Ma, Z., Gao, K. Spiral breakage and photo-inhibition of Arthrospira platensis (Cyanophyta) caused by accumulation of reactive oxygen species under solar radiation. *Env. Exp. Bot.* 68(2). (2010) 208–213.
12. Parsons, T.R., Strickland, J.D.H. Discussion of spectrophotometric determination of marine plant pigments, with revised equation for ascertaining chlorophylls and carotenoid. *J. Mar. Res.* 21. (1963) 155–163.
13. Porra, R.J. The chequered history of the development and use of simultaneous equations for the accurate determination of chlorophylls a and b. *Photosynth. Res.* 7(1). (2002). 149–156.

14. Suaad, H. Ali, A.B., Abdul, W.R. Effect of UV-B radiation on content of pigments and mineral elements in *Cladophora graminea* and *Spirogyra Deadaleoides* algae. *Indian Journal of Ecology*. 12 (2020) 221–224.
15. Xiong, F., Komenda, J., Kopecky, J., Nedbal, L. Strategies of ultraviolet-B protection in microscopic algae. *Physiol. Plant.* 100(2), (1997) 378–388.

An Overview of Recent Advancements in the Irrigation, Fertilization, and Technological Revolutions of Agriculture

Neeraj Pandey, Nitin Kamboj, Arvind Kumar Sharma, and Amrit Kumar

Abstract Technological advancement in the agricultural sector for increasing the productivity rate of crops has become a must to fulfill the demand of the massive population. Though Inida is progressing at a faster rate in terms of agriculture the growth in horticultural and some of the major crops are very less in comparison with other development countries. It was estimated by Food and Agricultural Organization (FAO) that by 2050, agricultural production would grow by 70% globally and almost 100% in developing countries to feed the hiking population. To fulfill this demand, technological invention should become fast. The invention has occurred in the field of irrigation, fertilization, or technology. Thus, by incorporating these inventions in the agriculture sector, the output gets increased to a greater extent. The methods of farming such as hydroponics, aeroponics, bioremediation, biosorption, vermicomposting, etc. have made a greater change in the agriculture output. Not only this even the technologies like GPS, Ultrasound, smartphones, cameras, etc. helps in regular monitoring of the agricultural processes and make it easy giving little relax to the farmers. In a very lesser space and time, yield can be hiked up to the maximum thereby increasing the economy. This book chapter aims to highlight the recent advancement that has occurred in last few decades and their effect on productivity and crop yield.

Keywords Agriculture · Technologies · Advancements · Production · Yield

1 Introduction

Agriculture can be defined as the manners by which yielding plants and domestic animals endure the worldwide human populace by giving food and different items to fulfill the basic needs [1]. A revolution is an important tool for any sector of the environment to upgrade the previously occurred mythological activities. Since 1950 a lot

N. Pandey · N. Kamboj (✉) · A. K. Sharma · A. Kumar
Department of Zoology and Environmental Sciences, Gurukula Kangri (Deemed to be University), Haridwar, Uttarakhand 249404, India
e-mail: nkamboj@gkv.ac.in

K. D. Bahukhandi et al. (eds.), *Environmental Pollution and Natural Resource Management*, Springer Proceedings in Earth and Environmental Sciences,
https://doi.org/10.1007/978-3-031-05335-1_11

of changes have occurred in the agricultural segment. Previously conventional means of agriculture practices were used that takes a long time from sowing to harvesting. And, also these methods pose a serious threat to the environment by deteriorating soil, water, air, fungal quality. So, a time for the advancement is the need for the hour. Development is an essential tool that is required to upgrade the existing technical inventions without compromising the future of the upcoming generation. It was observed that over the last few decades a drastic change has occurred in the technological aspects of agriculture. Previously conventional means were used for agricultural enhancement but in the current scenario, modern tools and techniques are adopted [2]. With the increasing population load on earth, the demand for food is hiking up at a greater pace. The land is less and the demand is more, so people have started using up the innovative ideas to enhance the crop yield in less space, time, and money [3]. An approach towards sustainable agriculture is focused nowadays. The initial step towards sustainable agriculture is the use of technology and external supplements without affecting future traits [4]. Sustainable agriculture practice may lead to the production of efficiently high quality and quality products thus fulfilling the necessity of a huge number of consumers [5]. Upgradation in agricultural technologies is the major need of the hour.

In the context of Indian agriculture growth, an increase of only 3.59% was observed in the year 2004–2014 which is lower than the expected value of 4% per annum growth until 2020. The major reason behind this reduced percentage is the production of food grains. The per capita production of food grain gets reduced to 179 kg in 2014/17 which was 207 kg in 1991/1995 [6]. The decreased level of food grain production poses a threat to the Indian population therefore much attention is required for maintaining food and nutritional security. With the scarce water supply and land resources, the target of 2020 of 4% can be obtained by improving the per-unit production of natural resources and farm output by the use of the improved and advanced technologies in a judicial way. So, keeping this in mind we have included all the possible inventions that have occurred in the agricultural fields in the last few decades and what benefits it brings to the consumers and the environment as well.

2 Developments in Varied Processes of the Farming System

To add more flavor in the field of agriculture, more and more inventions have occurred by the researchers both in the technology and machinery area. These are applied for enhancing productivity and yield together with performing more work in a short period. Some of them are given below:

2.1 Hydroponics

Hydroponics is a farming method in which soil is not used as a growing medium for crops. Instead, a water solution is given to the roots of the plants. All the nutrient essential for the plant growth is provided by the aqueous solution used. It is a more environment-friendly technique and affects the climate to a minimum, so it can also be termed as Controlled Environmental Agriculture (CEA). As plants are grown in absence of soil, it may also be termed as the soil-less culture technique, where plants are grown in a water-based nutrient-rich medium under a greenhouse set-up [7–9].

The main purpose of hydroponics is to provide nutrients to the plants in a soil-less environment, through direct contact of the plant roots to the nutrient-rich solutions, having an excess of oxygen that is essential for plant growth [10]. In India, farmers are not that much aware of this technique as it requires a lot of skills for making it functionally operational. The main drawbacks for the farmers are the low literacy rate. To make them perfect in this field proper guidance is required with proper hands-on training so that they can further guide the rest of the farmers.

2.1.1 Hydroponics Techniques

Various techniques can be used to grow plants in soil-less conditions by providing required nutrients through nutrient solutions. Some of them are described below and represented in Fig. 1.

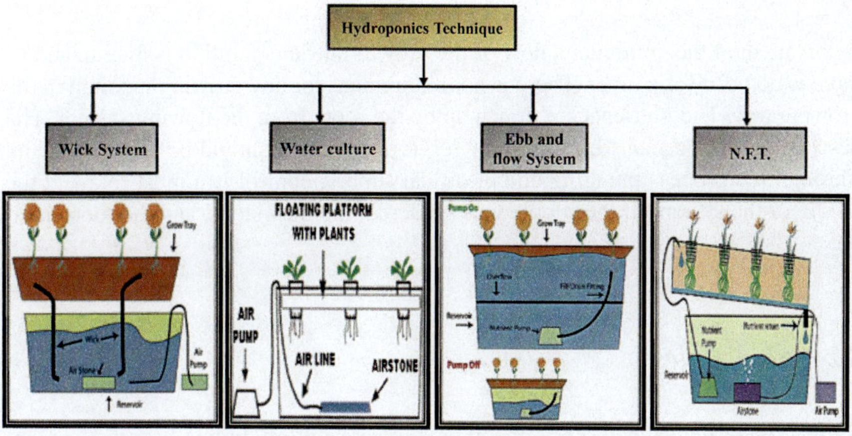

Fig. 1 Representing different types of hydroponics system

Wick System

It is one of the simplest methods of hydroponics which works passively without any moving part. The reservoir present at the bottom of the system are used to draw up the nutrient solution via several wicks present in the growing nutrient medium. This system uses a different variety of mediums, soil, perlite, and coco.

Water Culture

This system works actively having a moving part. In the water having the specific growth nutrient, the roots of the plants are immersed completely. The air pump which is joined with the water solution oxygenates the water thereby allowing the plant roots to breathe. This system works well for lettuce and some other plants.

Ebb and Flow System (Flood and Drain)

The hydroponics system that works on the flooding of the growing tray is known as the ebb and flow system or flood and drain system. The reservoir having the nutrient solution surrounds the roots and then starts draining. The whole system works on the automated water pump with a timer.

Nutrient Film Technique (N.F.T. System)

In this method the continuous flow of the growth nutrient solution is maintained so there is no need for a timer. There is no requirement for any growing medium in the growing tray. The nutrients are drawn up by the roots from the flowing system. The descending flow decants back into the reservoir where it should be recycled again. Through the process pumping and electrical supply, control is a must to avoid the failure of the system. If the supply gets hindered, the flow stops, and the roots dried up.

2.2 *Aeroponics*

Cultivation of plants in absence of soil or substrate culture. In this system, the plant grows in the air with the help of artificial support [11]. In other words, we can say, it is an air–water-associated cultivation technique, in which plants are hanged in the air inside the sealed pack container under the dark condition and the nutrient-rich solution is sprayed through the atomizer. The leaves and the exposed outwardly above the wet zone whereas the root zone is separated by the artificially made structure. The whole system utilizes the nutrient-rich droplets coming out from the pressurized

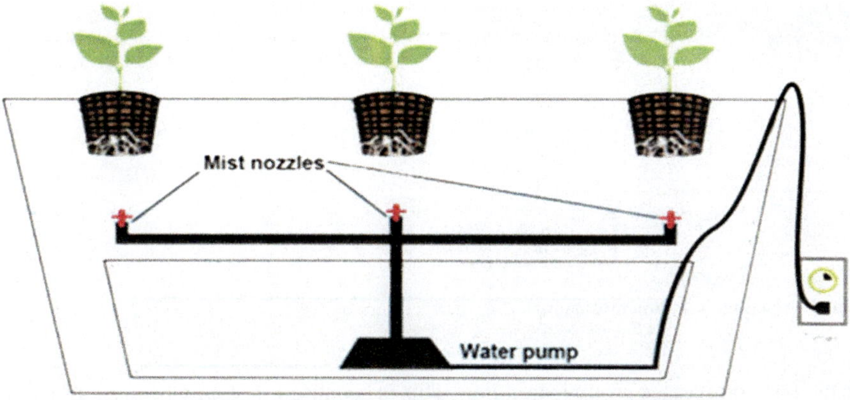

Fig. 2 Aeroponics system functioning

atomizer nozzles or froggers to sustain plant growth under the controlled environment [12]. The whole process of aeroponics is illustrated in Fig. 2.

3 Bioremediation

With the increasing pattern of agriculture practices, more and more supplements (i.e. pesticides and fertilizers) are being used to enhance the performance of the soil and plant growth. These supplements when directly or indirectly discharged into the water channels without prior treatment poses a serious to the aquatic body. The water contaminated with the supplements mainly contains a huge quantity of nitrates, sulfates, phosphates, etc. [13]. The intensification of these substances leads to severe abnormalities both in fauna and flora; on land and inside water. In the overall context, it leads to pollution on a wider scale and control becomes a tedious task [14].

To check the pollution level in water bodies through agriculture, a technique called bioremediation is being used. It can be defined as the process in which microorganisms degrade the harmful toxic substances into non-harmful ones likes water and carbon dioxide. Bioremediation is divided into different groups depending upon the substrate used for the remediation.

3.1 Mycoremediation

The type of remediation in which fungi and mushrooms are used as a substrate to remove the waste/pollutants from the environment is called mycoremediation. Mushrooms and fungi contain a set of enzymatic machinery that helps them to clean out the pollutants/waste products [15, 16]. In today's scenario, a mushroom belonging

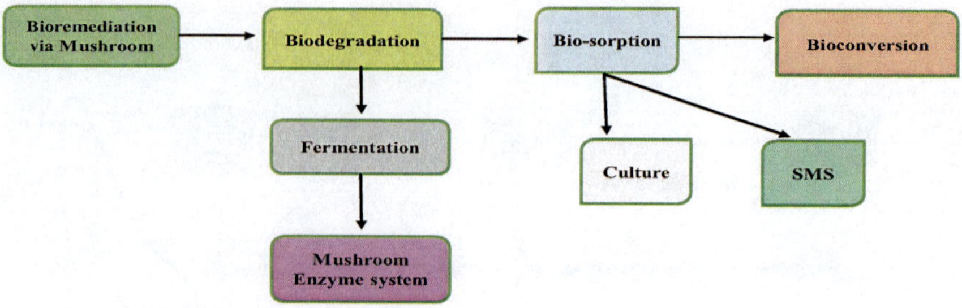

Fig. 3 Myco-remediation mechanism

to the basidiomycetes group is trending. This is because it not only be used as a tool for remediation but acts as a supplier of protein to the mycelium and the fruiting body. The efficacy of the fungus species to create food protein as biomass or fruiting bodies from different types of wastes depends upon the capacity to degrade waste through exudation of diversified hydrolyzing and oxidizing enzymes [17, 18]. Therefore, the researchers are now focusing on mushroom cultivation and remediation of waste through it (Fig. 3).

3.2 *Phytoremediation*

The process in which plants are used for remediating the contaminants present in water, wastewater, soil, sediments, sludge, etc. is called phytoremediation. Aquatic, semi-aquatic and terrestrial plants may be used in this process [19]. The whole process of phytoremediation is well illustrated in Fig. 4.

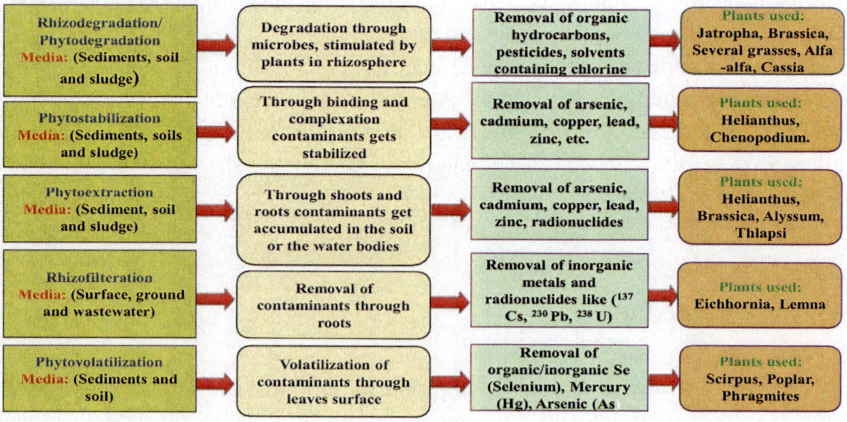

Fig. 4 The phytoremediation process [20]

3.3 *Vermiremediation*

Vermicomposting technology is recovering option for recycling the biodegradable waste into nutrient-rich manure which enhances the productivity of the soil by enhancing nutrients in the soil like NPK. Today rapid civilization and agricultural cultivation huge amount of biodegradable waste is produced from developed as well as developing countries. This organic biodegradable waste is a stinky smell which irritates and cause the various health problem to the mankind. By the vermicomposting techniques, we can easily convert this stinky smelly waste into the nutrients rich manure utilized in the agriculture field for the production of the crop. The vermicomposting enhances the productivity of the plants because the vermicompost is rich in nutrients who essential for the plant's development and growth. Vermicompost with microbial activities which present in the soil releases much of the nitrogen in the form of the nitrate while composting releases nitrogen into the soil in the formed ammonia, plant accumulates nitrogen in the form of nitrate, not ammonia the nitrate form of the nitrogen is easy to uptake by the. Vermicompost is an extreme organic biofertilizer, it initiates and propagates the rooting system in plants mung bean and positively elongates the shoot development in watercress and lettuce [21]. In vermicomposting to decompose the organic waste into the organic fertilizer (vermicompost) use various species of earthworms. Commonly used earthworm species are *Eisenia fetida* and *Eudrilus eugeniae* both species are epigeic, they feed on surface decaying organic matter and convert it into manure. Through vermicomposting improve the soil fertility vermicast improve the aeration in the soil.

Types of earthworm: there are three types of earthworms; epigeic, endogeic, and anecic. Epigeic earthworms are surface feeders of decaying organic wastes they consume approximately their weight of food per day also they do not construct the burrow live on the upper surface of the organic matter e.g.; *Eisenia fetida.* Hence this species of earthworms are most suitable for vermicomposting. Endogeic species of earthworms are live in horizontal borrow and feed minerals that present in the soil particles and decaying organic matter e.g.; *Aporrectodea calignosa.* Anecic species built vertical burrows and live up to 5–6 feet deep in the soil. They feed deep soil minerals and are called geophagous [22] (Fig. 5).

4 Bio-sorption

It is the process in which biological materials are used for the exclusion of contaminants from the water system through absorption, adsorption, ions exchange, complexation of surfaces, and precipitation [13]. The efficiency rate of the bio-absorbents is very high and they are easily available. They have a high tendency to bind heavy metals on their surface. Regeneration properties help it to create a favorable condition for the removal of contaminants. Regardless, if the concentration of the feed solution is remarkably high, the cycle of the process reaches its peak thus limiting

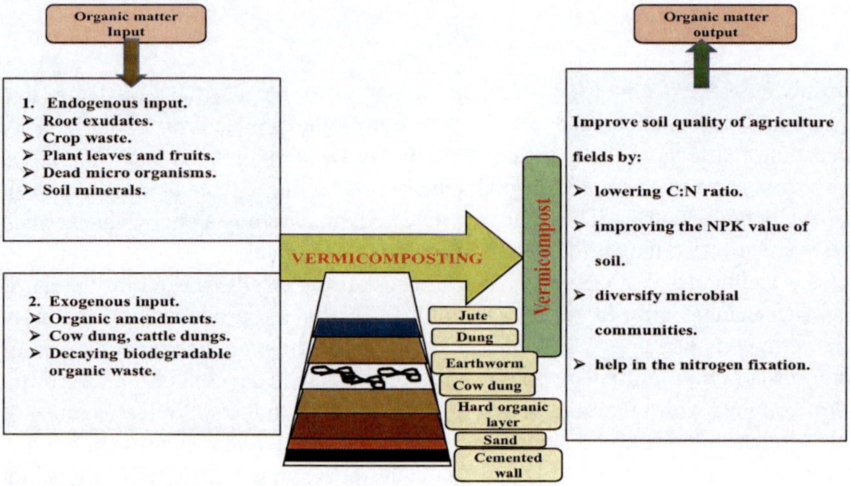

Fig. 5 Vermicomposting process

further removal of pollutants [23, 24]. The whole process of bio-sorption is shown below via a flowchart (Fig. 6).

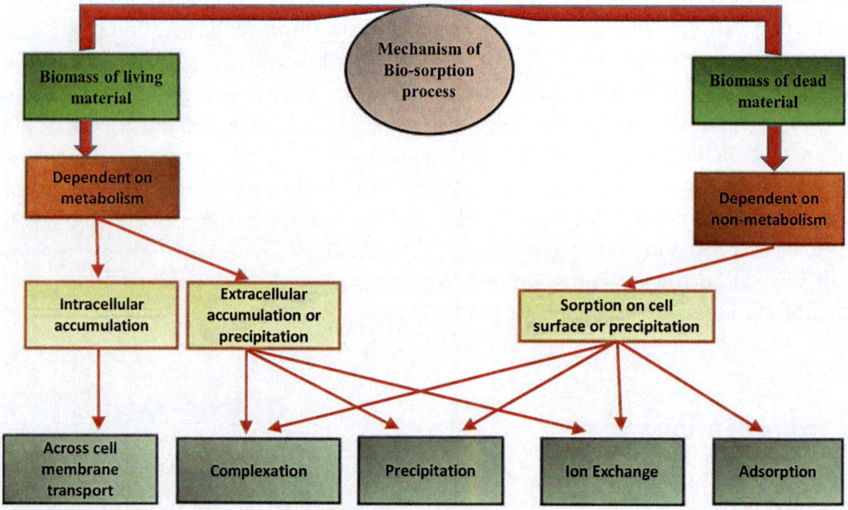

Fig. 6 Flowchart representing process of bio-sorption

5 Organic Farming

Organic farming is one of the agriculture practices in which the use of synthetic supplements (viz., fertilizers, weedicides, rodenticides, fungicides, additives in food, hormones, etc.) is avoided or excluded entirely. Instead of these supplements, organic products (viz., green manure, compost, vermicompost, crop residues, cow dungs, off-yard organic wastes, etc.) are used for enhancing productivity and crop yield. Organic farming works in favor of nature and depend on crop rotation, mobilization of nutrients and protection of plant in a natural way. Organic farming emphasizes chemical-less foodstuffs without the involvement of any chemical supplements, in a way to maintain the fertility of the soil for a longer period. Organic farming plays an important role in maintaining the soil micro and macro-diversity, maintains ecological balance, foster cycling of nutrients, etc. In today's time, this farming practice is developing at a greater pace and almost 170 countries globally produce organic agriculture products commercially [25].

Organic Farming is adopted these days globally because it does not involve expensive chemical supplements, thus it reduces the cost of production for the farmworkers and also the chance of pollution gets reduced. The fruits and vegetables grown through organic farming are nutrient-rich and are very tasteful as they are grown in the given time required for proper growth. So, organic farming can be considered an environment-friendly agriculture practice that helps in promoting the concept of sustainability [26]. Some of the basic steps of organic farming are illustrated in Fig. 7.

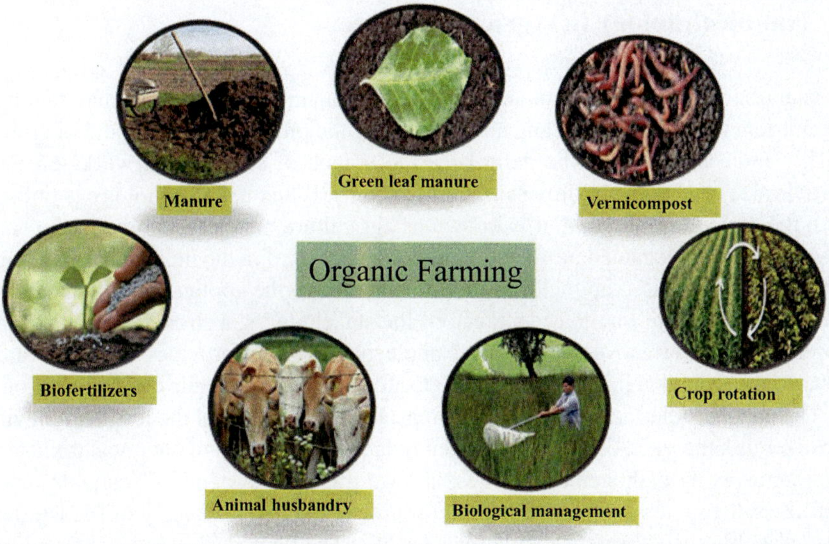

Fig. 7 Basic steps of organic farming

6 Biochar

When the biomass such as wood, manures, or leaves get undergoes a process called pyrolysis (i.e. heating of biomass in the complete absence or partially negligible oxygen and getting oil and different gases as a byproduct) leads to the formation of biochar. It is the highly enriched organic carbon that affects the sequestration of soil carbon by modifying its physicochemical and biological properties [27]. Due to the huge specific surface area of the biochar, porosity, functional groups such as hydroxyl, carbonyl, carboxyl, etc., it helps in eliminating contaminants from water bodies. These contaminants include heavy metals, dyes, antibiotics, which are toxic and affect the aquatic environment [28–31]. If the contaminants are not eliminated from the water bodies, it affects the soil properties when used in the field for irrigation, thus affecting the productivity and yield of crops. As it is a new kind of absorbent, it can be applied to improve the environment, energy, agricultural sustainability [32].

In recent scenarios, biochar provides a platform for the farmers to use it as a supplement in the fields in a way to increase the soil aeration, water holding capacity, enhanced activity of the soil organisms, maintaining the nutrient status of the soil, etc. Even some of the physicochemical properties of the soil get maintained to the equilibrium state, which is beneficial for increasing the productivity and yields of the crops [33]. The working process of the biochar, production methods, and its application is well demonstrated in the Fig. 8a–c.

7 Nanotechnology in Agriculture

In scientific advancements, nanotechnology holds an important place in maintaining the nutrient status of any soil and also increasing the productivity and yield of crops [34]. Nanotechnology can be defined as the branch of science that works on the particle size of nanometer dimensions (1–100 nm). When nanotechnology is linked with the agricultural sectors, it is known as agriculture nanotechnology. In the last few decades, it has gained momentum and is chiefly used in the fields for enhancing soil productivity and crop yield. In the Indian context, the application of nanotechnology is dependent on the acceptance of the stakeholders, a strong mechanism of governance, and effective regulations. Nanotechnology thus provides a step forward towards the second green revolution and also focuses on sustainable crop production.

The nanoparticles are required for increasing the efficiency of the fertilization via fertilizer use, the yield of crops, pesticidal reduction; detection of chemical toxins in food items, early pathogenic activities; shrewd transport systems of pesticides and fertilizers, the system and methods used for processing and packaging of foodstuffs; and agricultural food safety regulation [35, 36]. The degradation and reduction of the existing pollutants are achieved by the nano filters or the nano-catalysts. Thus, nanotechnology serves as a good resource management tool in the agricultural field, delivering drugs to the plants and helps in the maintenance of the fertility of the soil.

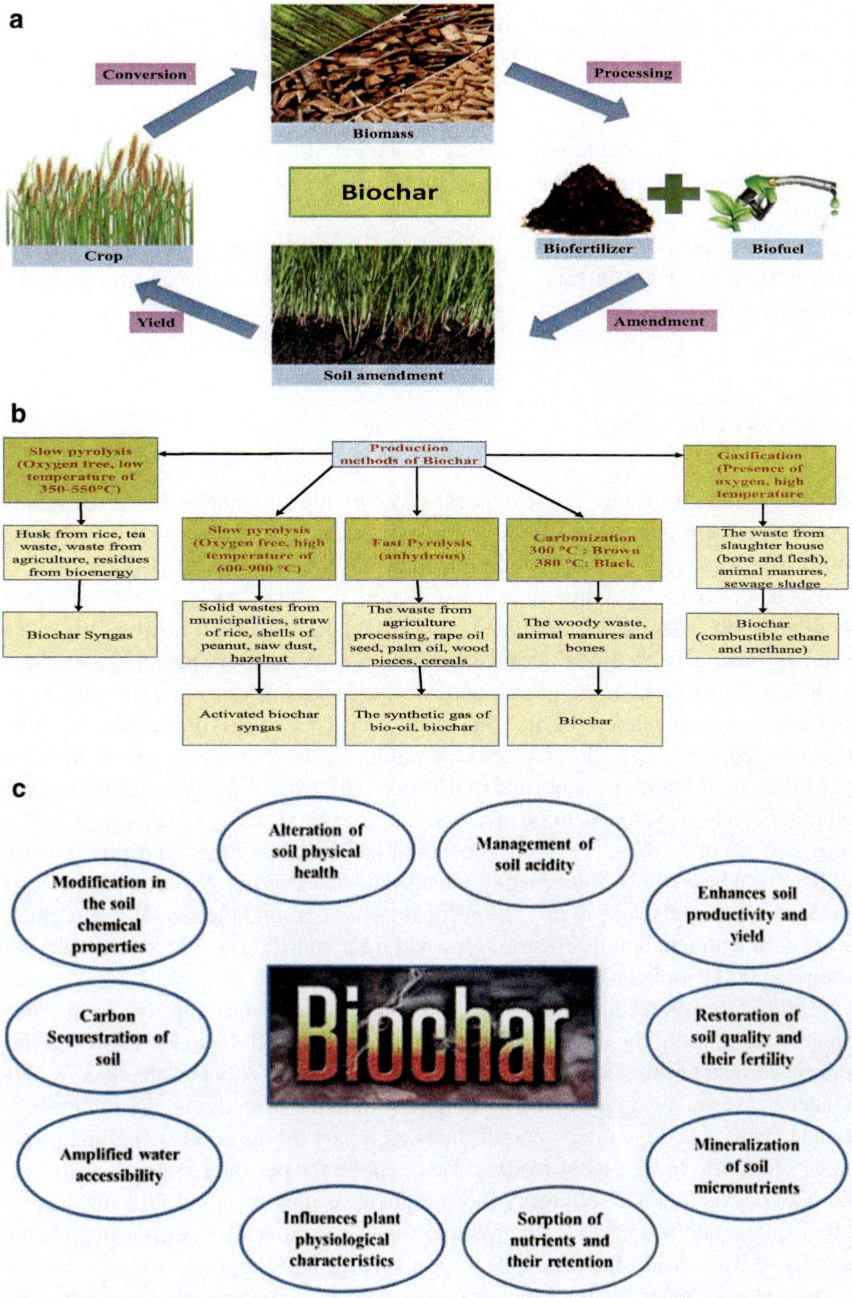

Fig. 8 **a** Working procedure of biochar, **b** production method, **c** applications of biochar

The role of this technology is also well observed in the usage of the biomass and waste from agriculture processes, food processing units, and packaging systems and also in the risk assessment [37]. Therefore, it can be easily said that the new and future advancements in nanotechnology that make a pathway for the food supply chain (from the agriculture field to the table: agrochemical usages such as nano-fertilizers, nano-pesticides, nano-fungicide, nano-herbicide, précised agricultural techniques, enrichment of food quality, quantity, and texture, feed in an intelligent way, nutrient standards and bioavailability, packaging, labeling, etc.) around the whole agricultural unit. More advancements are still to be faced in the way of nano-foods in the upcoming years [38].

8 Vertical Farming

Vertical farming is an innovative concept given by the eminent professor Despommier, that involves the use of basic methods of farming i.e. hydroponic and aeroponics, for faster yield. It can also be considered as the commercial farming method in which different life entities viz. plants, animals, fungi, etc. are cultivated for obtaining food, fiber, fodder, fuel, and other essential products. It involves the plantation vertically, stacked one above the other, in a very lesser space [39]. Vertical farming is taking momentum in megacities, where space is less and the requirement of products is much higher. It is in demand nowadays because (i) production capacity is more in comparison to the outdoor cultivation, (ii) water requirement is 70–95% lesser than that required for normal cultivation, (iii) do not get affected by natural calamities such as cyclones, flood, heavy rains, drought, etc. (iv) It helps grow pesticides free organic crops, (v) the occupational hazard is very less, and (vi) farmers are free from heavy farm equipment's and also the diseases like Malaria, exposure to poisonous chemicals, and so on. The vertical farm of around the size of a city block (9300 m^2), with a height of 30 stories, provides the nutrition of 200 kcal per day to around 15,000 people [40].

Vertical farming primarily involves three categories of farming: (a) In the first category of farming, the utopian concept of farming is used. It is also known as the "Phrase vertical farming". In this concept, vertical farming is done in an underground manner and is well observed in the Netherlands (b) In the second category of farming, vertical farming is done in open air or the skyscrapers for the control of climate and the consumption for the local feeders. This is done for personal or community use and not for commercial selling. A more innovative idea involved in it nowadays, that the plantation is done at the periphery of the skyscrapers, for getting an ambient quantity of light, (c) In the third category of vertical farming, cultivation is done in a closed sky scrapper for the large-scale production. This type of farming is well adopted in the countries like Singapore, Canada, London, etc. By 2050 it is expected that the population of the world gets increased by about 80% and most of the people live in urban areas. Due to this demand for food get hiked up, then vertical farming will be the best option to fulfill the food requirement [39, 40].

9 Modern Technological and Mechanical Inventions in the Field of Agriculture

Advanced technology and types of machinery that are employed in the current scenario for the enhancement of the crop yield and productivity is given below:

9.1 *Autopilot Tractors*

These are driverless tractors and sprayers worked on GPS in the agriculture fields. Previously, basic tractors are used in the fields to perform different works related to agriculture which is hectic for the farmers as they have to be continuously available in the fields. But, after the invention of autopilot tractors, this tedious task of farmers comes to nil. On the steering of the tractor, GPS is tied up that keeps it on the track, relaxing the driver from driving the vehicle. Even the spreading of fertilizers and pesticides is done in a controlled manner uniformly via a GPS tied on the tractors.

9.2 *Crop Sensors*

Crop sensors act as a means to apply the fertilizers in the field effectively for maximum uptake. Through sensing one can get information about how the crop is feeling, the amount of leaching that occurred, and also the amount of runoff in the underground water channel. This is taking variable rate innovation to a higher level. Crop sensors direct that how much quantity of fertilizers is to be applied in real-time into the field. This saves time in making the prescribed map for the fertilizer application in the field. The optical sensor senses what quantity of fertilizers a plant needs to be based on the light reflected in the sensor.

9.3 *Biotechnology*

Though biotechnology is not a new technique, it places an important role in having the potential to produce a product that is of more enhanced properties. Herbicidal resistance is most common in genetic engineering [41]. To make the crop resistant to pests, certain toxins are sprayed in the crop fields named herbicides that direct towards proper growth of the crops. Many of the farmers apply the same toxins that are found in certain organic pesticides, due to which they do not have to move to their fields for applying it [42]. This in turn saves the number of pesticides used, labor, fuel for operating machines, equipment, etc. Even there will be a cutback of the water used for irrigation by the farmers and also the crops yield does not suffer.

9.4 Documentation of Fields via GPS

In conventional methods, the documentation of the fields is done manually by moving one farm to another. But, with the invention of the GPS, the documentation work has become very easy and the data is obtained in a very precise way. Through the GPS data, farmers can figure out the possible changes that have to be made in the field for proper functioning. That means the data obtained via GPS technology are of very good quality. The yield map is the best form of documentation for the farmers by which they can figure out the years' cost of planning and the labor on a piece of a colorful paper sheet. The yield and moisture are being calculated when the harvesting equipment rolls through the field and binding it with the GPS coordinates. Then the map is printed, after the completion of tying. These maps are called the map of yield or the heat maps. Through these maps' farmers can find out that what varieties were the best, worst, or of steady yield in varied conditions. Even these maps can predict the drainage pattern working in that area.

9.5 Smartphones for Monitoring and Controlling the Irrigation System of Crops

Smartphones provide a new pathway to the farmers, thus playing a crucial role in maintaining and controlling the irrigation system in the agriculture fields [43]. Previously, farmers use to check their fields by driving themselves to each field individually, but after the launching of new software in smartphones and computers, the work becomes easy and is controlled by these gadgets. The moisture level in the soil is checked through the moisture sensors in the ground. These moisture sensors can transfer information about the availability of moisture at a certain deepness in the soil. Due to the flexibility, the water control and input of several supplements (such as fertilizers, pesticides, etc.) become precise when applied through the irrigation hinges. Farmers can also link it via Variable Rate Technology (VRT), for controlling the water rate applied precisely. Hence, it provides a way to use the resources effectively and efficiently.

9.6 Use of Mobile and Camera Technology

Mobile and cameras enable farmers to use social media sites effectively for multiple reasons. The apps like foursquare support to keep tabs on the workers. In order to keep an eye on the farm, nowadays cameras are placed around it. Even the livestock breeders put cameras around their barns, feedlot areas, meadows, etc. that sends the picture to the central location like home and office computers. Through this, they can able to put an eye on their livestock, when they are far away from their homes at night.

9.7 *Livestock Ultrasounds*

Ultrasounds are mostly used for checking out the status of the baby animal in the womb but in some cases, they can also be used to see the quality of the meat that has to be sent to the market. DNA testing enables the manufacturers to analyze the animals having good pedigrees and other necessary qualities. To improve the quality of the livestock, the information obtained from ultrasounds enables the farm keepers to improve the quality [44].

9.8 *Robotics*

An intelligent machine that is prepared with the collaboration of mechanical, electronics and computing field engineers to match the working outline of humans is called robotics [45]. In the agricultural sectors robotics is applied for various tasks such as plowing, seeding, planting the saplings, spraying supplements in the fields, irrigation of crops, harvest, and post-harvest operations, etc. [46]. Through this machine one can also work under the stressed environmental conditions and the farmers get relax from the long day's tiresome workload. Throughout the cycle of the production of crops, robotic can be applicable at three levels viz. seeding, plantation, and caring of plants and harvesting i.e. pre and post [5]. Though robotics reduced the stressful working environment and attracts the farming profession, its applicability is quite tough. Only the trained and knowledge-keeping manpower can handle it properly.

10 Conclusion

The population globally is increasing at a greater pace so the demand for food also gets increased. To provide an ample amount of feeding material advanced tools and techniques are to be adopted to fulfill the demand. Modernization in agricultural tools and techniques requires two most vital things: firstly, the higher yield of crops in lesser investment and secondly, high profit in the economy. Different Agriculture processes such as hydroponics, aeroponics, bioremediation, vermi-remediation, etc. help in enhancing crop productivity in a very lesser period and even have less impact on the environment. The technologies like GPS, Ultrasound, smartphones, cameras, etc. helps in regular monitoring of the agricultural processes and make it easy giving little relaxation to the farmers. In a very lesser space and time, yield can be hiked up to the maximum thereby increasing the economy.

Acknowledgements The authors are highly grateful to the Department of Zoology and Environmental Science, Gurukula Kangri (Deemed to be University), Haridwar for providing the lab facilities.

References

1. Harris, David, R. and Fuller, D. Q.: Agriculture: Definition and Overview. In Encyclopedia of Global Archaeology (Claire Smith, Ed.). *Springer,* New York. (2014). 104–113
2. Rehman, A., Jingdong, L., Khatoon, R., Hussain, I., Iqbal, M. S.: Modern agricultural technology adoption its importance, role and usage for the improvement of agriculture. *Life Science Journal*. Volume *14*(2). (2016)70–74
3. Islam, S. M. F., & Karim, Z.: World's demand for food and water: The consequences of climate change. In: *Desalination-Challenges and Opportunities*. IntechOpen. (2019)
4. Singh, R., Singh, H., Raghubanshi, A. S.: Challenges and opportunities for agricultural sustainability in changing climate scenarios: a perspective on Indian agriculture. *Tropical Ecology*. Volume 60(2). (2019) 167–185
5. Jyoti, B., Chandel, N.S., Agarwal, K. N.: Application of robotics in Agriculture: An Indian Perspective. (2020)
6. Pramanik, P., Krishnan, P., Maity, A., Mridha, N., Mukherjee, A., Rai, V. Application of nanotechnology in agriculture. In *Environmental Nanotechnology Springer, Cham*. Volume 4. (2020) 317–348
7. D. Despommier.: The rise of vertical farms. *Scientific American*. Volume 301 (5). (2009) 80–87
8. D. Despommier.: Farming up the city: The rise of urban vertical farms. *Trends Biotechnol.* Volume 31 (7). (2013) 388–389
9. D. Despommier.: *Encyclopedia of Food and Agricultural Ethics (Vertical Farms in Horticulture)*. Dordrecht: Springer Netherlands. (2014)
10. Putra, P. A., Yuliando, H.: Soilless culture system to support water use efficiency and product quality: a review. *Agriculture and Agricultural Science Procedia*. Volume *3*. (2015) 283–288
11. Osvald, J., Petrovic, N., Demsar, J.: Sugar and organic acid content of tomato fruits (*lycopersicon lycopersicum mill.*) grown on aeroponics at different plant density. *Acta Alimentaria*. Volume 30(1). (2001) 53–61. doi: https://doi.org/10.1556/AAlim.30.2001.1.6
12. Mbiyu, M. W., Muthoni, J., Kabira, J., Elmar, G., Muchira, C., Pwaipwai, P., Onditi, J.: Use of aeroponics technique for potato (Solanum tuberosum) minitubers production in Kenya. *Journal of Horticulture and Forestry*. Volume *4*(11). (2012) 172–177
13. Pandey, N., Kamboj, N., Bharti, M., Kamboj, V., Sharma, S., Bisht, A.: An overview on enormous effect of hazardous wastes on water components and their management. In: Advances in Environmental Pollution Management: Wastewater Impacts and Treatment Technologies. Volume 1. (2020) 158–173
14. Bharti, M., Kamboj, N.: Occurrence and diversity of mycofloral population in soil of two different land use types in Haridwar region (Uttarakhand), India. *Journal of Advanced Scientific Research*. Volume 11(1). (2020) 1–10
15. Purnomo, A. S., Mori, T., Putra, S. R., Kondo, R.: Biotransformation of heptachlor and heptachlor epoxide by white-rot fungus Pleurotus ostreatus. *International Biodeterioration & Biodegradation*. Volume *82*. (2013) 40–44
16. Kulshreshtha S, Mathur N, Bhatnagar P: In: *Fungi as Bioremediators: Soil Biology.* Goltapeh EM, Danesh YR, Varma A, editor. Springer Berlin, Heidelberg; Mycoremediation of paper, pulp and cardboard industrial wastes and pollutants. (2013) 77–116

17. Kuforiji, O. O., Fasidi, I. O.: Enzyme activities of Pleurotus tuber-regium (Fries) Singer, cultivated on selected agricultural wastes. *Bioresource Technology*. Volume *99*(10). (2008) 4275–4278
18. Zhu, M. J., Du, F., Zhang, G. Q., Wang, H. X., Ng, T. B.: Purification a laccase exhibiting dye decolorizing ability from an edible mushroom Russula virescens. *International Biodeterioration & Biodegradation*. Volume *82*. (2013) 33–39
19. Kumar, V., Singh, J., Kumar, P., Kumar, P.: Response surface methodology based electro-kinetic modeling of biological and chemical oxygen demand removal from sugar mill effluent by water hyacinth (Eichhornia crassipes) in a Continuous Stirred Tank Reactor (CSTR). *Environmental Technology & Innovation*. Volume *14*. (2019). 100327
20. Dhir, B.: *Phytoremediation: role of aquatic plants in environmental clean-up*. New Delhi: *Springer*. Volume 14. (2013)
21. Kenyangi, A., Blok, W.: Vermicompost as a component in potting mixes for growth promotion in ornamental plants. *Rwanda Journal*. Volume *28*. (2012). 53–63
22. Lee, K. E.: Earthworms: their ecology and relationships with soils and land use. Academic Press New York. (1985) 1–420
23. Silvas, F.P.C., Buzzi. D.C., Espinosa, D.C.R., Tenório, J.A.S.: Biosorption of AMD metals using *Rhodococcus opacus*. *Revista Escola de Minas*. Volume 64. (2011) 487–492
24. Adelodun, B., Ajibade, F.O., Abdulkadir, T.S., Bakare, H.O. and Choi, K.S.: SWOT analysis of agro waste-based adsorbents for persistent dye pollutants removal from wastewaters. In: Environmental Degradation: Causes and Reme-diation Strategies, Eds. Kumar, V., Singh, J. and Kumar, P. (2020) 88–103. https://doi.org/10.26832/aesa-2020-edcrs-07
25. Santhoshkumar, M., Reddy, G. C., Sangwan, P. S.: A review on organic farming-sustainable agriculture development. *International Journal of Pure & Applied Bioscience*. Volume *5*(4). (2017) 1277–128
26. Arrington, W.: Organic Farming-Great Alternative. Advance in agricultural Technology & Plant Sciences. Volume 3(2). (2020) 1–4
27. Lehmann, J., Joseph, S.: *Biochar for environmental management: an introduction*. Routledge. (2015) 33–46
28. Ahmad, M., Rajapaksha, A. U., Lim, J. E., Zhang, M., Bolan, N., Mohan, D., Ok, Y. S.: Biochar as a sorbent for contaminant management in soil and water: a review. *Chemosphere*. Volume 99. (2014) 19–33
29. Inyang, M. I., Gao, B., Yao, Y., Xue, Y., Zimmerman, A., Mosa, A., Cao, X.: A review of biochar as a low-cost adsorbent for aqueous heavy metal removal. *Critical Reviews in Environmental Science and Technology*. Volume *46*(4). (2016) 406–433
30. Rajapaksha, A. U., Vithanage, M., Zhang, M., Ahmad, M., Mohan, D., Chang, S. X., Ok, Y. S.: Pyrolysis condition affected sulfamethazine sorption by tea waste biochars. *Bioresource technology*. Volume *166*. (2014) 303–308
31. Tan, X., Liu, Y., Zeng, G., Wang, X., Hu, X., Gu, Y., Yang, Z.: Application of biochar for the removal of pollutants from aqueous solutions. *Chemosphere*. Volume *125*. (2015) 70–85
32. Laird, D. A.: The charcoal vision: a win–win–win scenario for simultaneously producing bioenergy, permanently sequestering carbon, while improving soil and water quality. *Agronomy journal*. Volume *100*(1). (2008) 178–181
33. Laghari, M., Naidu, R., Xiao, B., Hu, Z., Mirjat, M. S., Hu, M., Fazal, S.: Recent developments in biochar as an effective tool for agricultural soil management: a review. *Journal of the Science of Food and Agriculture*. Volume *96*(15). (2016) 4840–4849
34. Mukhopadhyay, S. S.: Nanotechnology in agriculture: prospects and constraints. *Nanotechnology, science and applications*. Volume *7*. (2014) 63
35. Moraru, C. I., Panchapakesan, C. P., Huang, Q., Takhistov, P., Liu, S., Kokini, J. L.: Nanotechnology: a new frontier in food science understanding the special properties of materials of nanometer size will allow food scientists to design new, healthier, tastier, and safer foods. *Nanotechnology*. Volume *57*(12). (2003) 24–29
36. Chau, C. F., Wu, S. H., Yen, G. C.: The development of regulations for food nanotechnology. *Trends in Food Science & Technology*. Volume 18(5). (2007) 269–280

37. Floros, J. D., Newsome, R., Fisher, W., Barbosa-Cánovas, G. V., Chen, H., Dunne, C. P., Ziegler, G. R.: Feeding the world today and tomorrow: the importance of food science and technology: an IFT scientific review. *Comprehensive Reviews in Food Science and Food Safety*. Volume *9*(5). (2010) 572–599
38. Prasad, R., Bhattacharyya, A., Nguyen, Q. D.: Nanotechnology in sustainable agriculture: recent developments, challenges, and perspectives. *Frontiers in microbiology*. Volume *8*. (2017) 1014
39. Rashmi, Royston, M., Pavithra, M.P.: Vertical Farming: A Concept. *International Journal of Engineering and Techniques*. Volume 4(3). (2018) 500–506
40. Garg, A., Balodi, R.: Recent trends in agriculture: vertical farming and organic farming. *Advances in Plants & Agriculture Research*. Volume 1(4). (2014) 142–144
41. Kniss, A. R.: Genetically engineered herbicide-resistant crops and herbicide-resistant weed evolution in the United States. *Weed Science*. Volume *66*(2). (2018) 260–273
42. Damalas, C. A., Koutroubas, S. D.: Farmers' exposure to pesticides: toxicity types and ways of prevention. *Toxics*. Volume 4. (2016) 1–10
43. Saiz-Rubio, V., & Rovira-Más, F.: From smart farming towards agriculture 5.0: A review on crop data management. *Agronomy*. Volume 10(2). (2020) 207
44. Torloni, M. R., Vedmedovska, N., Merialdi, M., Betran, A. P., Allen, T., Gonzalez, R., Platt, L. D.: Safety of ultrasonography in pregnancy: WHO systematic review of the literature and meta-analysis. *Ultrasound in Obstetrics and Gynecology: The Official Journal of the International Society of Ultrasound in Obstetrics and Gynecology*. Volume *33*(5). (2009) 599–608
45. Perez, J. A., Deligianni, F., Ravi, D., Yang, G. Z.: Artificial intelligence and robotics. *arXiv preprint* arXiv:1803.10813. (2018) *147*
46. Xia, C., Wang, L., Chung, B. K., Lee, J. M.: In situ 3D segmentation of individual plant leaves using a RGB-D camera for agricultural automation. *Sensors*. Volume *15*(8). (2015) 20463–20479

Dynamics of Soil Cationic Micronutrients in Different Land Use Systems in Lower Shiwalik Region of Uttarakhand, India

Manisha Bharti, Nitin Kamboj, Vishal Kamboj, Aditi Bisht, and Amrit Kumar

Abstract Due to the low nutrient content, high erosion rate, poverty, increasing urbanization and increasing agricultural intensity, micronutrients problem in ecosystem are widespread. Topography, soil texture and land use types are key factors affecting the soil nutrient availability. The current study provides information on the micronutrient status of soil in different land use patterns (forest, industrial, residential, agro-residential and agricultural area) in lower shiwalik region of Uttarakhand, India. Results indicated that significant differences in soil micronutrients among different land uses were observed. Iron content was found low in industrial area and medium to high in other selected land use patterns. Soil of selected land use patterns has high range of copper content except only forest area. Zinc content was found low in selected land uses and showing the negative correlation ($p < 0.05$) with organic matter in soil. Further observations suggested that type of land use patterns was the dominant factor altering the nutrient dynamics vertically as well as horizontally.

Keywords Land use · Micronutrient status · Nutrient dynamics · Soil properties · Urbanization

1 Introduction

Soil fertility is one of the most significant factors for crop yield and development. Soil fertility evaluation is an important aspect in sustainable agriculture production [1]. A total of 16 soil nutrients are occurring as macronutrients and micronutrients which are essential for plant growth. Macronutrients (C, H, O, N, P, K, Ca, Mg and S) are needed in large quantity and micronutrients (Fe, Zn, Cu, Mn, Bo, Cl, and Mo) are needed in small quantities. Five additional elements (Na, Co, Ve, Ni and Si) are also considered as nutrients in soil system. On the basis of nutrient utility and requirement,

M. Bharti · N. Kamboj (✉) · V. Kamboj · A. Bisht · A. Kumar
Department of Zoology and Environmental Science, Gurukula Kangri Vishwavidyalaya, Haridwar 249404, India
e-mail: nkamboj@gkv.ac.in

K. D. Bahukhandi et al. (eds.), *Environmental Pollution and Natural Resource Management*, Springer Proceedings in Earth and Environmental Sciences,
https://doi.org/10.1007/978-3-031-05335-1_12

minerals nutrients are classified into different classes i.e. macronutrient and micronutrient. Further, soil micronutrients are playing an effective role in plant metabolism, nutrient regulation, chlorophyll synthesis, reproductive growth, production of carbohydrates, development of seed and fruits [2]. Micronutrients are declining due to various abiotic and biotic stresses. One of the most responsible factors for micronutrient deficiency is uncontrolled use of NPK fertilizers that substantially contributed to the worldwide problem in devastating human health consequences like malnutrition and under nutrition [3]. Agriculture is the main source of nutrition in India. Government of India introduced many policies to this respect for high agricultural production with food security goals and nutrition security that are most important for plants and human being [4–6].

Land use may be defined as a part of land used by human for various purposes such as industrialization, settlements and agricultural activities [7]. Changes in extent and composition of forests, urbanized areas, residential areas and other agricultural areas have large impacts on provision of soil nutrient dynamics. Different type of land use has different type of soil structure and their nutrient properties. Soil nutrients concentration are different in each and every type of land use which directly affect the crop productivity, ground water quality, soil fertility, soil microbial behaviour, soil respiration etc. Unfortunately, many soils do not have adequate levels of all necessary nutrients due to contamination and leaching of pollutants from anthropogenic activities in particular type of land use. Then conditions in the soil are unfavourable for plant uptake of certain nutrients [8–10]

2 Materials and Method

2.1 Description of Study Area

Ranipur Rao is situated between 29° 56′ 9′ N to 78° 4′ 50′ E, south-western part of Shiwalik foot hills range. Some of the study area falls on the Rajaji National Park which is selected for forest area. The rest of the area is assortment of wasteland, residential, agricultural land under different crops, plantations and villages. This seasonal stream originates from Shiwalik hills of Rajaji National Park passes from the SIDCUL, BHEL industrial area, Shivalik Nagar and Tehri colony residential area and reaches and merges upto the agricultural land of Bhagtantpur Abidpur village in district Haridwar. The total length of Ranipur Rao seasonal stream is about 12.83 km. Out of the its total length, approximately 2.15 km is under forest area of Rajaji National Park and remaining 10.68 km part is under industrial, residential and agricultural areas. However, for the study point of view the total length of Ranipur Rao is divided into 5 zones and which mentioned Table 1.

Haridwar is humid and sub humid agro climatic region with three different cropping seasons. The summer season is called zaid (April–May), rainy season is called kharif (June–September) and winter season is called Rabi (October–March). The

Table 1 Details of sampling sites along with their locations and types

S. No.	Sampling zone	Location	Type of land use area	Geo-coordinates	Elevation (ft)
1	Zone-A	Sureshwari temple trail sector	Forest area	29° 58′ 10.66″ N, 78° 06′ 27.73″ E	1057
2	Zone-B	between the SIDCUL and BHEL industrial area	Industrial area	29°57′11.31″ N, 78° 05′ 10.62″ E	984
3	Zone-C	Tehri colony and Shivalik Nagar	Residential area	29° 55′ 29.60″ N, 78° 04′ 43.57″ E	931
4	Zone-D	Alipur Ibrahim village	Agro-residential area	29° 54′ 20.93″ N, 78° 04′ 30.14″ E	898
5	Zone-E	Bhagtanpur Abidpur village	Agricultural area ending point of the Ranipur Rao seasonal river	29° 52′ 35.18″ N, 78° 04′ 36.43″ E	872

average annual rainfall of the district is about 1074 mm and distribution is variable recording high in northern part and low in southern part. Ganges is the main and perennial stream flowing through the district and some Seasonal streams (Solani, Ratmau, Ranipur, and Pathari) are also occurs which originating from Shiwalik ranges.

2.2 *Topography*

The topography of the study area mainly consisted of the shiwalik hills and alluvial deposits. The upland area is subjected to heavy erosion because the hills and lowland plains are mainly composed of the sandstone and sand rocks, due to which there is a rapid and fast runoff from heavy rains during monsoon season. The bottom of the streams, rivulets and valleys consist mainly of large and small-sized rounded stones and pebbles which are used as shelter by many wild animals such as reptiles and amphibians. The formation occurring of Shiwalik hills are alluvial fan deposits of recent age. Ranipur Rao is also comprised of two hydrogeo-morphological units i.e. shiwalik and upper piedmont or bhabbar zone.

2.3 Soil Classification

The hills and sub-montane areas have sandy loam to loamy sand and dry soil, in the adjoining and moist areas the soil is covered by humus and is dry. The soil of study area shows wide variations in soil physico-chemical characteristics, soil microbial biomass, macro and micro nutrient availability due to many variations in topography, intensity of soil erosion, parent weathered material and other environmental factors. Agricultural soils of study area are suitable for sugarcane, wheat, rice and oil seeds cultivation. Study areas have Orthent (Entisol) type of soil except RNP. In RNP is characterized by Ultisol type of soil which is the brown hill soil.

These piedmont soils are highly fertile which consists boulders, pebbles, sand, silt and clay. It is relatively logical to use the factors like colour of soil layers, thickness, soil texture, density of soil, maturity of soil profile and the nature of sediments for the classification of soil in relation to time and physiographic divisions.

2.4 Atmospheric Temperature

There is a thrilling variation in climate of the study area, giving rise to three different seasons namely winter, summer and monsoon. Winter commences from mid-November and lasts till March. Summer starts by mid of March and lasts till the onset of monsoon in July. During summer there is occasional short duration sand and dust storms accompanied by rain or sometimes hail. Onset of monsoon varies but usually July witnesses the first shower and rains last till October. People do many activities for living. Most of them are laborers, working in the nearby factors and newly established industrial estate. About 25 to 30% of the population is cultivator.

2.5 Agriculture

Farmers grow varied crops including maize, sorgum, pear millet, foxtail millet, rice, ground nut and dhaincha. Sugar cane and tomato are also grown. These crops are grown either as sole crops or intercropped in different combinations. About 20–40% of the population is dependent on the forests especially Rajaji National Park for daily fuel wood needs.

2.6 Collection of Soil Sample

Soil samples were collected from the five different selected Land Use Pattern (LUP) followed by composite soil sampling method during 2016–17. These sites were distributed throughout the catchment according to different land use and topography. Each sampling site was surveyed for analyze the spatial characteristics of particular land use types. For collection of soil samples five 10 m × 10 m sampling plots were laid down in each of the land use pattern. These collected soil samples will brought to the laboratory in sterilized polythene bags and further stored at 4 °C temperature until further analysis [11].

2.7 Soil Analysis

Soil samples were analyzed for selected physic-chemical and micronutrient characteristics in soil. Particle size distribution was analyzed by the international pipette method [11], soil temperature was recorded by soil thermometer, soil pH was analyzed by digital pH meter (ESICO-1012), moisture content (gravimetrically method) and organic matter [12] were estimated [13].

Some selected micronutrients were analyzed in soil samples by using atomic absorption spectroscopy (AAS; model no. AAS4129, ECIL, India). Operate AAS according to the instructions provided for the instrument. Run the selected micronutrient standards and draw a celebration curve. Measure the concentration if elements in soil extracts by using appropriate lamp for each element. Finally, calculate the micronutrient concentration according to the calibration curve.

$$Micronutreintscation(ppm) = \text{ppm MC (from calibration curve)} \times \frac{\text{V}}{\text{Wt}}$$

where:

MC Micronutrient cation
V Total volume of the extract (mL)
Wt Weight of air-dry soil (g)

3 Results and Discussion

3.1 Physico-Chemical Properties of Soil

Physico-chemical and micronutrient characteristics of soil and their correlation with each other in different land use patterns were depicted in Tables 2, 3 and 4.

Particle size distribution: The highest sand concentration was found in agro-residential area (79.11 ± 1.16%) followed by the other land uses, whereas the silt content was found maximum in forest soil (44.60 ± 0.76%). Clay content was found maximum in agro-residential area (6.13 ± 0.94%). Generally, there was a significant difference in soil particle size distribution in different land use patterns. The soil of Forest, industrial, residential and agriculture area were classified as sandy loam type

Table 2 Seasonal variation in soil texture of Ranipur Rao watershed under different land uses patterns during 2016–2017 (Mean ± S.D.)

Type of land use	Particle type	Summer	Monsoon	Winter	Mean ± S.D	Textural classes
Forest area	Sand	49.72 ± 1.546	51.94 ± 2.07	51.72 ± 2.33	51.13 ± 1.00	Sandy loam
	Silt	44.74 ± 2.71	43.59 ± 1.25	45.45 ± 1.48	44.60 ± 0.76	
	Clay	5.53 ± 3.58	4.47 ± 1.18	2.83 ± 1.80	4.28 ± 1.11	
Industrial area	Sand	52.55 ± 1.63	56.39 ± 2.11	55.00 ± 2.28	54.65 ± 1.59	Sandy loam
	Silt	44.19 ± 2.12	37.47 ± 1.44	42.95 ± 1.94	43.35 ± 0.59	
	Clay	3.26 ± 2.20	6.14 ± 2.21	2.05 ± 1.18	2.00 ± 1.05	
Residential area	Sand	53.13 ± 1.37	54.33 ± 0.98	53.34 ± 0.56	53.60 ± 0.52	Sandy loam
	Silt	43.86 ± 2.25	43.33 ± 2.24	43.11 ± 1.34	42.78 ± 0.32	
	Clay	3.01 ± 1.20	2.34 ± 1.82	3.56 ± 1.76	2.75 ± 0.50	
Agro-residential area	Sand	78.87 ± 4.52	80.64 ± 2.92	77.83 ± 2.01	79.11 ± 1.16	Loamy sand
	Silt	15.36 ± 2.91	14.17 ± 3.18	14.76 ± 2.66	14.76 ± 0.49	
	Clay	5.78 ± 2.25	5.19 ± 3.29	7.42 ± 1.40	6.13 ± 0.94	
Agricultural area	Sand	52.55 ± 1.63	56.61 ± 2.16	55.00 ± 2.28	54.72 ± 1.67	Sandy loam
	Silt	46.00 ± 2.43	42.10 ± 2.24	43.45 ± 1.43	43.85 ± 1.62	
	Clay	1.46 ± 1.82	1.29 ± 0.60	1.55 ± 1.45	1.43 ± 0.11	

of soil. Due to maximum concentration of sand and clay, soil of agro-residential area was classified as loamy sand.

Soil temperature (ST): Maximum soil temperature was recorded in industrial area (23.48 ± 0.17 °C) during the summers. The average soil temperature was found maximum in agro-residential area (19.88 ± 3.71 °C). Minimum soil temperature was found in forest area (11.45 ± 0.43 °C) during the winter season. The average soil temperature was found minimum in residential area (19.43 ± 2.71 °C) (Fig. 1). There was a significant positive correlation ($p < 0.05;0.01$) was found between soil temperature and all selected cationic micronutrients.

Moisture Content (MC): Maximum moisture content was recorded in forest area (14.98 ± 0.55%) during the monsoon. The average moisture content was found maximum in residential area (13.11 ± 1.48%). Minimum moisture content was found in industrial area (8.34 ± 0.14%) during the summer season. Soils of industrial area also have average minimum moisture content (11.53 ± 1.47%) among the different land use patterns in watershed area (Fig. 2).

pH: Maximum pH was recorded in soil of agricultural area (8.17 ± 0.18) during the monsoon and also having maximum average pH value (7.27 ± 0.40) in the study area. Minimum soil pH was found in industrial area (6.13 ± 0.21) during the winter season. Soils of forest area have minimum average pH value (6.63 ± 0.29) among the different land use patterns in watershed area (Fig. 3).

Organic matter (OM): Maximum organic matter was recorded in soil of forest area (3.57 ± 0.41%) during the monsoon and also having maximum average organic

Table 3 Range and mean ± S.D. of some physico-chemical and DTPA-extractable micronutrients in soils in different land use patterns of Ranipur Rao watershed

Parameters	Forest area		Industrial area		Residential area		Agro-residential area		Agricultural area	
	Range	mean ± S.D	Range	Mean ± S.D	Range	Mean ± S.D	Range	Mean ± S.D	Range	Mean ± S.D
Soil temperature (°C)	11.45–23.37	19.46 ± 3.86	11.62–23.01	19.56 ± 2.53	11.64–23.25	19.44 ± 3.79	11.62–23.27	19.88 ± 3.71	17.44–23.07	19.87 ± 1.79
Moisture content (%)	9.03–14.98	12.97 ± 1.75	8.34–13.94	11.53 ± 1.47	10.68–16.15	13.12 ± 1.49	11.02–12.98	12.45 ± 0.74	11.22–14.79	12.75 ± 2.04
pH	6.15–7.03	6.63 ± 0.29	6.13–7.25	6.64 ± 0.49	6.89–7.48	7.13 ± 0.15	6.92–7.85	7.25 ± 0.26	6.49–8.17	7.27 ± 0.40
Organic matter (%)	2.75–3.57	3.07 ± 0.26	1.02–1.37	1.35 ± 0.45	1.38–1.92	1.60 ± 0.17	1.17–2.37	1.88 ± 0.31	1.82–2.79	2.07 ± 0.26
Fe (mg/kg)	4.376–5.693	4.91 ± 0.49	3.21–4.086	3.60 ± 0.29	5.10–5.92	5.43 ± 0.25	5.23–6.87	6.13 ± 0.48	5.25–6.53	6.04 ± 0.42
Cu (mg/kg)	0.163–0.191	0.184 ± 0.008	0.420–2.56	1.11 ± 0.72	0.21–0.56	0.37 ± 0.11	0.331–0.543	0.47 ± 0.06	0.72–1.12	0.79 ± – 0.28
Zn (mg/kg)	0.123–0.187	0.16 ± 0.05	0.117–0.246	0.17 ± 0.036	0.116–0.334	0.249 ± 0.061	0.319–0.392	0.35 ± 0.02	0.119–0.443	0.26 ± 0.11
Mn (mg/kg)	1.542–1.991	1.79 ± 0.127	0.221–0.816	0.57 ± 0.19	0.67–1.01	0.892 ± 0.10	1.63–2.02	1.83 ± 0.10	1.52–1.86	1.66 ± 0.17

Table 4 Correlation between selected soil properties under different land use patterns of Ranipur Rao watershed

	ST	MC	pH	OM	Fe	Cu	Zn	Mn
ST	1.0							
MC	−0.158	1.000						
pH	0.670*	0.388*	1.000					
OM	−0.117	0.543**	−0.271*	1.000				
Fe	0.606*	0.661**	0.878**	0.211	1.000			
Cu	0.295*	−0.840**	−0.094	−0.666*	−0.471*	1.000		
Zn	0.714**	0.174	0.868**	−0.268*	0.792**	−0.149	1.000	
Mn	0.523**	0.505*	0.346*	0.736**	0.720**	−0.559*	0.418*	1.000

*Significant at p 0.05; **significant at p 0.01

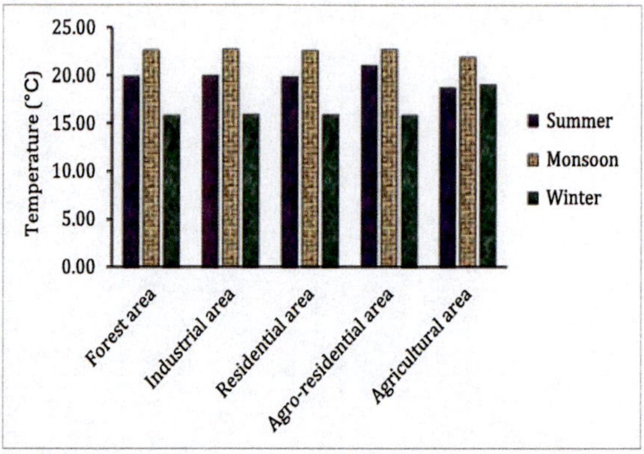

Fig. 1 Seasonal variation in soil temperature at selected zones during April 2016 to March 2017

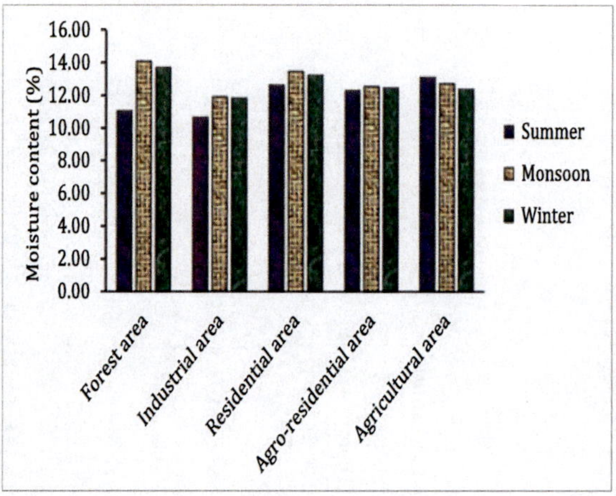

Fig. 2 Seasonal variation in moisture content at selected zones during April 2016 to March 2017

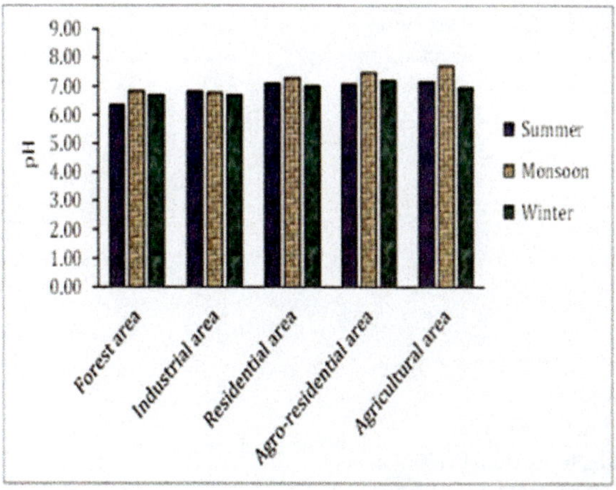

Fig. 3 Seasonal variation in pH at selected zones during April 2016 to March 2017

content (3.07 ± 0.26%) among the other type of land uses. Minimum organic matter was found in industrial area (1.02 ± 0.11%) during the summer season and also having minimum average organic content (1.35 ± 0.45%) in watershed area (Fig. 4).

Iron (Fe): Maximum iron content was recorded in soil of agro-residential area (6.87 ± 1.21 mg/kg) during the monsoon and also having maximum average iron content (6.13 ± 0.48 mg/kg) among the other type of land uses. Minimum iron content was found in industrial area (3.21 ± 0.06 mg/kg) during the summer season and also having minimum average iron content (3.60 ± 0.29 mg/kg) in the watershed area (Fig. 5).

Copper (Cu): Maximum copper content was recorded in soil of industrial area (2.56 ± 0.05 mg/kg) during the winters and also having maximum average copper content (1.11 ± 0.72 mg/kg) among the other type of land uses. Minimum copper content was found in forest area (0.163 ± 0.002 mg/kg) during the summer season

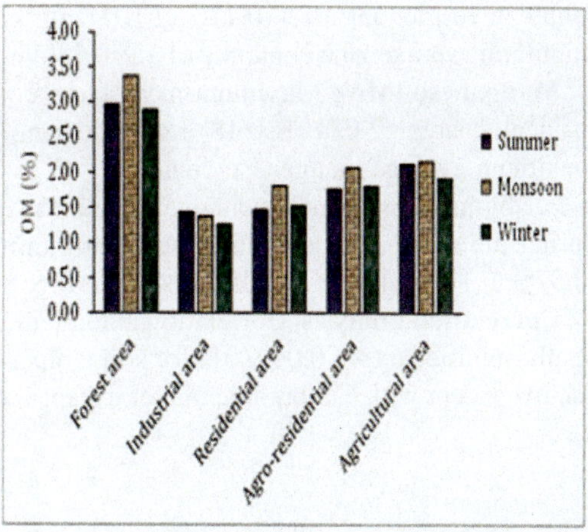

Fig. 4 Seasonal variation in organic matter at selected zones during April 2016 to March 2017

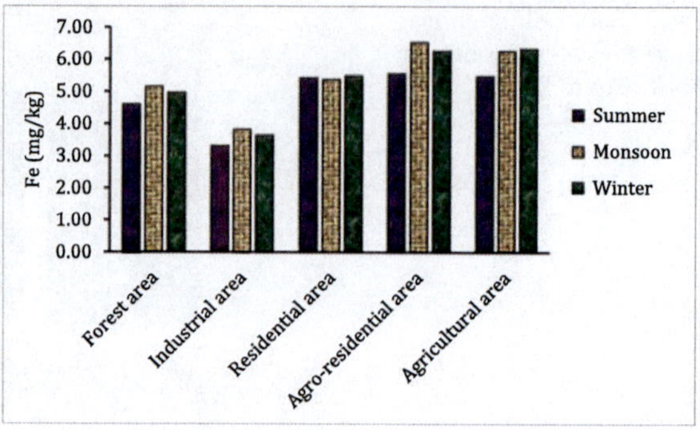

Fig. 5 Seasonal variation in iron content at selected zones during April 2016 to March 2017

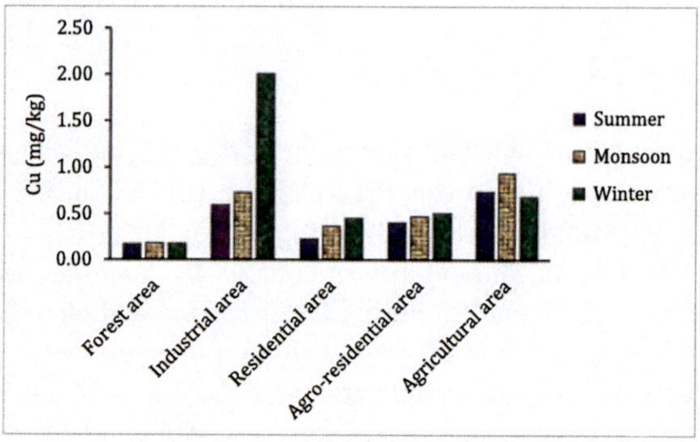

Fig. 6 Seasonal variation in copper content at selected zones during April 2016 to March 2017

and also having minimum average copper content (0.184 ± 0.008 mg/kg) in the watershed area (Fig. 6).

Zinc (Zn): Maximum zinc content was recorded in soil of agriculture area (0.443 ± 0.01 mg/kg) during the monsoon and maximum average zinc content (0.35 ± 0.02 mg/kg) was found in soil of agro-residential area. Minimum zinc content was found in residential area (0.116 ± 0.04 mg/kg) during the summer season and minimum average zinc content (0.16 ± 0.246 mg/kg) in soil of forest area (Fig. 7).

Manganese (Mn): Maximum manganese content was recorded in soil of agro-residential area (2.024 ± 0.18 mg/kg) during the monsoon and also found the maximum average manganese content (1.83 ± 0.10 mg/kg) among the other land uses. Minimum manganese content was found in industrial area (0.221 ± 0.02 mg/kg) during the summer season and also having minimum average manganese content (0.57 ± 0.19 mg/kg) in soil of forest area (Fig. 8).

Correlation analysis: Correlation among micronutrients were found to be statistically significant ($p < 0.05;0.01$) for soil temperature, moisture content except with Zn, pH except with Cu, organic matter except with Fe and Copper except Zn (Table 4).

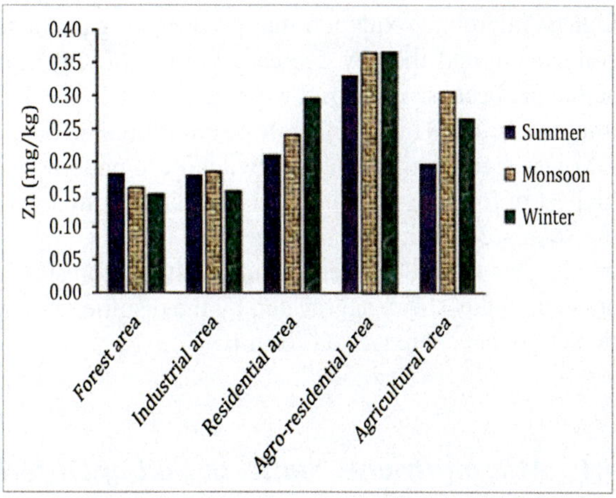

Fig. 7 Seasonal variation in zinc content at selected zones during April 2016 to March 2017

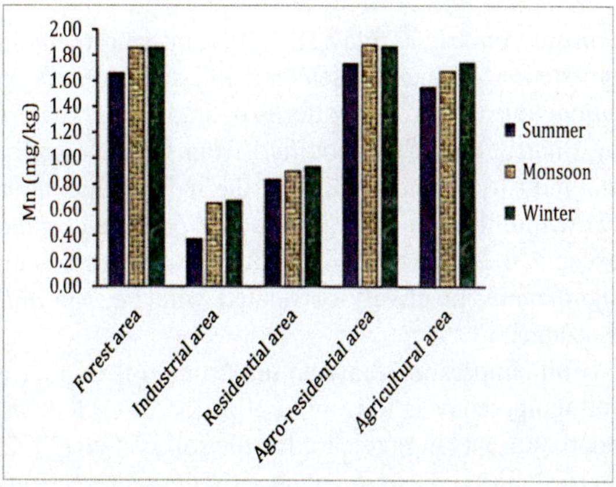

Fig. 8 Seasonal variation in manganese content at selected zones during April 2016 to March 2017

4 Discussion

Dynamics of soil cationic micronutrients were significantly affected by land use systems in lower shiwalik region of Uttarakhand, India. Soil particle size distribution is primarily a sand fraction. Slightly a variation in soil texture was found among the all-land use patterns except agro-residential area. Soil texture is mostly affected by parent materials. Similar study was done by Kaur et al. [14], they also found the sandy loam texture of soils under different land use systems of lower shiwalik foothills of Himalaya, India.

High pH in soil of agricultural area might be due to attributed to the periodic use of lime (dolomite) in the agricultural fields to better productivity [15]. So, significant difference was observed in pH of different land use systems. The lowest pH in the forest area due to soil acidification resulting from litter decomposition of tree leaves, stems, barks, logs, flowers and fruits etc. [7, 16, 17]. It may be due to its highest microbial oxidation that produces organic acids which afford H ions to the soil system and thereby decrease the pH of soil. Industrial area also showed the acidic pH because industrial waste and waste water have the acidic constituents and diminishes the pH and mineralization of organic acid containing elements [18, 19]. OM was significantly affected by land use patterns in the study area. Soil OM was highest in forest area and lowest in industrial area. Urbanization process increasing the socio-economic and demographic activities in a particular area but their land quality has been degraded significantly. Deforestation and conversion to built-up area might have been aggravated by the insufficient inputs of nutrient substrate, OM reduction and decrease soil fertility.

4.1 Micronutrients Status in Soil of Different Land Use Patterns

Soil micronutrients play an important role in Indian sustainable agricultural/crop production. Values of selected micronutrients were found significantly different among the all land use patterns of study area. According to soil nutrient index given by Bharti et al. [20], modified from [19, 21] soil pH found to be acidic in forest and industrial area. According the index value, another all type of land use has the pH within the safe limit. There was positive correlation of pH with Fe, Mn and Zn except Cu. Similar study was done by Brajendra et al. [22], they found that pH was significantly positively correlated with Fe, Mn and Zn in of Mount Kilimanjaro, Tanzania.

Soil temperature plays an important role in micronutrient availability [23]. When soil temperature is low, mineralization of OM slows down resulting less amount of micronutrient being released in the soil system [24, 25]. Fe is essential in chlorophyll and protein synthesis, enzyme activities and electron transfer chain reactions [26]. Its deficiency causes the chlorosis in newly young leaves due to reduction in chlorophyll synthesis. Iron content was found in the following ascending order in different land uses: IA < RA < FA < AA < ARA, respectively. Lowest iron content was found in industrial area due to low organic matter concentration in cultivated soils of industrial area. Only contaminated site and areas surrounding the industries like steel factories, chemical factories etc. have high Fe because of heavy metal contamination from various industrial waste sources. Fe status in other all land use patterns in the study area was sufficient to meet the crop recruitments.

Copper plays an important role in various enzyme activities and for chlorophyll and seed production. There was a significantly negative correlation was found between organic matter and copper content. Cu content was found in the following ascending order in different land uses: FA < ARA < AA < RA < IA, respectively. Soil of selected land use patterns has high range of copper content except only forest area. Cu content was high in industrial area due to copper toxicity from various copper containing metals dumped into the soils and also from accumulation soils have been contaminated. In the residential area, the sources of Cu content were sewage sludge, dumping of solid waste and industrial effluent passing through the residential area [27]. Continuous application of copper containing fertilizers can develop copper toxicity in soil. It affects seed germination, development of root system, and plant strength [28].

Zinc occurs naturally in rocks and their availability depends on the parent material [29]. Zn is an important component in many metabolic reaction and enzyme activities for crop growth and development [30, 31]. Low Zn content was found in soil of all land use patterns. It was found in the following ascending order in different land uses: FA < IA < RA < ARA < AA, respectively. Carbohydrate, protein and chlorophyll formation are significantly affected by Zn deficiency in soil. Therefore, continue application of zinc fertilizer is needed for optimum growth of crop and maximum yield.

Manganese is a major contributor to various biological processes like photosynthesis, respiration and nitrogen assimilation [32]. It is also an important for pollen germination, growth of pollen tube, root cell elongation and resistance to root pathogens. Mn values found to be high in all land use patterns and take place in the following trend: IA < RA < AA < FA < ARA. Reddish brown spots, burning of their tips and margins have occurs due to the manganese toxicity.

5 Conclusion

Above results of this study concluded that micronutrient cations (Fe, Cu, Zn and Mn) are widely associated with crop productivity, animals and human health. Adoption of intensive and new cropping practices for high yield and imbalanced fertilizer used resulted in emergence of soil micronutrient deficiency and toxicity. Industrial activities and domestic waste generation also cause the same problems due to soil contamination in the study area. Therefore, micronutrient management depending on the soil types, land use types, soil properties and time, rate source and frequency of the application needs for the suitable agricultural production and maintenance of human health. Soil micronutrients analysis would be helpful for understanding regarding nature and extent of micronutrient deficiencies and toxicity in soil, plants, animals and human being.

Acknowledgements Authors highly thankful to the Department of Zoology and Environmental Sciences, Gurukula Kangri (Deemed to be University), Haridwar for providing necessary facilities during my research work. This research work was financially supported by the University Grants Commission (UGC), New Delhi, India, [RGNF: 201718-RGNF-2017-18-SC-UTT-30512] to Ms Manisha Bharti.

References

1. Prasad, P.N.S., Subbarayappa, C.T., Sathish, A.: Spatial Distribution and Mapping of Available Micro Nutrient Status in Relation to Soil Properties of Tomato Grown Soils of Chikkaballapura District, Karnataka. International Journal of Plant & Soil Science. volume 32(10). (2020) 1–10
2. Tripathi, D.K., Singh, S., Singh, S., Mishra, S., Chauhan, D.K., Dubey, N.K.: Micronutrients and their diverse role in agricultural crops: advances and future prospective. Acta Physiologiae Plantarum. volume 37. (2015) 139 https://doi.org/10.1007/s11738-015-1870-3
3. Thompson, B., Cohen, M.J., Meerman, J.: World Food Insecurity and Malnutrition: Scope, Trends, Causes and Consequences. In: The Impact of Climate Change and Bioenergy on Nutrition. Springer Netherlands. (2012) 21–41
4. White, P.J., Brown, P.H.: Plant nutrition for sustainable development and global health. Ann Bot-London. volume 105 (7). (2010) 1073–1080
5. Marschner, H.: In: Marschner's mineral nutrition of higher plants. P. Marschner (Ed.). Academic press. volume 89. (2012) 1–89
6. Maathuis, F.J., Diatloff, E.: Roles and functions of plant mineral nutrients. In: Plant mineral nutrients. Humana Press. (2013) 1–21
7. Kumar, M., Jha, A.K., S. Hazarika, S., Verma, B.C., Choudhary, B.U., Ramesh, T., Moiranhthem, P., Kumar, R., Brajendra, Rajkhowa, D.J., Kumar, A, Devi, M.H.: Micronutrients (B, Zn, Mo) for improving crop production on acidic soils of Northeast India. National Academy Science Letters. volume 39(2). (2016) 85–89
8. David, M.M., Power, M.E., Comin, F.A., Yockteng, R.: Structural and functional loss in restored wetland ecosystems. Plos Boilogy. volume 10. (2012) e1001247, https://doi.org/10.1371/journal.pbio.1001247
9. Calhoun, A.J.K., Arrigoni, J., Brooks, R.P., Hunter, M.L., Richter, S.C.: Creating successful vernal pools: A literature review and advice for practitioners. Wetlands. volume 34. (2014) 1027–1038 https://doi.org/10.1007/s13157-014-0556-8
10. Gao, J., Zhang, X., Lei, G., Wang, G.: Soil organic carbon and its fractions in relation to degradation and restoration of wetlands on the Zoige Plateau, China. Wetlands. volume 34. (2014) 235–241 https://doi.org/10.1007/s13157-013-0487-9
11. Behera, P.K.: Soil and solid waste analysis. A laboratory Manual. New Delhi: Dominant Publishers and Distributors Pvt Ltd. (2014) 1–128
12. Walkley, A. and Black, I.A.: An examination of the Degtjareff method for determining soil organic matter, and a proposed modification of the chromic acid titration method. In: Soil Science. volume 37(1). (1934) 29–38 https://doi.org/10.1097/00010694-193401000-00003
13. Trivedy, R.K. and Goel, P.K.: In: Chemical and Biological Methods for Water pollution studies. Environmental Publication. In, Carad, India. (1998) 1–10
14. Kaur, T., Sehgal, S.K.,Singh, S., Sharma, S., Dhaliwal, S.S., Sharma, V.: Assessment of Seasonal Variability in Soil Nutrients and Its Impact on Soil Quality under Different Land Use Systems of Lower Shiwalik Foothills of Himalaya, India. Sustainability. volume 13. (2021) 1398 su13031398
15. Mahmood, F., Khan, I., Ashraf, U., Shahzad, T., Hussain, S., Shahid, M., Abid, M., Ullah, S.: Effects of organic and inorganic manures on maize and their residual impact on soil physico-chemical properties. Journal of Soil Science and Plant Nutrition. volume 17. (2017) 22–32
16. Bharti, M., Kamboj, N.: Impact of different land uses on soil characteristics in Ranipur Rao watershed in Haridwar district, Uttarakhand. International Journal of Creative Research Thoughts. volume 6(1). (2018) 1147–1154
17. Moharana, P.C., Sharma, B.M., Biswas, D.R.: Changes in the soil properties and availability of micronutrients after six-year application of organic and chemical fertilizers using STCR-based targeted yield equations under pearl millet-wheat cropping system. Journal of Plant Nutrition. volume 40. (2017) 65–176
18. Saha, S, Hazra, G.C., Saha, B.N., Mandal, B.: Assessment of heavy metals contamination in different crops grown in long-term sewage-irrigated areas of Kolkata, West Bengal. India

Environmental Monitoring and Assessment. volume 187. (2015) 4087 https://doi.org/10.1007/s10661-014-4087-9
19. Bharti, M., Kamboj, N., Kamboj, V.: Effect of untreated wastewater on soil quality: A case study in Ranipur Rao watershed in Haridwar region (Uttarakhand), India. In Advances in Environmental Pollution Management: Wastewater Impacts and Treatment Technologies. volume 1. (2020) 145–157 https://doi.org/10.26832/aesa-2020-aepm-010
20. 21. Parker, F.W., Nelson, W.L., Winters, E., Miller, L.E.: The broad interpretation and application of soil test information. Agronomy Journal. volume 43. (1951) 105-112
21. Brajendra, A., Latha, P.C., Surekha, K., Prasad Babu, M.B.B., Ghosh, T.J., Bhadana, V.B.: Soil Health Testing Manual. Parmar Publishers and Distributors, India. (2014)
22. Mathew, M.M., Majule, A.E., Merchant, R., Sinclair, F.: Variability of Soil Micronutrients Concentration along the Slopes of Mount Kilimanjaro, Tanzania. Applied and environmental science, volume 16. (2016) 1–8 https://doi.org/10.1155/2016/9814316
23. Foth, H.D.: Fundamentals of Soil Science, John Wiley & Sons, New York, NY, USA, 8th edition (1990)
24. Temu, A., Waized, B., Henson, S.: Mapping value chains for nutrient-dense foods in Tanzania in Reducing Hunger and Under nutrition. E.r. 76, Institute of Development Studies, Brighton, UK, 2014.
25. Klopfenstein, S.T., Hirmas, D.R., Johnson, W.C.: Relationships between soil organic carbon and precipitation along a climosequence in loess-derived soils of the Central Great Plains, USA. Catena. volume 133. (2015) 25–34
26. Shukla, A.K., Tiwari, P.K., Pakhare, A., Prakash, C. Zinc and iron in soil, plant, animal and human health. Indian Journal of Fertisers. volume 12. (2016) 133–149
27. Saha, S., Saha, B., Seth, T., Dasgupta, S., Ray, M., Pal, B., Pati, S., Mukhopadhyay, S.K., Hazra, G.: Micronutrients Availability in Soil–Plant System in Response to Long-Term Integrated Nutrient Management Under Rice–Wheat Cropping System. Journal of Soil Science and Plant Nutrition. volume 19. (2019) 712–724 https://doi.org/10.1007/s42729-019-00071-6
28. Benke, M.B., Indraratne, S.P., Hao, X., Chang, C., Goh, T.B.: Trace element changes in soil after long–term cattle manure applications. Journal of Environmental Quality. Volume 37. (2008) 798–807
29. Shiwakoti, S., Zheljazkov, V.D., Gollany, H.T., Kleber, M, Xing, B., Astatkie, T.: Micronutrients in the Soil and Wheat: Impact of 84 Years of Organic or Synthetic Fertilization and Crop Residue Management. Agronomy. volume 9 (464). (2019) 1–13 https://doi.org/10.3390/agronomy9080464
30. Rudani, K., Patel, V., Prajapati, K.: The importance of zinc in plant growth- A review. International research journal of natural and applied sciences. volume 5(2). (2018) 38–48
31. Shukla, A.K., Srivastava, P.C., Tiwari, P.K., Prakash, C., Patra, A.K., Singh, P., Pachauri, S.D.: Mapping current micronutrients deficiencies in soils of Uttarakhand for precise micronutrient management, Indian Journal of Fertilisers. Volume 11(7). (2015) 52–63
32. Khaliq, A., Zafar. M., Abbasi, M.K., Hussain, I.: Soil-plant micronutrients dynamics in response to integrated fertilization under wheat–soybean cropping system at Rawalakot, Pakistan. Archives of Agronomy and Soil Science. volume 64. (2017) 640–653 https://doi.org/10.1080/03650340.2017.1376245

Monitoring of Channel Morphology of Ganga River Using Remote Sensing and GIS Data

Vishal Kamboj, Aditi Bisht, and Nitin Kamboj

Abstract The present study deals with the dynamics of channel morphology of river Ganga in Haridwar region during the last two decades (1992–2019). Land Use/Land cover classification and channel geometry characteristics of river Ganga were assessed using remote sensing and GIS analysis approach. The LULC classification was divided into three classes: River, Dry Riverbed and River Ganga catchment or floodplain covering an area of 8366.22 ha. Moreover, channel geometry was calculated using sinuosity index and geomorphological features. The study area is divided into 10 segments showing the different types of channel patterns. It was found that LULC classification of river banks depends on water flow and seasonal ecological and anthropogenic activities. The channel pattern of river Ganga is modified by natural processes such as heavy rainfall, floods, bank erosion and anthropogenic processes such as dams, riverbed mining and agricultural activities near the banks.

Keywords Channel morphology · Ganga river · Remote sensing and GIS · Sinuosity index · Haridwar

1 Introduction

Rivers serve as geological agents that transfer material from one ecosystem to another and are also responsible for the formation of various landforms on the earth's surface [1]. Flowing water serves as a constant erosional factor and shapes the channel pattern of a river [2, 3]. Moreover, the process of erosion and deposition of river sediments creates, destroys and recreates the channel pattern, landforms and finally habitat [1, 3]. The shape of the riverbed depends on some geomorphological features such as point bars, bars, islands and natural dams [1, 4]. Basically, a river system is a combination of corridors such as channel patterns, bars, levees, floodplains,

V. Kamboj · A. Bisht · N. Kamboj (✉)
Department of Zoology and Environmental Science, Gurukula Kangri (Deemed to be University), Haridwar 249404, India
e-mail: kambojgurukul@gmail.com; nkamboj@gkv.ac.in

K. D. Bahukhandi et al. (eds.), *Environmental Pollution and Natural Resource Management*, Springer Proceedings in Earth and Environmental Sciences,
https://doi.org/10.1007/978-3-031-05335-1_13

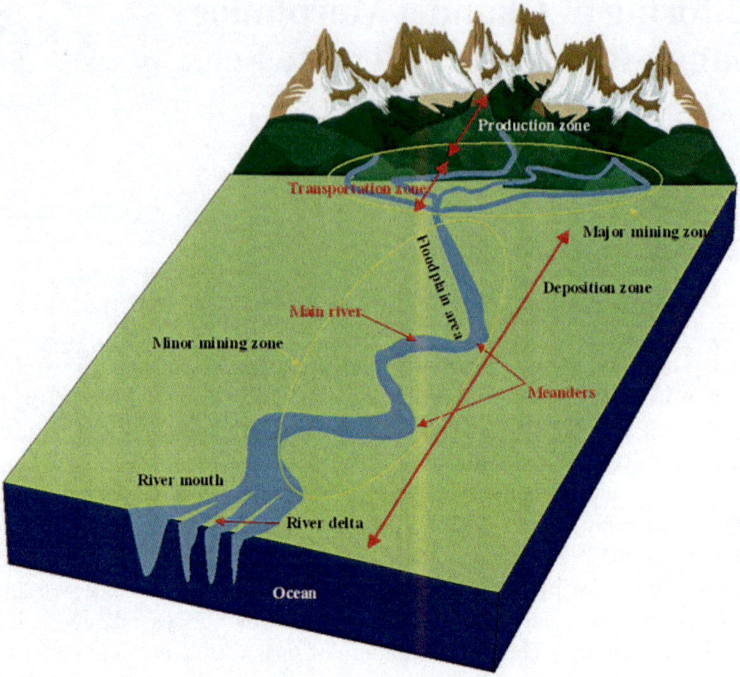

Fig. 1 Showing the river corridors from upland to lowland area

riparian zones, etc., extending from the uplands or mountains to the lowlands or plains (Fig. 1).

2 Geological Characteristics of the River System

The river bed consists of a combination of stones, boulders, gravel, sand and silt known as bed material and depends on the depositional areas [5]. A fluvial water system is divided into three zones based on its function and structure: Production zone, Transport zone and Deposition zone. In all three zones, the most affected part of the fluvial system is the channel pattern [1, 6]. The production zone is very steep and eroded. The eroded material is transported by flowing water through the transport zone and later deposited in the depositional zone under favourable conditions [1].

2.1 Substrate Structure and River Channelization

According to [7, 8], the composition of sediments in fluvial systems is classified according to the size of boulder (>256 mm), cobble (64–256 mm), pebble (16–64 mm), gravel (2–16 mm), sand (0.0625–2 mm), silt (0.0039–0.0625 mm) and clay (0.000244–0.0039 mm). Sediments are non-cohesive, granular particles composed mainly of a mixture of rocks and minerals that move with the flow in quartz form [3, 9, 10]. The availability of larger sediment particles is higher in the upper reaches than in the lower reaches. The small particles are deposited in the lower reaches due to their size and gravity and form the floodplain [4]. The amount of sediment depends on the erosion rate and water flow of the river and forms and maintains the channels and floodplains [11]. The river channels are continuously damaged due to natural and anthropogenic disasters all over the world.

In the present era, the changes in the rivers are studied with the help of remote sensing images and GIS applications. These scientific tools are inexpensive and efficient for mapping and monitoring the geomorphological features of river systems [12, 13]. These tools and techniques provide spatial and temporal data for evaluating the detection of changes in river channels [14].

In Northern India, most of the rivers are glacier fed rivers like Ganga and Yamuna. Ganga is also affected by natural and anthropogenic disasters in the form of floods, bank erosion, high runoff, mining, agricultural and industrial activities. Due to these disasters, the river bed of the rivers is destroyed and altered. The objective of the present study is to map and monitor the changes in the river flow of Ganga during the last two decades at Haridwar District. The results obtained indicate that the changes in channel pattern and floodplain are important for the conservation of river Ganga.

2.2 Description of the Study Area

Haridwar district (29° 56′ N and 78° 09′ E) is located in the southwestern part of Uttarakhand and covers an area of about 2360 km^2. Haridwar is the most famous holy place and also known as Gangadwara because the river Ganga touches the plains after a distance of 267 km in the mountains [15]. Due to the geological features of the river system, Haridwar is located in both the transport and deposition zones of the river Ganga. River Ganga provides a large amount of raw material such as sand, boulders, gravel and paving stones for construction purposes in Haridwar district [5, 16]. In recent decades, excessive extraction of these materials has altered the channel pattern and degraded the aquatic habitat, which has affected the biotic community of river Ganga.

Table 1 Satellite data used to analyse the LULC of the study area

S.No.	Data/Satellites name	Path:Row	Spatial resolution (m)	Years
1.	Landsat 4–5 TM C1	146:39	30	1992
2.	Landsat 4–5 TM C1	146:39	30	2001
3.	Landsat 7 ETM + C1	146:39	15	2011
4.	Landsat 8 OLI/TIRS	146:39	15	2019

Source Earth explorer (USGS)

3 Data and Methodology

3.1 Data Sources Used for Preparation of LULC Maps

Land Use/Land Cover (LULC) and channel pattern was done using various data sources i.e. satellite imagery, toposheet and ground surveys over two decades (1992–2019). The cloud-free satellite image was obtained from the official website of the United States Geological Survey (USGS; Earth Explorer) for the years 1992, 2001, 2011, and 2019 (see Table 1).

3.2 Methodology Adopted for Preparation of the LULC Maps

LULC change detection was performed using Arc GIS 10.5 software with a supervised classification method. First, the area of interest (AOI) was cropped from the downloaded satellite data of the respective years 1992, 2001, 2011 and 2019. For this classification, training samples were obtained during a field visit using Garmin GPS MAP Model No. 76CSx. During the study, the area is classified into three classes: Water body, river bed and river basin or floodplain. The classified image was validated using Arc GIS (10.5) accuracy assessment function (see Fig. 2).

3.3 Methodology Applied for the Channel Pattern

Channel pattern refers to the water flow paths of a river system. The study of channel pattern was carried out using sinuosity index and geomorphological features of the river [17, 18].

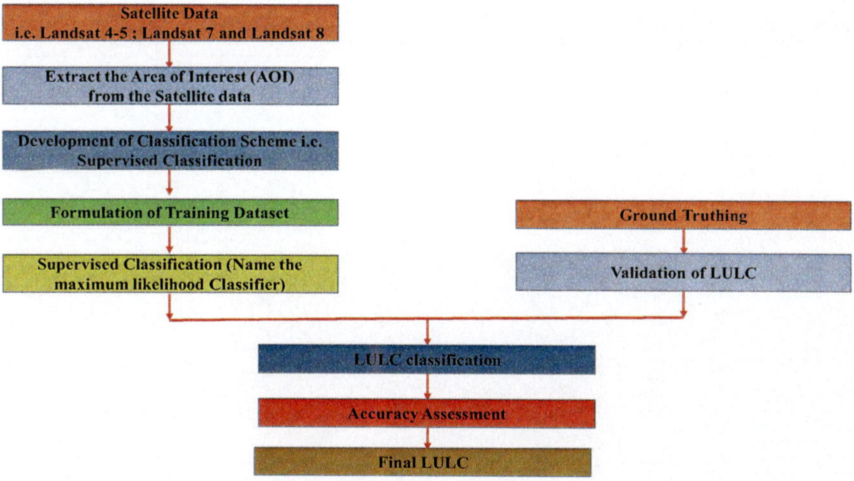

Fig. 2 Flow chart for preparation the land use/land cover

3.3.1 Identification of Channel Pattern Based on Sinuosity Index

Sinuosity index is the length of the watercourse divided by the shortest distance between two points (Fig. 3). The sinuosity index is calculated according to the following formula.

$$Sinuosity\,index\,(SI) = \frac{lenght\,of\,water\,course\,(Lw)}{shortest\,length\,of\,the\,river\,(SLw)}$$

Based on the sinuosity index, the river channel is divided into three types: straight channel, sinuous channel and meandering channel (Table 2).

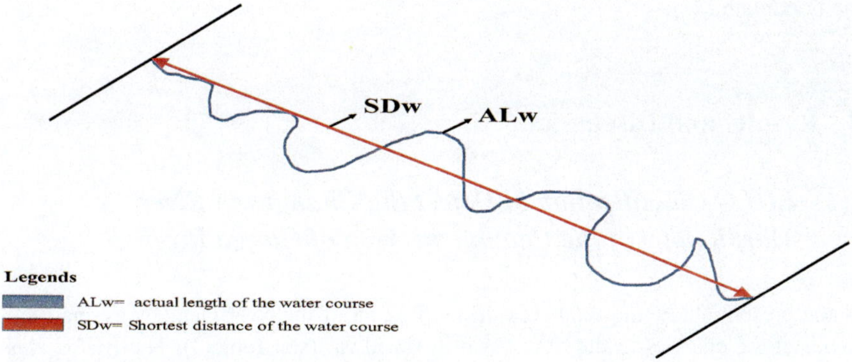

Fig. 3 Methods for calculating the sinuosity index [19]

Table 2 Types of the river channel based on sinuosity index

S.No.	Types of river	Sinuosity index
1.	Straight channel	<1.05
2.	Sinuous channel	1.05–1.50
3.	Meandering channel	>1.50

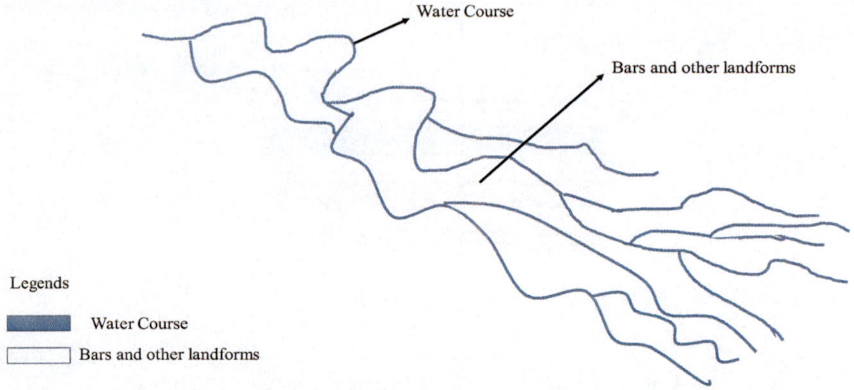

Fig. 4 Sketch map showing the braided channel condition [19]

3.3.2 Identification of Channel Patterns Based on Geomorphological Features

On the basis of geomorphological features, channel patterns are classified into two types, namely braided channel and Anabranching. Braided channel is a fragmentation of the river channel by permanent and temporary bars and islands, which are shown in Fig. 4 [20]. Moreover, Anabranching channel is a separation of the river into different channels by vegetation, an alluvial island in the active channel and not in the floodplain [21].

4 Results and Discussion

4.1 LULC Classification to Detect the Changes in River Morphology in the Catchment Area of Ganga River

In the present study, the total area of 8366.22 ha of the river Ganga was analysed to detect the changes in the river basin in Haridwar (see Table 3). The LULC was classified into three classes: River or Water, Dry Riverbed and Basin or Floodplain of the selected years 1992–2019 (Figs. 5, 6, 7 and 8). The river or water covers an area of 781.92 ha, 1463.49 ha, 856.89 ha and 934.92 ha in the respective years 1992,

Table 3 LULC classes of Ganga river basin for river morphology in selected study zones

S.No.	Classes name	1992		2001		2011		2019	
		Area (ha)	Area (%)	Area (ha)	Area (%)	Area (ha)	Area (%)	Area (ha)	Area (%)
1.	River	781.92	9.35	1463.49	17.49	856.89	10.24	934.92	11.17
2.	Dry river bed	2324.43	27.78	2109.60	25.22	4025.52	48.12	2978.55	35.60
3.	River basin	5259.87	62.87	4793.13	57.29	3483.81	41.64	4452.75	53.22
	Total area (ha)	8366.22	100.00	8366.22	100.00	8366.22	100.00	8366.22	100.00

ha = hectares, % = percentage

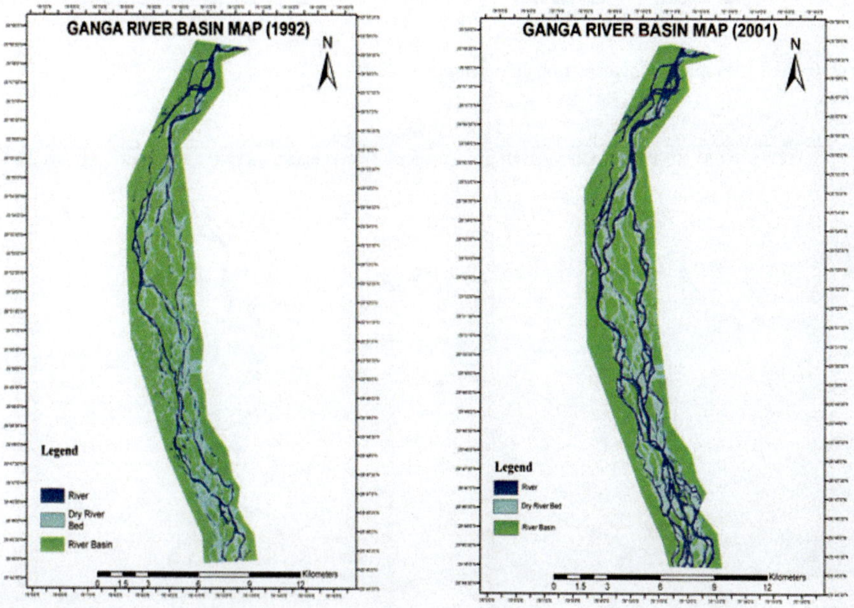

Fig. 5 LULC map of Ganga river basin during 1992–2001

2001, 2011 and 2019. In addition, the dry riverbed covers an area of 2324.43 ha, 2109.60 ha, 4025.52 ha and 2978.55 ha in the years 1992, 2001, 2011 and 2019. In addition, the floodplain or river basin covers an area of 5259.87 ha, 4793.13 ha, 3483.81 ha and 4452.75 ha in the respective years 1992, 2001, 2011 and 2019.

The LULC results of Ganga Basin show that the selected classes, i.e., water or river, dry riverbed and floodplain or river basin, depend on the flow discharge of the river and other environmental factors. Water covers a large area in the monsoon

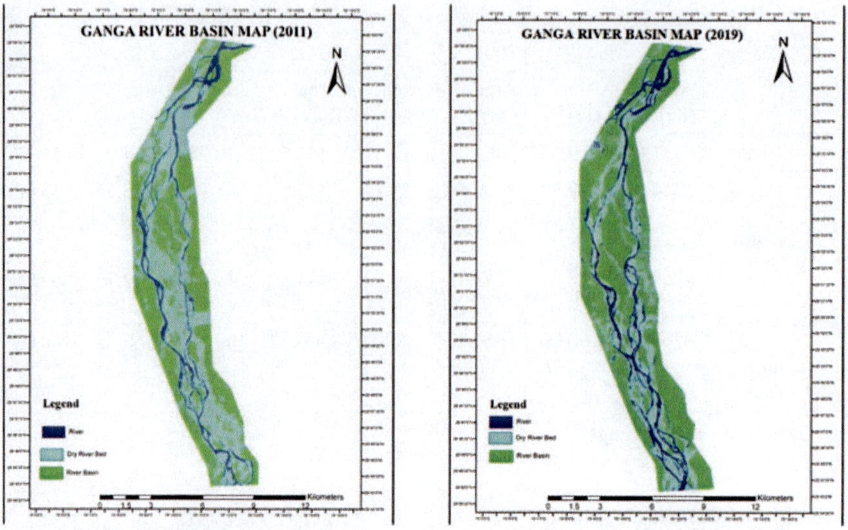

Fig. 6 LULC map of Ganga river basin during 2011–2019

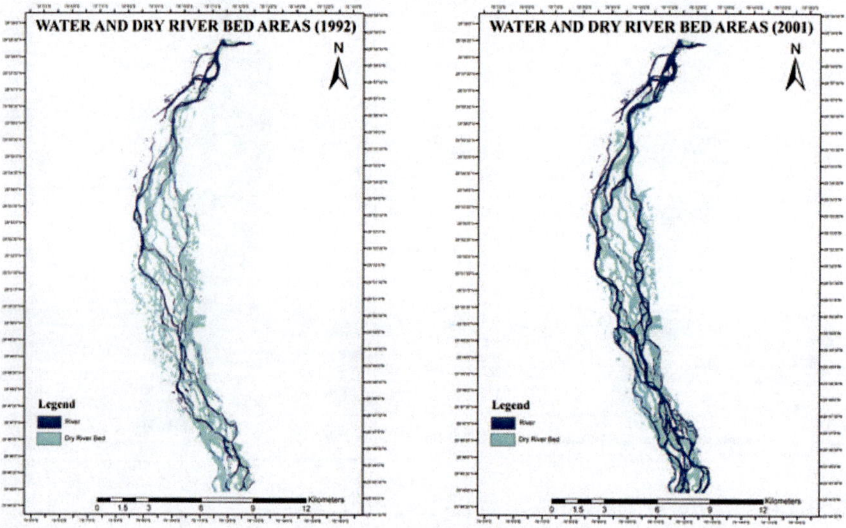

Fig. 7 Water and dry river bed areas during 1992–2001

season due to heavy rainfall and high runoff. The dry riverbed and floodplain or river basin cover a large area in summer and winter as the discharge of river water is less.

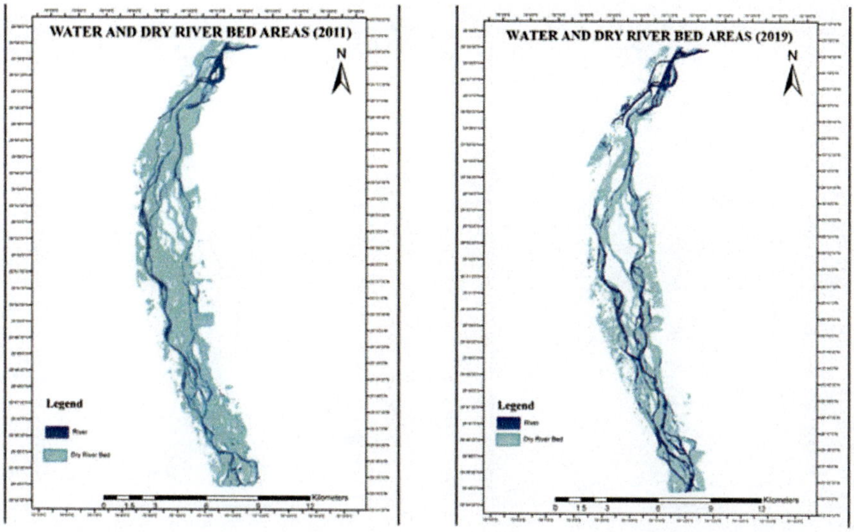

Fig. 8 Water and dry river bed areas during 2011–2019

4.2 Detection of Changes in Channel Morphology

Channel morphology refers to the shape, size, length, depth, width, slope, cross-sectional area, and presence of bed material. The characteristics of channel morphology depend on the geologic setting and discharge of a river. Most rivers have a cross-sectional area, also known as traverse section, which shows unique features of the river. At the cross-section site, the channel forms in the form of valleys and terraces and increases the width, depth, sediment character, gradient and floodplain area of the channel.

During the study, satellite data of selected years from two decades were used to analyse the changes in the channel, discharge pattern and river flow (see Figs. 9 and 10). Then, the current channel pattern was analysed using sinuosity index and geomorphological features of the river.

To study the current channel morphology, geometry and stream type, the selected stretch of Ganga was divided into 10 segments based on the channel pattern shown in Table 4 and Figs. 11 and 12. In the selected 10 segments, segment I shows the straight channel with sinuosity index of 0.89 while the six segments II, III, IV, VI, VII and X show the sinuous channel. Moreover, the left segment namely V, VIII and IX, shows the braided channel. The result shows that the formation of straight channel is possible due to the short length of the segment I. However, the formation of sinuous and meandering channel depends on the cross-sectional areas. In this area, riffles are mostly present in the cross section and pools in the bands. Many researchers also reported the higher number of pools and riffles in the cross section and bands of a river system [1, 4, 10, 14]. However, based on the geomorphological features,

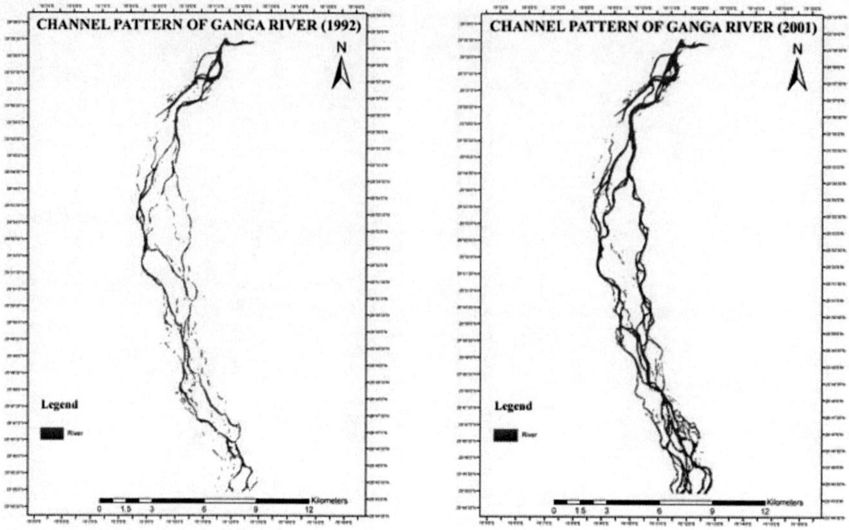

Fig. 9 Channel pattern of Ganga river during 1992–2001

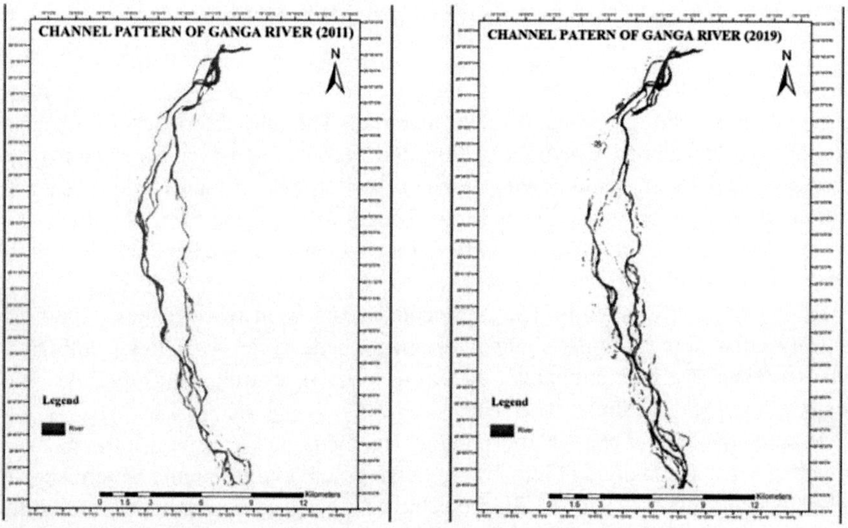

Fig. 10 Channel pattern of Ganga river during 2011–2019

braided conditions were found in the downstream sections. The braided condition is due to the unstable bars and numerous channels. The braided condition depends on the natural and anthropogenic activities and changes from day to day, month to month and season to season [4, 10]. In the present study, it was found that the braided channel formed in the river Ganga may be due to the unplanned mining of river bed

Table 4 Channel geometry of the Ganga river in selected stretch

S.No.	Segment name	Segment code	Width (m)	Depth (ft) (range)	Length of stream channel (m)	Length of straight-line distance (m)	Sinuosity index value	Type of channel
1	Bhimgoda barrage to Chandi bridge	I	517.42–594.54	5–8	1653.5	1853	0.89	Straight channel
2	Chandi bridge to upstream of Kangri village	II	547–1072.13	6–12	3158	2822.90	1.12	Sinuous channel
3	Kangri village to Gajiwala (left bank)	III	274.54–711.71	4–9	3089.06	2,858.19	1.08	Sinuous channel
4	Gajiwala to Shyampur (left bank)	IV	180.55–798.37	5–8	2241.86	1909.25	1.17	Sinuous channel
5	Shyampur to Sajanpur Peeli	V	775.53–991	4–9	2761.58	2511.23	1.09	Sinuous and braided channel
6	Kangri village to Ajitpur (right bank)	VI	311.05–654.92	4–8	3941.28	3757.06	1.05	Sinuous and braided channel
7	Ajitpur to Bisanpur	VII	654.92–805.47	4–12	3778.89	3282.22	1.15	Sinuous and braided channel
8	Bisanpur to Tanda	VIII	805.47–991.37	4–11	2403.23	2264.19	1.06	Sinuous and braided channel
9	Tanda to Bhogpur	IX	991.37–1145	4–13	2683.13	2530.55	1.06	Sinuous and braided channel
10	Sajanpur to Bhogpur	X	991–1145	4–13	3192.33	2849.77	1.12	Sinuous and braided channel

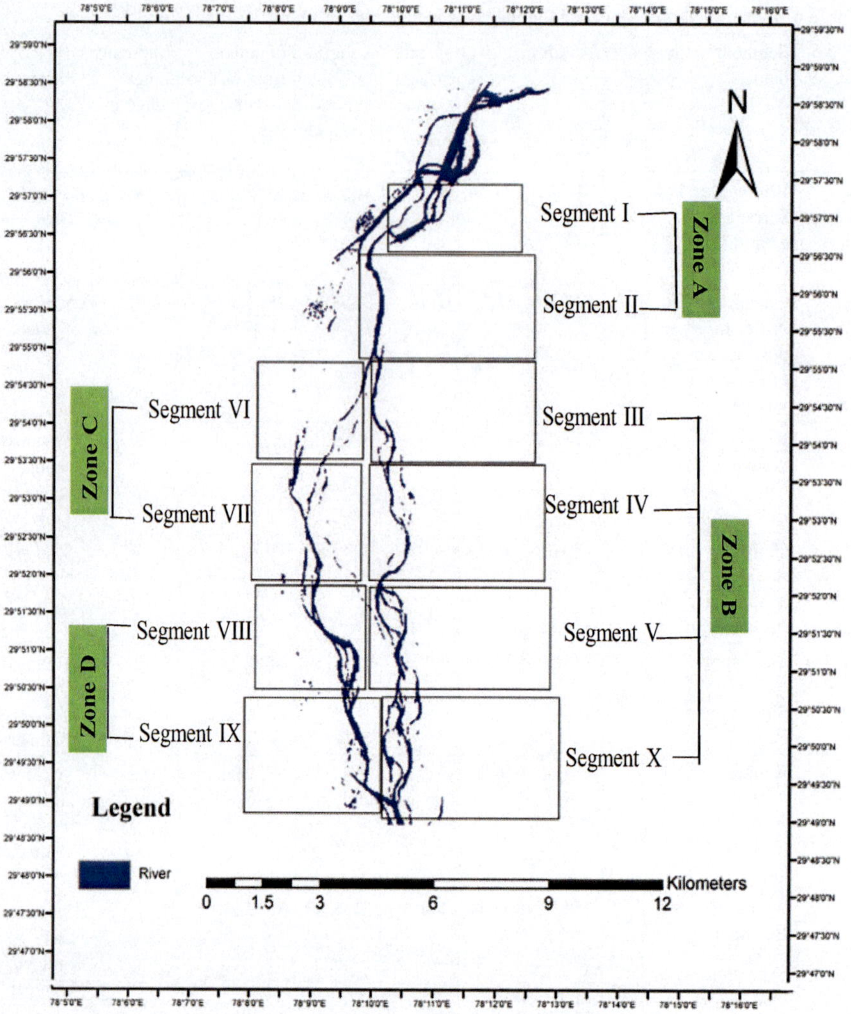

Fig. 11 Map showing the channel pattern with segment number I–X of river Ganga

in this area as well as natural flooding and flood discharges in the monsoon season. The study conducted by [22, 23] showed the conditions of river Ganga due to mining activities in the area (see Fig. 13) and reported that river Ganga is changing its actual water course and water quality and aquatic biodiversity due to mining activities.

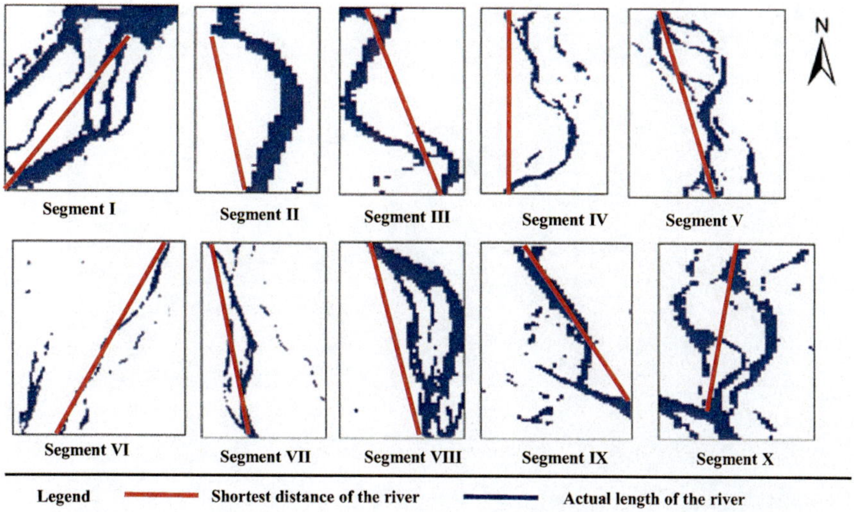

Fig. 12 Segments of river Ganga

5 Conclusion

The present study focused on the changes in LULC and channel pattern in the catchment area of River Ganga in Haridwar during the last two decades (1992–2019). The present study covers a total area of 8366.22 ha in the catchment area of river Ganga. It is to be noted that the three classes selected for LULC mapping of Ganga catchment mainly depend on the discharge of the river. Moreover, the study of channel pattern of river Ganga is divided into ten segments on the basis of morphological characteristics. Out of the 10 river segments, segments I have straight channel, II–IV have sinuous channel and segments V–X have both sinuous and braded channel. The reasons for channel alteration may be natural processes i.e. bed load and sedimentation, floods and anthropogenic activities like unscientific mining in the selected catchment.

Fig. 13 Mining activity at selected stretch of Ganga river at Haridwar

Acknowledgements The authors highly grateful to the Department of Science and Technology, New Delhi for financial support through INSPIRE fellowship (IF160805) to Vishal Kamboj and also thanks to Department of Zoology and Environmental Science, Gurukula Kangri (Deemed to be University), Haridwar for providing the lab facilities.

Conflict of Interest There is no conflict of interest.

References

1. Padmalal, D., Maya, K.: Sand mining. environmental impacts and selected case studies. *Environmental Science and Engineering*, (2014) 1–161. https://doi.org/10.1007/978-94-017-9144-1
2. FAO. Rehabilitation of rivers for fish, food and agriculture. *United Nations Organization,* (1998) 260
3. Kamboj, V., Kamboj, N., Sharma, S.: Environmental impact of riverbed mining-a review. *International Journal of Scientific Research and Reviews,* 7(1) (2017) 504–520
4. Kondolf, G.M.: Hungry water: effects of dams and gravel mining on river channels. *Environmental Management,* 21 (1997) 533–551
5. Kamboj, N., Pandey, A., Shoaib, Moh., Kumar, R.: Environmental impact assessment of illegal Ganga mining at Kangri village, district Haridwar (Uttarakhand) India. *Journal of Sustainable Environmental Research,* 1 (2012) 67–71
6. Schumm, S.A.: The fluvial system. *John Wiley and Sons, New York,* (1977) 338
7. Wentworth, C.K.: A Scale of Grade and Class Terms for Clastic Sediments. *The Journal of Geology* (1922)
8. Lane, E.W.: Report of the subcommittee on sediment terminology. *Trans. A.G.U.*, 28 (1947) 125
9. Jensen, M.L., Bateman, A.M.: *Economic mineral deposits.* John Wiley & Sons., New York, 3 (1979) 593
10. Kondolf, M.G.: Geomorphic and Environmental Effects of In stream Gravel Mining. *Landscape and Urban Planning,* 28 (2007) 2–3
11. Whiting, P.J.: Floodplain maintenance flows. *Rivers*, 6 (1998) 160–170
12. Batalla, R., Iroumé, A., Hernández, M., Llena, M., Mazzorana, B., Vericat, D.: Recent geomorphological evolution of a natural river channel in a Mediterranean Chilean basin. *Geomorphology,* 303 (2018) 322–337
13. Wang, B., Xu, Y.J.: Dynamics of 30 large channel bars in the Lower Mississippi River in response to river engineering from 1985 to 2015. *Geomorphology*, 300 (2018) 31–44
14. Langat, P.K., Kumar, L., Koech, R.: Temporal Variability and Trends of Rainfall and Streamflow in Tana River Basin, Kenya. *Sustainability,* 9(11) (2017) 1963
15. Sajnani, M.: *Encyclopedia of tourism resources in India (Vol.1).* New Delhi: Kalpaz publishers (2001)
16. Sharma, S., Kamboj, N.: Impact of river bed mining on floral diversity in Mohand Rao Watershed of Haridwar region (Uttarakhand), India. *Journal of Emerging Technologies and Innovative Research,* 6(1) (2019) 13–26
17. Leopold, L.B., Wolman, M.G., Miller, J.P.: Fluvial processes in geomorphology. *WH Freeman, San Francisco,* (1964) 522
18. Miall, A.D.: Architectural-element analysis: A new method of facies analysis applied to fluvial deposits. *Earth Science Reviews*, 22 (1985) 261–308
19. Kamboj, V., Kamboj, N., Sharma, A. K.: A review on general characteristics, classification and degradation of river systems. *Environmental degradation: Causes and remediation strategies,* 1 (2020) 47–62
20. Schumm, S., Kahn, H.: Experimental study of channel patterns. *Bulletin of the Geological Society of America,* 83(6) (1972) 1755–1770. https://doi.org/10.1130/0016-7606(1972)83
21. Nanson, G.C., Knighton, A.D.: Anabranching rivers: their cause, character and classification. *Earth Surface Processes and Landforms,* 21(3) (1996) 217–239.https://doi.org/10.1002/(SICI)1096837(199603)21:3%3C217::AIDSP611%3E3.0.CO;2-U
22. Kamboj, N., Kamboj, V.: Riverbed mining as a threat to in-stream agricultural floodplain and biodiversity of Ganges River, India. Contaminants in Agriculture and Environment: Health Risks and Remediation. 25 (2019) 1:250–63
23. Kamboj, N., Kamboj, V.: Water quality assessment using overall index of pollution in riverbed-mining area of Ganga-River Haridwar, India. *Water Science* 33,1 (2019) 65–74

Identification of Aspect Impact of Construction Site: A Case Study

Mukesh Ruhela, Vishal Dixit, and Faheem Ahamad

Abstract Now-a-days the project implementation period is getting further compressed resulting in more intense activity of construction and much more increase in pollution level in the shorter period of construction which may lead to very prolonged adverse effect on the public health in general. The present study was carried out at the construction sites of metro and elevated viaduct to identify the aspects and their impacts and to perform the risk analysis of the identified aspects impacts. Different aspects such as consumption of natural resources, usage of new spare parts, generation of food, sewage, metal, and E-waste, noise and vibration, fire and explosion, damage of material due to rusting, generation of toxic fumes, water and land contamination, gas leakages, spillage of transformer oil, and biomedical waste generation, etc. are identified. Depletion of natural resource, water, air and noise pollution, odour, and land contamination are the impacts of the identified aspects. In case of operation of plant and machinery, all the identified aspects and impacts fall in legal concern category which results the risk in significant category. In case of form work and finished of concrete process, all the identified aspects and impacts didn't need legal concern category which results the risk in non-significant category. In all the other processes a mix trend was observed where the risk in both significant and non-significant category.

Keywords Aspect impact · Metro tunnel · Elevated viaduct · Risk analysis · Severity · Probability · Scale

---

M. Ruhela · V. Dixit · F. Ahamad (✉)
Department of Environmental Engineering, Swami Vivekanand Subharti University, Meerut 250005, Uttar Pradesh, India
e-mail: faheem.rs@gkv.ac.in

Keral Verma Subharti College of Science (KVSCOS), Swami Vivekanand Subharti University, Meerut 250005, Uttar Pradesh, India

K. D. Bahukhandi et al. (eds.), *Environmental Pollution and Natural Resource Management*, Springer Proceedings in Earth and Environmental Sciences,
https://doi.org/10.1007/978-3-031-05335-1_14

1 Introduction

Ever increasing population exerts a pressure on cities, especially in developing countries, are rendering the existing infrastructure facilities totally inadequate and pushing the pollution level to a new high [1]. To mitigate the problem of city dwellers, infrastructure projects are undertaken on a large scale including the construction of metro system to relieve the city from problem of transportation and consequent reduction in pollution [2]. The existing level of pollution in such cities for most of the parameters like suspended particulate matter (SPM), CO_2, CO, SOX, NOX, etc. is either already on higher side or above the acceptable level causing serious concern to the health of city dwellers [3–6]. Major construction activity, by its very nature cannot be environmental friendly and is known to grossly aggravate the existing problems due to pollution especially when such activities are underway in a dense habitation [7, 8]. Small rise in pollution for a short duration at places where the pollution is already at threshold level is bound to leave adverse effect for a long period to a large cross section of people [9–11]. Along with asthmatics and old people, air pollution also impacts the young people which results in different diseases throughout their remaining life [12]. Increase in pollution level also affects the heath of pregnant ladies along with the child in womb [13]. Apart from human health in general, the effect on certain historical structures, patients in hospitals and children in the school in the area where there is sudden increase in the pollution level are likely to be much worse [14, 15]. Thus before starting any construction activity, assessment of existing pollution level, traffic density, socio-economic consideration and density of population etc. and then prediction in the increase in pollution level is necessary to safeguards the people [16–19]. This will enable the metro planning to have an overall view of the problem of pollution as well as the project's requirements (Fig. 1).

New Delhi having a population of almost 14 million people and being one among the most polluted cities of the world. The city is considered as the fourth-most

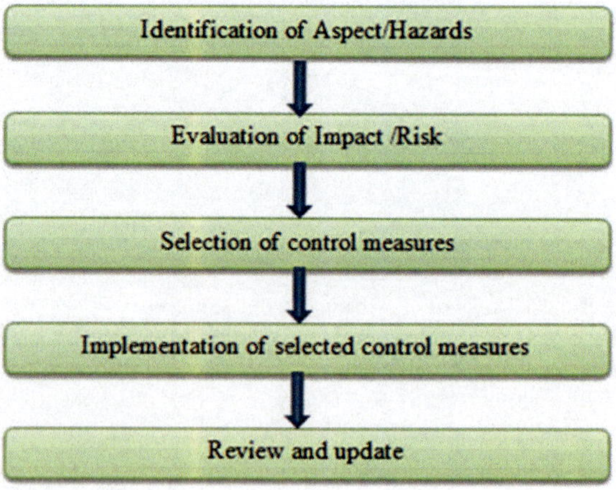

Fig. 1 Showing the procedure followed at the construction site during the study period

Table 1 Showing the study sites

S. No.	Location	Station	CH (M)
1	Maujpur–Yamuna Vihar	Yamuna Vihar	0.000–1466.179
2	Yamuna Vihar–Bhajanpura	Bhajanpura	1466.179–2903.877
3	Bhajanpura–Khajuri Khas	Khajuri Khas	2903.877–3742.748
4	Khajuri Khas–Sonia Vihar	Sonia Vihar	3742.748–4984.848
5	Sonia Vihar–Soorghat (Yamuna River Bed)	Soorghat	4984.848–6673.148 (5644.659–6204.659)
6	Soorghat–Jagatpur Village	Jagatpur Village	6673.148–8039.343
7	Jagatpur Village–Jharoda	Jharoda	8039.343–9098.572
8	Jharoda–Majlis Park	Burari	9098.572–12,098.520

polluted city in the world due to Suspended Particulate Matter content in Delhi, which is much higher than the acceptable limits of MOEF and other the concentration of other pollutants is also higher but within acceptable level of MOEF which is amply. Periodical assessment of air quality is mandatory before, during and after the construction activity in any area [19]. Project Implementation Agencies should carry out an assessment with major sources for other relevant pollution parameters and population density not only along the locations where environment is likely to be affected because of construction but on the alternative routes for movement of materials [20]. The above information can form a key input for further planning, design, execution and monitoring of metro rail system. The construction industry of India accounted about 40% of the development investment in the last 50 years. The Indian construction industry provides employment to about 30 million people (20% of the India's working peoples) and creates assets worth over Rupee 200 billion contributing more than 5% to the nation's GDP [21] (Table 1).

2 Materials and Methods

2.1 *Study Area*

For the present study, chosen site was the construction site of Elevated Viaduct from Maujpur (excluding) to Majlis Park (excluding) including Bridge across river Yamuna, Integrated Elevated Viaduct with PWD flyover at lower deck and metro line at upper deck from Yamuna Vihar to Bhajanpura and 08 Elevated Stations viz Bhajanpura, Yamuna Vihar, Khajuri Khas, Sonia Vihar, Soorghat, Jagatpura, Jharoda, and Burari. The present study was carried out to find out the aspect impact of construction at the selected sites and to perform the risk analysis in the study area (Table 2).

Table 2 Showing the numerical value assigned to scale (Sc), severity (Se), probability (P) and duration (D) against activities

Scale of impact		
No action required		1
Low impact	No action required—unless escalation of impact is possible	2
Medium impact	Process need to be modified	3
High impact	Process cannot be done in this form	4
Severity of environmental impact		
Slight effect	Remediable immediately	1
Minor effect	Remediable within 3 days	2
Localized effect	Remediable within 1 month	3
Major effect	Not remediable	4
Probability of occurrence		
Rare	Occur rarely	1
Once a year	Occurs any once in a year	2
Once a month	May occur once in a month	3
Once a week	Duration of occurrence is weekly	4
Duration of impact		
Unlikely to last for 24 h		1
Unlikely to last for more than 1 week		2
Unlikely to last beyond 1 month		3
Unlikely to last beyond 12 months		4

2.2 Steps for the Identification of Aspects—Impacts

a. Listing of all the activities to be undertaken during execution of project work.
b. Identification of environmental aspect from those activities.
c. Control/mitigation measures were put in place against each significant aspect.
d. Once the impact was reduced by implementing control measures again the process is carried out to evaluate the residual impacts.
e. Assigning of numerical value to each impact against Scale (Sc), Severity (Se), Probability (P) and Duration (D).
f. The sum value of Scale (Sc), Severity (Se), Probability (P), and Duration (D) impacts is compared with the significant impacts and hence significant aspects.

3 Results and Discussion

All the abbreviation used in the tables of results is given in Table 3. The identified aspect, impact and risk assessment of operation of plant and machinery, running of project site office and facilities, piling and excavation work (including Yamuna

Table 3 Showing abbreviation used in the tables of results

Abbreviations	
RA	Routine activity
NRA	Non-routine activity
AP	Air pollution
NP	Noise pollution
WP	Water pollution
NRD	Natural resource depletion
LC	Land contamination
Lc	Legal concern
BC	Business concern
IPC	Interested party concern
Sc	Scale of impact
Se	Severity of impact
P	Probability of occurrence
D	Duration of impact
SI	Significant impact
NSI	Non-significant impact

Bridge), casting yard work, natural resource depletion (NRD), site storage, FORM work and finished of concrete, and first aid activity was given in Table 4. Mitigation measures of significant environmental impacts (Air quality impacts of casting yard and viaduct and station, noise and vibration impacts, surface water quality impacts, and soil and ground water quality impacts) are given in Table 5a, 5b.

3.1 Operation of Plant and Machinery

In the operation of equipment's process (routine activity-RA), the aspect observed was consumption of natural resources and its impact observed was natural resource depletion (NRD), noise pollution (NP), and air pollution (AP) with an impact scale (Sc) of 3 and severity scale (Se) of 2. All the impacts observed falls in legal (Lc), business (BC) and interested party concern (IPC). The scale of probability of occurrence and duration of the impact was 2 and 1 respectively. The risk scale of the identified impact was calculated as 8 which is found as significant as the aspect falls in legal concern category. In the maintenance of plant and machinery (RA), the aspect observed was lubrication of mechanical parts, usage of new spare parts, and used oil, oily rags and contaminated material/discarded batteries and its impact observed was NRD, NP, LC, WP, and AP) with an Sc of 2 and 1 and Se of 3 and 2. All the impacts observed falls in Lc, BC and IPC. The scale of probability of occurrence and duration of the impact was 2 and 3 and 2 respectively. The risk scale of the identified

Table 4 The identified aspect, impact and risk assessment of operation of plant and machinery, running of project site office and facilities, piling and excavation work (including Yamuna Bridge), casting yard work, natural resource depletion (NRD), site storage, FORM work and finished of concrete, and first aid activity

S. No.	Activity/process	RA/NRA	N/AB/E	Aspect	Impact	Lc	BC	IPC	Sc	Se	P	D	Risk = S + S + P + D	SI/NSI
Operation of plant and machinery														
1	Operation of equipment	RA	N	Consumption of natural resources	NRD, NP, AP	Y	Y	Y	3	2	2	1	8	SI
2	Maintenance of plant and machinery	RA	N	Lubrication of mechanical parts	NRD, LC, WP	Y	Y	Y	2	3	2	2	9	SI
				Usage of new spare parts	NRD, NP, LC,	Y	Y	Y	1	2	2	3	8	SI
				Used oil, oily rags and contaminated material/discarded batteries	LC	Y	N	N	1	2	2	3	8	SI
Running of project site office and facilities														
1	Office refreshment	RA	N	Food waste Sewage water	LC, WP, Odour	Y	Y	Y	1	2	3	2	8	SI
2	Site office ablution facility	RA	N	Domestic sewage	LC, WP	Y	Y	Y	2	3	2	2	9	SI
3	Consumption of office stationery	RA	N	Refuse generation	NRD, LC	N	Y	N	1	2	2	1	6	NSI
4	Use of electronic equipment's	RA	N	Generation of E-waste	LC, AP	N	Y	N	1	3	2	2	8	NSI
Piling and excavation work (including Yamuna Bridge)														
1	Piling and excavation	RA	N	Sediments	WP, NP, LC	N	Y	Y	2	2	2	3		

(continued)

Table 4 (continued)

S. No.	Activity/process	RA/NRA	N/AB/E	Aspect	Impact	Lc	BC	IPC	Sc	Se	P	D	Risk = S + S + P + D	SI/NSI
2	Disposal of excavated soil	RA	N	Muck and excavated soil	LC, WP, AP	Y	Y	Y	2	2	3	2	9	SI
3	Land dumping of wallgrip (polymer) slurry	NR	N	Used and discarded slurry	LC	Y	Y	Y	2	2	3	2	9	SI
4	Land dumping of muck slurry (Yamuna Bed)				WP, LC	Y	Y	Y	2	2	3	2	9	SI
Casting yard work														
1	Rebar cutting and bending work	RA	N	Rebar cage fabrication	NRD, AP, LC	N	Y	Y	1	2	2	3	8	NSI
2	Material loading and unloading work	RA	N	Noise and vibration	NP	N	Y	Y	3	2	2	1	8	NSI
3	Stacking of material	NR	N	Fall and damage	LC, AP	N	N	Y	2	2	3	2	9	NSI
4	Storage of cylinder	NR	N	Fire and explosion	AP	Y	Y	Y	2	2	3	2	9	SI
5	Form work	RA	N	Generation of noise Damage of material due to rusting	LC, NP	N	Y	Y	2	3	3	2	10	SI

(continued)

Table 4 (continued)

S. No.	Activity/process	RA/NRA	N/AB/E	Aspect	Impact	Lc	BC	IPC	Sc	Se	P	D	Risk = S + S + P + D	SI/NSI
6	Concreting work	RA	N	Slurry and chemical generation	LC, WP	N	Y	Y	2	3	3	2	10	SI
7	Erection of plant	NR	N	Generation of toxic fumes and noise	AP, NP	Y	Y	Y	2	3	3	2	10	SI
8	Storage of raw material	NR	N	Generation of dust and noise	AP, NP	N	Y	Y	2	2	3	2	9	NSI
9	Loading of concrete batch	RA	N		LC, WP	N	Y	Y	1	2	3	1	7	NSI
10	Cleaning of plant	RA	N	Water and land contamination	WP, LC	Y	Y	Y	2	2	3	1	8	SI
11	Gas cutting	RA	N	Gas leakages, fire and explosion	NRD, AP, NP	Y	Y	Y	2	2	3	2	9	SI
12	Welding	RA	N	Toxic fumes generation, metal waste	AP, LC	Y	Y	Y	2	2	3	2	9	SI
13	Grinding	RA		Dust and noise generation	AP, NP	N	Y	Y	2	2	3	2	9	NSI
NRD														
1	Transmission of electricity	RA	N	Spillage of transformer oil, con	NRD, LC	N	Y	Y	2	2	2	2	8	NSI

(continued)

Table 4 (continued)

S. No.	Activity/process	RA/NRA	N/AB/E	Aspect	Impact	Lc	BC	IPC	Sc	Se	P	D	Risk = S + S + P + D	SI/NSI
2	Production of electricity by DG sets	RA	N	Generation of noise, stack emission, consumption of fuel, spillage of oil	NRD, AP, LC	Y	Y	Y	2	2	3	1	8	SI
3	Running of diesel generator	RA	N	Generation of used oil	LC	Y	N	N	2	3	2	2	9	SI
				Discharged parts (filters, batteries etc.)	LC	Y	N	N	2	2	2	2	8	SI
				Noise generation	NP	Y	N	N	3	2	2	1	8	SI
Site storage														
1	Storage of batteries	NR	N	Spillage of acid	LC	N	Y	N	2	2	2	2	8	NSI
2	Storage of gas cylinders	NR	N	Leakage of gas	LC	N	Y	N	2	2	3	2	9	NSI
3	Storage of hazardous waste	NR	AB	Dust generation, fire in oil storage area	LC, AP	Y	N	N	2	2	3	1	8	SI
4	Storage and handling of oil and fuel	RA	AB	Spillage of oil, waste oil lubricants etc., fire in oil storage	LC, AP	N	Y	Y	2	2	3	1	8	NSI

(continued)

Table 4 (continued)

S. No.	Activity/process	RA/NRA	N/AB/E	Aspect	Impact	Lc	BC	IPC	Sc	Se	P	D	Risk = S + S + P + D	SI/NSI
5	Transportation and storage of material	RA	N	Oil leakage	NRD, LC	N	Y	Y	2	3	2	2	9	NSI
FORM work and finished of concrete														
1	Consumption of material as a form work such as timber/metals	RA	N	Damage and corrosion of material	NRD, LC	N	Y	Y	2	3	2	2	9	NSI
2	Consumption of cement mortar	RA	N	Consumption of material	NRD, LC	N	Y	Y	2	3	2	2	9	NSI
3	Consumption of nuts, bolts, screw etc. for form work	RA	N	Corrosion of material	NRD	N	Y	Y	2	3	2	2	9	NSI
First aid activity														
1	First aid activity	NR	E	Biomedical waste generation, sewage water generation	LC, WP	N	Y	Y	2	3	3	2	10	SI

impact was calculated as 9 and 8 which is found as significant as the aspect falls in legal concern category.

Table 5a Mitigation measure for significant environmental impacts

Pollutant	Source	Impact	Rating	Required mitigation measures
Air quality impacts (casting yard)				
Dust and particulate matter	Vehicular movement loadingand hauling of materials Excavation work	Air pollution and health impacts to workers	Medium	1. Peripheral barricade to be erected at casting yard 2. Wheel wash facility with high pressure water jets near exit gate, surface from wheel wash to exit gate to be paved (either bitumen or concrete) 3. Storage facility for dust generation material covered with tarpaulin or hessian cloth sprinkling to be carried out 3 times a day or as directed if done manually 4. No construction material will be stored outside barricaded area 5. Soil, sand, aggregate, debris of any kind and all dust prone material will be fully covered with tarpaulin or hessian cloth 6. Mist sprinkling system to be used, during delivery and handling of material 7. Casting yard and batching plant, where regular movement of vehicles PCC will be done 8. Speed limit of vehicles at 10 kmph 9. No usage of star batching plant, cement to be purchased through bulkers, wind breaking system of 6 m barricades in surrounding of RMC plant, conveyor belt will be covered, provision of exhaust fans with ducts in cement go-down

(continued)

Table 5a (continued)

Pollutant	Source	Impact	Rating	Required mitigation measures
Gaseous pollutants NOx, SOx, CO, HC, etc	Emission from plant and machinery, transport vehicle, diesel generator and vessel operation, stack emissions	Air pollutionand deterioration of health of workers	Medium	1. Use of generators prohibited during implementation of GRAP from 15 Sept to 15 March 2. PUC certification of vehicles 3. Air monitoring will be conducted as planned and mitigation measures will be taken in its accordance
Green house gases	Burning of fossil fuel during construction work	Global warming	Low	1. Burning of any waste is not allowed at site
Air quality impacts (viaduct & station)				
Dust and particulate matter	Excavation, breaking of concrete, vehicular movement loading and hauling of materials	Air pollution and health impacts to workers	Medium	1. Barricades to be erected on either side 2. No gap between bottom of road and barrier 3. Pre wet area of excavation or during breaking of concrete structure 4. Dust suppression through sprinkling of water will be carried out 5. Where there is building or other important structures, green net will be provided above barricades 6. Storage facility for dust generation material covered with tarpaulin or hessian cloth and 0.5 m below barricade board 7. If storage is for long period, dust control measures like growing grass to be done 8. Air monitoring will be conducted as planned and mitigation measures will be taken in its accordance
Noise and vibration impacts and its mitigation measures				

(continued)

Table 5a (continued)

Pollutant	Source	Impact	Rating	Required mitigation measures
	Running of plant and machinery, diesel generator, compressor, compactor, jack hammer, drilling, grinding and chipping work	Increased noise level of the surrounding causing annoyance to living organism, occupational health impacts (reduced hearing, deafness, numbness, sleeplessness, depression, high blood pressure, irritation and headache) to worker	Medium	1. Early identification and assessment of potential problem areas 2. Monitoring of ambient noise and mitigation measures will be taken in its accordance 3. Acoustic barriers to be placed near influence zone at viaduct and station 4. Minimize usage of impact device, instead pneumatic tools to be used 5. DG complying with noise regulations to be used 6. Equipment's to adhere with noise standards 7. Schedule and deploy machinery in manner to reduce noise pollution as simultaneous nearby noise generating activity not to be taken, plant and equipment not kept idle when not in use, noisy operation during highest ambient noise levels, well maintained plant and equipment will be used hauling of heavy load dump trucks away from residential area to reduce vibration, avoid impact piling
Surface water quality impacts and its mitigation measures				
Waste water	Waste water from ablution facility, canteen, pantry and kitchen	Water pollution leads to the deterioration of water quality and makes it unfit for drinking, irrigation, domestic use and for the survival of aquatic animals	Medium	1. Baton wash system for RMC plant 2. Storm water drainage and wheel wash system's sedimentation tank with oil and grease separator 3. Provision of bio-toilets at viaduct and stations
Fuel and lubricants	Hydrocarbon spill			Weather proof shed to stop entering storm water inside storage area
Hazardous chemical	Chemical spill			No storage of chemicals near drain/bank of river

(continued)

Table 5a (continued)

Pollutant	Source	Impact	Rating	Required mitigation measures
Waste water	Surface runoff in construction side containing contaminants			1. Polymer slurry to be collected and reused if possible if not to be disposed at landfill site (Clause No. 52.5.4) 2. No storage of chemicals and hydrocarbons near any drain/river
Soil and ground water quality impacts and its mitigation measures				
	Discharge of wastewater to ground from sanitation facility, wheel washing, concrete washout and other maintenance area	Soil and ground water contamination	Medium	Reuse for dust suppression, and gardening
	Spill of chemicals and hydrocarbons			1. Drip pans of suitable size provided 2. Hazardous waste is segregated and stored in designated area, later send to authorized dealer for disposal
	Dumping of waste			1. Processing of C and D waste is done as govt. designated site 2. Material is procured as per the guidelines given C and D waste management rules 2016
	Dumping of excavated soil and reclaimed sediments			Excavated material is used in the dumping areas designated in the drawings

3.2 *Running of Project Site Office and Facilities*

In the office refreshment process (RA), the aspect observed was food waste and sewage generation and its impact observed was LC, WP, and odor with and Sc of 1 and Se of 2. All the impacts observed falls in BC and IPC category. The P and D of the impact was 3 and 2 respectively. The risk scale of the identified impact was calculated as 8 which is found as significant as the aspect falls in legal concern category. In the site office ablution facility process (RA), the aspect observed was generation of domestic sewage and its impact observed was LC and WP with and Sc of 2 and Se of 3. All the impacts observed falls in Lc, BC and IPC category. The P and D of the impact was 2 and 2 respectively. The risk scale of the identified impact was calculated as 9 which is found as significant. In the consumption of office stationery process (RA), the aspect observed was refuse generation and its impact observed was NRD and LC with and Sc of 1 and Se of 2. All the impacts observed falls in BC category. The P and D of the impact was 2 and 1 respectively. The risk scale of the identified impact was calculated as 6 which is found as non-significant. In the use of electronic equipment's process (routine activity-RA), the aspect observed was generation of E-waste and its impact observed was LC and AP with and Sc of 1 and Se of 3. All the impacts observed falls in business concern (BC) category. The P and D of the impact was 2 and 2 respectively. The risk scale of the identified impact was calculated as 9 which is found as non-significant.

3.3 Piling and Excavation Work (Including Yamuna Bridge)

In the piling and excavation process (routine activity-RA), the aspect observed was sediments and its impact observed was LC, WP and NP with and Sc of 2 and Se of 2. All the impacts observed falls in BC and IPC category. The P and D of the impact was 2 and 3 respectively. The risk scale of the identified impact was calculated as 8 which is found as non-significant. In the disposal of excavated soil process (routine activity-RA), the aspect observed was muck and excavated soil and its impact observed was LC, WP and AP with and Sc of 2 and Se of 2. All the impacts observed falls in Lc, BC and IPC category. The P and D of the impact was 3 and 2 respectively. The risk scale of the identified impact was calculated as 8 which is found as non-significant. In the land dumping of wall grip (Polymer) slurry process (non-routine activity-NRA), the aspect observed was used and discarded slurry and its impact observed was LC with and Sc of 2 and Se of 2. All the impacts observed falls in Lc, BC and IPC category. The P and D of the impact was 3 and 2 respectively. The risk scale of the identified impact was calculated as 9 which is found as significant as the aspect falls in legal concern category. In the land dumping of muck slurry (Yamuna Bed) process (NRA), the aspect observed was used and discarded slurry and its impact observed was WP and LC with and Sc of 2 and Se of 2. All the impacts observed falls in Lc, BC and IPC category. The P and D of the impact was 3 and 2 respectively. The risk scale of the identified impact was calculated as 9 which is found as significant as the aspect falls in legal concern category.

3.4 Casting Yard Work

In the rebar cutting and bending work process (RA), the aspect observed was rebar cage fabrication and its impact observed was NRD, AP and LC with and Sc of 1 and Se of 2. All the impacts observed falls in BC and IPC category. The P and D of the impact was 2 and 3 respectively. The risk scale of the identified impact was calculated as 8 which is found as non-significant. In the material loading and unloading process (RA), the aspect observed was noise and vibration and its impact observed was NP with and Sc of 3 and Se of 2. All the impacts observed falls in BC and IPC category. The P and D of the impact was 2 and 1 respectively. The risk scale of the identified impact was calculated as 8 which is found as non-significant. In the stacking of material process (NRA), the aspect observed was fall and damage and its impact observed was LC and AP with and Sc of 2 and Se of 2. All the impacts observed falls in IPC category. The P and D of the impact was 3 and 2 respectively. The risk scale of the identified impact was calculated as 9 which is found as non-significant. In the storage of cylinder process (NRA), the aspect observed was fire and explosion and its impact observed was AP with and Sc of 2 and Se of 2. All the impacts observed falls in Lc, BC and IPC category. The P and D of the impact was 3 and 2 respectively. The risk scale of the identified impact was calculated as 9 which is found as significant as the aspect falls in legal concern category. In the form work process (RA), the aspect observed was generation of noise and damage of material due to rusting and its impact observed was LC and AP with and Sc of 2 and Se of 3. All the impacts observed falls in BC and IPC category. The P and D of the impact was 3 and 2 respectively. The risk scale of the identified impact was calculated as 10 which is found as significant.

In the concreting work process (RA), the aspect observed was slurry and chemical generation and its impact observed was LC and WP with and Sc of 2 and Se of 3. All the impacts observed falls in BC and IPC category. The P and D of the impact was 3 and 2 respectively. The risk scale of the identified impact was calculated as 10 which is found as significant. In the erection of plant process (NRA), the aspect observed was generation of toxic fumes and noise and its impact observed was AP and NP with and Sc of 2 and Se of 3. All the impacts observed falls in Lc, BC and IPC category. The P and D of the impact was 3 and 2 respectively. The risk scale of the identified impact was calculated as 10 which is found as significant. In the raw material storage process (NRA), the aspect observed was generation of dust and noise and its impact observed was AP and NP with and Sc of 2 and Se of 2. All the impacts observed falls in BC and IPC category. The P and D of the impact was 3 and 2 respectively. The risk scale of the identified impact was calculated as 9 which is found as non-significant. In the loading of concrete batch process (RA), the aspect observed was generation of dust and noise and its impact observed was LC and WP with and Sc of 1 and Se of 2. All the impacts observed falls in BC and IPC category. The P and D of the impact was 3 and 1 respectively. The risk scale of the identified impact was calculated as 7 which is found as non-significant. In the cleaning of plant process (RA), the aspect observed was water and land contamination and its impact observed was LC and WP with and Sc of 2 and Se of 2. All the impacts observed falls in Lc, BC and IPC category. The P and D of the impact was 3 and 1 respectively. The risk scale of the identified impact was calculated as 8 which is found as significant as the aspect falls in legal concern category. In the gas cutting process (RA), the aspect observed was gas leakages, fire and explosion and its impact observed was NRD, AP and NP with and Sc of 2 and Se of 2. All the impacts observed falls in Lc, BC and IPC category. The P and D of the impact was 3 and 2 respectively. The risk scale of the identified impact was calculated as 9 which is found as significant as the aspect falls in legal concern category. In the welding process (RA), the aspect observed was toxic fumes generation and metal waste and its impact observed was AP and LC with and Sc of 2 and Se of 2. All the impacts observed falls in Lc, BC and IPC category. The P and D of the impact was 3 and 2 respectively. The risk scale of the identified impact was calculated as 9 which is found as significant as the aspect falls in legal concern category. In the grinding process (RA), the aspect observed was dust and noise generation and its impact observed was AP and NP with and Sc of 2 and Se of 2. All the impacts observed falls in BC and IPC category. The P and D of the impact was 3 and 2 respectively. The risk scale of the identified impact was calculated as 9 which is found as non-significant.

3.5 NRD

In the electricity transmission process (RA), the aspect observed was spillage of transformer oil and its impact observed was NRD and LC with and Sc of 2 and Se of 2. All the impacts observed falls in BC and IPC category. The P and D of the impact was 2 and 2 respectively. The risk scale of the identified impact was calculated as 8 which is found as non-significant. In the production of electricity by DG sets (RA), the aspect observed was generation of noise, stack emission, consumption of fuel, spillage of oil and its impact observed was NRD, AP and LC with and Sc of 2 and Se of 2. All the impacts observed falls in Lc, BC and IPC category. The P and D of the impact was 3 and 1 respectively. The risk scale of the identified impact was calculated as 8 which is found as significant as the aspect falls in legal concern category. In the running of diesel generator (RA), the aspect observed was generation of used oil, discharged parts (filters, batteries etc.), noise generation and its impact observed was AP and LC with and Sc of 2 and Se of 2. All the impacts observed falls in Lc category. The P and D of the impact was 3 and 2 respectively. The risk scale of the identified impact was calculated as 8 which is found as significant as the aspect falls in legal concern category.

3.6 Site Storage

In the storage of batteries (NRA), the aspect observed was spillage of acid and its impact observed was LC with and Sc of 2 and Se of 2. All the impacts observed falls in BC category. The P and D of the impact was 2 and 2 respectively. The risk scale of the identified impact was calculated as 8 which is found as non-significant. In the storage of gas cylinders (NRA), the aspect observed was leakage of gas and its impact observed was LC with Sc of 2 and Se of 2. All the impacts observed falls in BC category. The P and D of the impact was 3 and 2 respectively. The risk scale of the identified impact was calculated as 9 which is found as non-significant. In the process of hazardous waste storage (NRA), the aspect observed was dust generation, fire in oil storage area and its impact observed was LC and AP with Sc of 2 and Se of 2. All the impacts observed falls in Lc category. The P and D of the impact was 3 and 1 respectively. The risk scale of the identified impact was calculated as 8 which is found as significant as the aspect falls in legal concern category. In the storage and handling of oil and fuel (RA), the aspect observed was spillage of oil, waste oil lubricants etc. fire in oil storage and its impact observed was LC and AP with Sc of 2 and Se of 2. All the impacts observed falls in BC and IPC category. The P and D of the impact was 3 and 1 respectively. The risk scale of the identified impact was calculated as 8 which is found as non-significant. In the transportation and storage of material (RA), the aspect observed was oil leakage and its impact observed was LC and NRD with Sc of 2 and Se of 3. All the impacts observed falls in BC and IPC category. The P and D of the impact was 2 and 2 respectively. The risk scale of the identified impact was calculated as 9 which is found as non-significant.

3.7 FORM Work and Finished of Concrete

In the consumption of material as a form work such as timber/metals (RA), the aspect observed was damage and corrosion of material and its impact observed was LC and NRD with Sc of 2 and Se of 3. All the impacts observed falls in BC and IPC category. The P and D of the impact was 2 and 2 respectively. The risk scale of the identified impact was calculated as 9 which is found as non-significant. In the consumption of material as a form work such as timber/metals (RA), the aspect observed was damage and corrosion of material and its impact observed was LC and NRD with Sc of 2 and Se of 3. All the impacts observed falls in BC and IPC category. The P and D of the impact was 2 and 2 respectively. The risk scale of the identified impact was calculated as 9 which is found as non-significant. In the process of consumption of cement mortar (RA), the aspect observed was consumption of material and its impact observed was LC and NRD with Sc of 2 and Se of 3. All the impacts observed falls in BC and IPC category. The P and D of the impact was 2 and 2 respectively. The risk scale of the identified impact was calculated as 9 which is found as non-significant. In the process of consumption of nuts, bolts, screw etc. for form work (RA), the aspect observed was corrosion of material and its impact observed was NRD with Sc of 2 and Se of 3. All the impacts observed falls in BC and IPC category. The P and D of the impact was 2 and 2 respectively. The risk scale of the identified impact was calculated as 9 which is found as non-significant.

3.8 First Aid Activity

In the process of first aid activity (RA), the aspect observed was biomedical waste generation, Sewage Water generation and its impact observed was LC and WP with Sc of 2 and Se of 3. All the impacts observed falls in BC and IPC category. The P and D of the impact was 3 and 2 respectively. The risk scale of the identified impact was calculated as 10 which is found as significant.

4 Conclusion

The present study was carried out for the identification of aspect impact of construction sites of metro and elevated viaduct. For the present study, chosen site was the construction site of elevated viaduct from Maujpur (excluding) to Majlis Park (excluding) including dridge across river Yamuna, integrated elevated viaduct with PWD flyover at lower deck and metro line at upper deck from Yamuna Vihar to Bhajanpura and 08 other elevated stations. Various processes of construction industry was elaborated and consumption of natural resources, usage of new spare parts, food waste, sewage generation, E-waste generation, muck and excavated soil, used and discarded slurry, noise and vibration, fire and explosion, damage of material due to rusting, generation of toxic fumes, water and land contamination, gas leakages, metal waste, spillage of transformer oil, corrosion of material, and biomedical waste generation, etc. are the identified aspects of the activities running at construction sites. The impacts of the identified aspects were natural resource depletion, water, air and noise pollution, odour, and land contamination. In case of operation of plant and machinery, all the identified aspects and impacts fall in legal concern category which results the risk in significant category. In case of form work and finished of concrete process, all the identified aspects and impacts didn't need legal concern category which results the risk in non-significant category. In all the other processes a mix trend was observed where the risk in both significant and non-significant category.

Table 5b Mitigation measure for significant environmental impacts

S. No.	Activity/process	Aspect	Required control	Additional control	Legal app
Operation of plant and machinery					
1	Operation of equipment	Consumption of natural resources	1. All plant and equipment's shall be regularly inspected and maintained to reduce 2. Use of techniques such as enclosure and acoustical treatment to reduce engine noise 3. PUC where necessary to be maintained at site	Noise barrier shall be used to control any unwanted noise, exhaust system/height of exhaust to be maintained as per regulatory requirements	Center Motor Vehicle Act Amendment May 2019
2	Maintenance of plant and machinery	Lubrication of mechanical parts Usage of new spare parts Used oil, oily rags and contaminated material/discarded batteries	1. All plants and equipment's must be regularly inspected and maintained to reduce spillages and leakages 2. Oil spillage collection trays/spill kits shall be used to prevent wastage and contamination of land and water 3. Routinely inspect all fuel, oil or fluid containing fittings, hoses and seals, and hydraulic lines 3. Environmental training and awareness programs	Oil spills kits will be used remedial measure and control in any spilled oil situation, further will be handed over to authorized facility for disposal	Hazardous and other wastes (Management and Transboundary Movement) Rules, 2016
			1. Optimum utilization of scrap at work site 2. Developing a proper channel for collection and storage of scraps to reduce excess-wastage 3. Recycling of scrap by giving it to authorized vendors	Handling and end disposal process of refused material will be done as per regulatory requirements	
			1. Returning of old/refused/discarded material to agencies supplying new one 2. Recycling of used engine oil and batteries 3. Use oil collection tray to prevent contamination of land	Storage of old/discarded will be done as per regulatory requirements	

(continued)

Table 5b (continued)

S. No.	Activity/process	Aspect	Required control	Additional control	Legal app
Running of project site office and facilities					
1	Office refreshment	Food waste Sewage water	1. Avoid over buying stock 2. Train educate employee to reduce waste 3. Donate leftovers foods to needy people 4. Use waste food for generating compost which will be further used for gardening purpose 5. Reducing, reusing and recycling of generated wastewater		The Water (Prevention and Control of Pollution) Act, 1974 (Amended 1988) and Amended Rules, 2011
2	Site office ablution facility	Domestic sewage	1. Proper sewage channel connectivity channel shall be prepared seeking approvals from concerned authority 2. Septic tank shall be setup for treatment of waste generated		The Water (Prevention and Control of Pollution) Act, 1974 (Amended 1988) and Amended Rules, 2011
3	Consumption of office stationery	Refuse generation	1. Efforts shall be made to make all office work digitized 2. Recycling of refuse by handing over to authorized vendors		
4	Use of electronic equipment's	Generation of E-waste	1. Efforts shall be taken to generate less E- waste 2. Proper E-waste collection system shall be developed 3. E-waste shall be carefully disposed of by handing over to authorize vendors for recycling purpose		

(continued)

Table 5b (continued)

S. No.	Activity/process	Aspect	Required control	Additional control	Legal app
Piling and excavation work (including Yamuna bridge)					
1	Piling and excavation	Sediments	1. Muck collection system shall be developed to avoid its spillage 2. Generated muck shall be analyzed, and it shall be disposed accordingly in safe area designated for this purpose only		
2	Disposal of excavated soil	Muck and excavated soil	1. Wet process shall be adopted to prevent dust generation 2. Dust mask shall be used while performing such operation	Will be disposed to authorized vendor as per regulatory requirement, 4 copy challan system and monitoring of process will be done	C&D Waste Management Rules 2016
3	Land dumping of wallgrip (polymer) slurry	Used and discarded slurry	1. The slurry materials shall be analyzed prior to dumping/disposal in the identified locations 2. Approvals shall be taken from competent authority for dumping of slurry so that it does not cause any other impact to the environment	1. Generated muck shall be used, for the purpose of back filling of low-lying areas at stations and viaduct 2. Drip tray and spill kit provided 3. Color bins for waste segregation and collection	
4	Land dumping of muck slurry (Yamuna Bed)				Waste Management Rules 2016

(continued)

Table 5b (continued)

S. No.	Activity/process	Aspect	Required control	Additional control	Legal app
Casting yard work					
1	Rebar cutting and bending work	Rebar cage fabrication	1. Working with approved construction drawings only 2. Cut pieces shall be adequately collected to avoid spreading of scrap in entire site 3. Collected scrap shall be reused to optimum level 4. The waste left out after optimum utilization shall be sold to authorized vendors 5. Trained workforce shall be engaged to perform the job		
2	Material loading and unloading work	Noise and vibration	1. Mechanical material handling shall be performed 2. All mechanical machinery and equipment shall be well maintained to be free from noise and smoke generation		
3	Stacking of material	Fall and damage	1. Store cool and dry place under the shed to avoid direct exposure of sun 2. Sufficient fire extinguisher shall be provided to mitigate the cause 3. Loading and unloading shall be done using ramp and rubber matt arrangement and transportation of cylinders should be done using trolley to avoid dust generation		

(continued)

Table 5b (continued)

S. No.	Activity/process	Aspect	Required control	Additional control	Legal app
4	Storage of cylinder	Fire and explosion	1. Store cool and dry place under the shed to avoid direct exposure of sun 2. Sufficient fire extinguisher shall be provided to mitigate the cause 3. Loading and unloading shall be done using ramp and rubber matt arrangement and transportation of cylinders should be done using trolley to avoid dust generation	1. Test and inspection certificate of cylinder will be maintained 2. Empty cylinder to be segregate from the filled ones and care shall be taken that al the valves are tightly shut 3. Cylinder will not be stored along with any combustible material	The Gas Cylinders Rules, 2004
5	Form work	Generation of noise damage of material due to rusting	1. Care shall be taken to avoid spillage of concrete 2. Proper collection system shall be maintained, and excess concrete generation shall be avoided		
6	Concreting work	Slurry and chemical generation	1. Care shall be taken to avoid spillage of concrete 2. Proper collection system shall be maintained, and excess concrete generation shall be avoided	Will be sent to authorized vendor for disposal and reuse, challan system will be used for monitoring	C&D Waste Management Rules 2016
7	Erection of plant	Generation of toxic fumes and noise	1. Plant erection should be done in isolated place 2. Filters shall be installed	Bulker will be used instead of loose bags of cement	The Air (Prevention and Control of Pollution) Act, 1981 (Amendment in 1987) and Rules, 1982

(continued)

Table 5b (continued)

S. No.	Activity/process	Aspect	Required control	Additional control	Legal app
8	Storage of raw material	Generation of dust and noise	1. Plant erection should be done in isolated place 2. Filters shall be installed 3. Ducting for collection of dust generated		
9	Loading of concrete batch		Care shall be taken to avoid concrete wastage		
10	Cleaning of plant	Water and land contamination	1. Proper collection of wastewaters shall be carried out 2. Wastewater treatment shall be carried out 3. Sediments shall be cleared before releasing water freely in land	RO system will be used for treatment and then water will be reused	The Water (Prevention and control of pollution) Act, 1974 (Amended 1988) and Amended Rules, 2011
11	Gas Cutting	Gas leakages, fire & Explosion	1. Cylinder shall be always in vertical position stacked on a trolley which chain locking arrangement to arrest its fall 2. Flashback arresters shall be used on both cylinder and torch sides 3. Entire gas cutting area shall be free from flammable materials 4. Adequate fire extinguishers shall be installed to avoid fire and explosion hazard	1. Inspection and marking of cylinder to be done 2. A safe distance of 10 m will be maintained from combustible material 3. Empty and filled cylinder to be kept separately	The Gas Cylinders Rules, 2004
12	Welding	Toxic fumes generation, metal waste	1. Electrodes with less toxic emission shall be preferred 2. LEV system shall be preferred 3. Face shield and filtered nose mask shall be used 4. Waste to be disposed at waste collection bins		The Air (Prevention and Control of Pollution) Act, 1981 (Amendment in 1987) and Rules, 1982
13	Grinding	Dust and noise generation	1. Electrodes with less toxic's emission shall be preferred 2. LEV system shall be preferred 3. Filtered nose mask shall be used		

(continued)

Table 5b (continued)

S. No.	Activity/process	Aspect	Required control	Additional control	Legal app
NRD					
1	Transmission of electricity	Spillage of transformer oil, con	1.Drip pans shall be used to protect oil spillage 2.Regular maintenance of transformer 3.Monitoring of fuel consumption shall be carried out on monthly basis		
2	Production of electricity by DG Sets	Generation of noise, stack emission, consumption of fuel, spillage of oil	1. Regular maintenance of DG set to be done 2. Maintain proper stack height of DG as per the guidelines given by CPCB/NGT/DPCC 3. Drip pans shall be used to protect oil spillage 4. Monitoring of fuel consumption shall be carried out on monthly basis	Only acoustic DG sets are used	Notification on Control of Noise from Diesel generator (DG Sets), 2002
3	Running of diesel generator	Generation of used oil Discharged parts (filters, batteries etc.) noise generation	1. Spill collection trays shall be used as required 2. Arrangements shall be made required to avoid spillage of used oil 3. Used oil shall be collected properly and must be disposed off to authorized vendors only for further treatment	Oil contaminated cloths shall also be treated as hazardous waste and must be disposed of separately to authorized vendors only	Hazardous and other Wastes (Management and Transboundary Movement) Rules, 2016
			1. All the dischargeable parts shall be considered as hazardous waste 2. Proper collection of all the parts shall be carried out and must be disposed off to authorized vendors only	Handing and end disposal process of refused material will be done as per regulatory requirements	Hazardous and other Wastes (Management and Transboundary Movement) Rules, 2016
			Acoustic barriers wall enclosures fencing, anti-vibration mounts, exhaust silencer attenuators shall be used to reduce noise		Notification on Control of Noise from Diesel Generator (DG Sets), 2002

(continued)

Table 5b (continued)

S. No.	Activity/process	Aspect	Required control	Additional control	Legal app
			While working in close vision of DG set ear plugs shall be used		
Site storage					
1	Storage of batteries	Spillage of acid	Proper handling shall be carried out		
2	Storage of gas cylinders	Leakage of gas	1. Store cool and dry place under the shed to avoid direct exposure of sun 2. Sufficient fire extinguisher shall be provided to mitigate the cause 3. Loading and unloading shall be done using ramp and rubber matt arrangement 4. Filled and empty cylinders shall be stacked separately		
3	Storage of Hazardous waste	Dust generation, fire in oil storage area	1. All hazardous waste material disposals shall be done as per hazardous waste management rules 2016 2. Hazardous waste shall be collected and stacked properly as stated in the rules 3. It shall be handed over to authorized vendors for its recycling 4. Adequate fire extinguishers shall be provided	1. Loading of wastes in storage sheds should only be done under the supervision of the well trained and experienced staff 2. Minimum of 1-m clear space should be left between two adjacent rows of drums in pair for inspection	Hazardous and Other Wastes (Management and Transboundary Movement) Rules, 2016

(continued)

Table 5b (continued)

S. No.	Activity/process	Aspect	Required control	Additional control	Legal app
4	Storage and handling of oil and fuel	Spillage of oil, waste oil lubricants etc., fire in oil storage	1. Use drip tray or absorbent materials under the oil drums 2. Maintain good housekeeping and Provide concrete floor 3. No electric connection shall be present inside the storage yard, only battery-operated hand lamps shall be preferred 4. Firefighting arrangement shall be provided		
5	Transportation and storage of material	Oil leakage	1. The vehicles and equipment shall be checked for leakage and if found shall be attended 2. Regular maintenance of plants and equipment's shall be done 3. Provide hearing protection wherever found necessary		
FORM work and finished of concrete					
1	Consumption of material as a form work such as timber/metals	Damage and corrosion of material	1. Foam work material stacking area must be isolated keep in dry place 2. Material stacking height should not be above 2 m and shall be according to shape and size 3. Optimum utilization of materials shall be carried out 4. Damaged materials shall be properly collected and sold to scrap agency if not to be used again		

(continued)

Table 5b (continued)

S. No.	Activity/process	Aspect	Required control	Additional control	Legal app
2	Consumption of cement mortar	Consumption of material	1. Optimum utilization of the material shall be carried out 2. Wastage shall be reduced 3. Only trained persons should be allowed to perform the job		
3	Consumption of nuts, bolts, screw etc. for form work	Corrosion of material	1. Stacking of material in dry places 2. Regular oiling/greasing of material 3. Handling of materials shall be specially taken care of		
First aid activity					
1	First aid activity	Biomedical waste generation, sewage water generation	1. Biomedical waste shall be sent to the local Govt. Hospital for further treatment and disposal 2. Biomedical waste shall be collected in different bin with color codes for clear identification 3. Reducing, reusing and recycling of generated wastewater	1. Waste will be collected daily or as frequently as required 2. Good lightning and least passive ventilation will be done	Bio Medical Waste Management Rules 2016

References

1. Sakr, D.A., Sherif, A., El-Haggar, S.M.: Environmental management systems' awareness: an investigation of top 50 constractors in Egypt. Journal of cleaner production (2010) 18(3): 210–218
2. Macedo, A. T., *et* Martins, M.F.: A Sustentabilidade urbana na perspectiva das empresas construtoras em Campina Grande – PB", artigo apresentado no XII ENGEMA: Encontro Nacional sobre Gestão Empresarial e Meio Ambiente, São Paulo, SP, 5–7 de Dezembro, (2011)
3. Vechi, N.R.G., Gallardo, A.L.C.F., Teixeira, C.E.: Environmental aspects of the construction industry: framework for the adoption of environmental management system for small and medium enterprises of pro viding services. Systems & Management (2016) 11: 17–30
4. Cheng, H.: Correlation and measures associated with construction dust and haze. Journal of Construction Safety (2014) 29(4): 50–52
5. Rui, D.M., Chen, J.J., Feng, Y.C.: Study on source apportionment for ambient air PM10 in Nanjing. Journal of Environmental Science and Management (2008) 33(4): 56–58
6. Huang, G.J. Emissions and effects of PM2.5 resulting from construction dust. Journal of Environment and Life ((2014) 16): 141–142

7. Guan, J., Zhang, Z.: Study on an environmental impact assessment of a construction process based on the life cycle assessment approach. Proceedings from Smart Construction and Management in the Context of New Technology, Kunming, China, (2014) 404–413
8. Chou, J.S., Yeh, K.C.: Life cycle carbon dioxide emissions simulation and environmental cost analysis for building construction. Journal of Cleaner Production (2015) 101: 137–147
9. Mandal, A., Suva, P.: Haematological changes among construction workers exposed to cement dust in west Bengal, India. Progress in Health Sciences (2014) 4(1): 88–94
10. Yang, J.C., Zhou, Z.H., Zhang, G.J., Li, H.J., Ma, Q., Mo, Z.Y., Chen, Z.M., Mao, J.Y., Huang, J.Y.: Atmospheric particulate matter emissions of road and building construction dust estimation and prevention measures in Baise. Journal of Popular Science and Technology (2016) 18(8): 29–31
11. Bishnoi, P., Tripathi, K.N., Malhotra, A.: Air and Noise control and Analysis in Construction Industry. International Journal of Engineering Research & Technology (IJERT) (2020) 9(01): 414–417
12. Johncy, S.S., Ajay, K.T., Dhanyakumar, G., Raj, N.P., Samuel, T.V.: Dust exposure and lung function impairment in construction workers. Journal of Physiological and Biomedical Sciences (2011) 4(1): 9–13
13. Li, X.D., Su, S., Huang, T.J.: Quantitative evaluation on health loss due to construction dust. Journal of Tsinghua University (Science and Technology) (2015) (1): 50–55
14. Guo, E.G., Wang, C., Peng, Z.H., Fang, C.: Advances in physical and chemical character of urban airborne suspended particulate matter and its health effect. Journal of Ecology and Environment (2008) 17(2): 851–857
15. Liu, X.Y., Xie, P., Liu, Z.R., Li, T., Zhong, L.: Economic assessment of acute health impact due to inhalable particulate air pollution in the Pearl River delta. Journal of Acta Scientiarum Naturalium Universitatis Pekinensis (2010) 46(5): 829–834
16. Huang, Y.S.: An analysis of economic losses and influencing factors from environmental pollution of Beijing and Shanghai. Lanzhou University, Lanzhou, China, (2016) 12–14
17. Shah, J.J., Kiang, C.S., Lo, C., Judith, C.C., Watson, J.G., Loh C., L.M., Lents, J.M., Mollna, M.J., Mollna, L.T.: Megacities and atmospheric pollution. Journal of the Air & Waste Management Association (2004) 54(10): 1226–1235
18. Fadiya, O.O., Georgakis, P., Chinyio, E.: Quantitative Analysis of the Sources of Construction Waste. Journal of Construction Engineering, Article ID 651060, 9 pages, http://dx.doi.org/ https://doi.org/10.1155/2014/651060 (2014)
19. Viana, E.T., Ribeiro, L.C.L.J., Lintz, R.C.C., Gachetbarbosa, L.A., Pires, M.S.G.: Implementation of environmental management on a construction site. Applied Mechanics & Materials (2013) 441–444
20. Yao, F., Liu, G., Ji, Y., Tong, W., Du, X., Li, K., Shrestha, A. and Martek, I. (2020). Evaluating the Environmental Impact of Construction within the Industrialized Building Process: A Monetization and Building Information Modelling Approach. Int. J. Environ. Res. Public Health, 17, 8396; doi:https://doi.org/10.3390/ijerph17228396
21. https://en.wikipedia.org/wiki/Construction_industry_of_India#:~:text

Inventory of Glaciers in the Himalayan Region: A Study Through Field Survey and Sentinel 2B Based Imagery Analysis

Arvind Pandey, Deepanshu Parashar, Bidyutjyoti Baruah, Sarita Palni, J. S. Rawat, Ajit Pratap Singh, Pankaj Kumar Bhatt, and Aditi Bisht

Abstract Glaciers play a major role in the human ecosystem. Since they are the primary source of our freshwater reserve, glaciers are very much crucial to sustaining life on the planet. Moreover, its highly sensitive nature to the changes in the environment and climatic conditions, glaciers also act as a major indicator for climate change. Hence, the inventory and monitoring of glaciers have significant importance in today's world. Rapid deglaciation primarily caused by the global increase in temperature can lead to a rise in the sea level to an extreme level. Remote sensing techniques help the researchers a lot to identify and monitor the changes that occurred to the most physically inaccessible glaciers. Normalized Differential Indexes like Normalised Differential Snow Index (NDSI) can help the researchers to identify the snow cover regions and also to make an inventory of the glaciers supported by other techniques. While monitoring snow cover and glacier-covered regions, researchers need to understand different characteristics such as knowledge of reflectance characteristics, spatial resolution requirement, repeat data acquisition duration, sensor saturation, discrimination toward snow clouds, and monitoring of snow cover under mountain shadow.

A. Pandey (✉) · D. Parashar · B. Baruah · S. Palni
Department of Remote Sensing and GIS, Soban Singh Jeena University, Almora 263601, India
e-mail: pandeyarvind02@gmail.com

J. S. Rawat
Department of Geography, Soban Singh Jeena University, Almora 263601, India

A. P. Singh
Civil Engineering Department, Birla Institute of Technology and Science, Pilani 333031, India
e-mail: aps@pilani.bits-pilani.ac.in

P. K. Bhatt
Department of Geography, L.S.M. PG College, Pithoragarh, India

A. Bisht
Department of Zoology and Environment Science, Gurukula Kangri (Deemed to be University), Haridwar, Uttarakhand, India

K. D. Bahukhandi et al. (eds.), *Environmental Pollution and Natural Resource Management*, Springer Proceedings in Earth and Environmental Sciences,
https://doi.org/10.1007/978-3-031-05335-1_15

Keywords Glacier · Snow covered region · Kumaun Himalaya · Remote sensing · NDSI · Sentinel multi spectral satellite

1 Introduction

Mapping of glaciers is one of the key applications areas of remote sensing data which has been widely applied in particular after the launch of Landsat imageries for the public and offering orthorectified (i.e., terrain corrected) products [1]. Because glacier changes are key climate change indicators [2] and their outlines are mandatory for any glacier-specific measurements and modelling (e.g., changes in length, volume/mass, run-off, ice thickness, and potential glacier development), there has been a strong demand for a full global inventory of glaciers [3]. The classification of glaciers is based on a clear difference in the visible and near-infrared (VNIR) spectral reflectance of snow and glacier ice compared to shortwave infrared (SWIR). Since snow absorbs most of the incident radiation in the SWIR while clouds do not, Normalised Differential Snow Index (NDSI) can distinguish snow from clouds. It prepares a basis for snow/ice cover mapping and glacier monitoring applications [4]. Illumination effects due to topography are reduced by separating the raw digital numbers (DN's) of a VNIR band by those of the SWIR, and glaciers stand out bright against a dark background. The ratio image is transformed into a glacier map (clean ice only) with a simple threshold, and the corresponding glacier outlines [5]. Compared to other, much more complex and computationally intensive approaches, simple band ratios have proven to be fast, robust (results are insensitive to the thresholds) and precise [6, 7] or reference datasets [8]. The method has thus been widely adopted for a large-scale glacier inventory with medium resolution satellite datasets in many areas of the globe [9–13]. However, due to their spectral similarity to the surrounding landscape, it is not possible to map debris-covered glacier sections from spectral properties alone. Accordingly, a wide range of different combined approaches has been developed to separate debris-covered from glacier parts [14–17].

As per the report of IPCC 2014 [18], the glaciers are continuously shrinking very rapidly at the global level because of the ongoing phenomenon of climate changing. The recession of glacier in the Himalayan region has not only showed impact on the streamflow but also on the proglacial lakes [19, 20]. The shrinking of the Himalayan cryosphere and retreat occurrences in different glaciers resulted in decreasing the streamflows in the Himalayan rivers and impacting vegetation and grasslands [21–24] which directly affects the availability of water in downstream. Moreover, growing freshwater scarcity or inequitable availability of fresh and safe potable water leads to severe health issues, poverty, and degradation of the environment that might result in global hunger, civil unrest, and conflicts [25]. The overall land use land cover pattern system changes in the region of interest have been linked to depleting streamflows, unplanned land transformation and economic considerations [26].

In this research, 11 Glaciers have been identified and selected based on field survey and visual interpretation of satellite imagery observations. The locations of glaciers are highlighted with red marks. The five selected Glaciers are named Kafini Glacier, Chayanguchha Glacier, Pindari Glacier, Baljuri Base Camp Glacier and Sundardhung Glacier from east to west, respectively in Fig. 7. There were also Six unnamed Glaciers selected in the vicinity of the research area. Approximately Permanent snow cover is 213.77 km^2 area which is 54.61% of the study area. All Identified 11 glaciers are cover ~45 km^2 area which is 21.05% of the permanent snow cover area in the upper valley of the Pindari region. The Sundardhung Glacier is located at the highest elevation point among these glaciers, which is at ~6785 m above sea level, while the Pindari Glacier is located at the lowest point, i.e., 3800 m above sea level.

2 Study Area

Uttarakhand, which is located on the North of India, experiences great weather and climatic conditions. Most of Uttarakhand comes under the lesser Himalayas which belong to the Western Himalayan site. Uttarakhand Himalaya includes a 320 km stretch of the mountains between Kali river to Satluj river, which forms the Indo-Nepal border in the east and the Tons Pabar valleys (concerning Kali river) forming the eastern border of Himachal Pradesh (concerning Satluj river). The study area, viz., the Pinder watershed (Kumaun Himalaya) (Fig. 1) extends between 30°6′15.429″ N 30°19′15.45″ N latitudes and 79° 47′57.434″ E to 80°5′41.672″ E longitudes and encompasses an area of 391.05 km^2. Pindari Glacier is a prime glacier in the study area which is located in Bageshwar district of Uttarakhand State, India. It comes in Upper Himalayan range and is neighbour to Chhanguch glacier. Pindar river is the biggest tributary of the Ganga river, and the current snout point location of Pindar River is recorded at 80°00′29″ E and 30°16′28″ N, situated at 3847 m elevation. The study area has mixed weather conditions, ranging from sunny to heavy snowfall and a track of around 70 km on foot from Song village [27]. The average approximate temperature of the region lies between −10 and −15 °C during summers and −20 to −40 °C during winters. The tongue of Pindari Glacier is called 'Trail pass' which is the shortest way to Milam glacier. G.W. Trail, the first British Deputy Commissioner of Kumaun division, was the first to reach the pass in 1830. Prominent peaks near Pindari glacier from right to left are Baljuri (5922 m), Pawalidwar (6663 m), Nandakhat (6611 m), Changuch (6322 m, Unclimbed Peak) and Nandakot (6860 m). Pindari Kanda Trail's Pass (el. 5300 m or 17,400 ft) is a mountain pass through the Himalaya located between Nanda Devi and Nanda Kot peaks in the Uttarakhand state in the districts of Pithoragarh and Bageshwer.

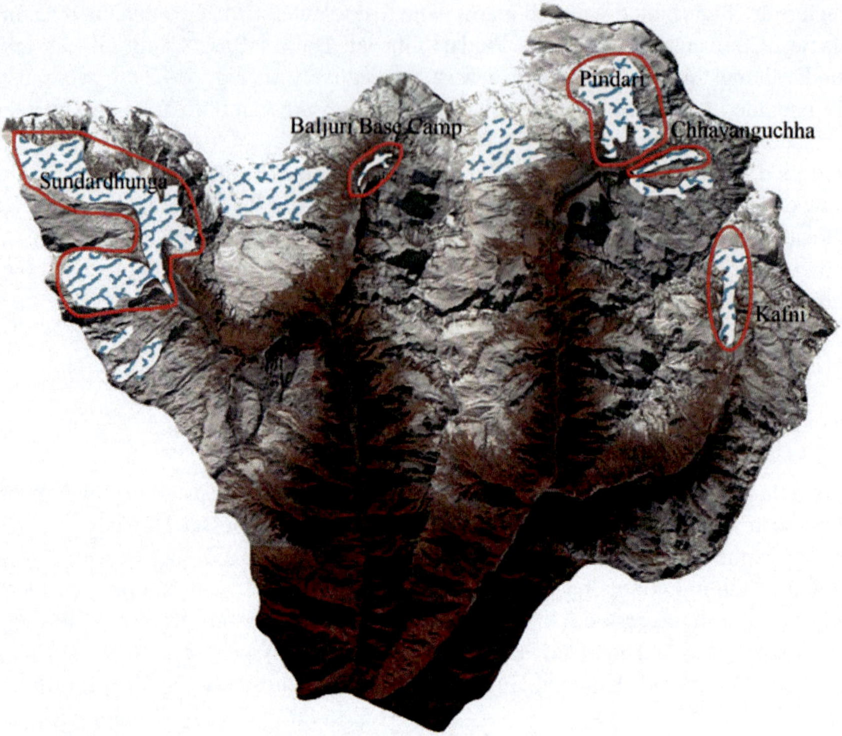

Fig. 1 Study area map

3 Data and Methodology

In this study, we used Sentinel multi-resolution satellite data downloaded from the copernicus for extraction the glacier covers over the study sites. The sentinel multi resolution data provides us with resolution varying from 10, 20 and 60 m. Visual image interpretation techniques were used to identify the glacier boundary. ERDAS imagine and ArcGIS softwares were used to pre-process the images to reduce the atmospheric noises and identifying the glaciers respectively (Fig. 2 and Table 1).

4 Results

4.1 Kafini Glacier

Kafini Glacier is one of the main glaciers in the study area and the source of the Kafini River, Pindar River's main tributary, Dwali confluence, 30°10′36.55′ N River, Dwali

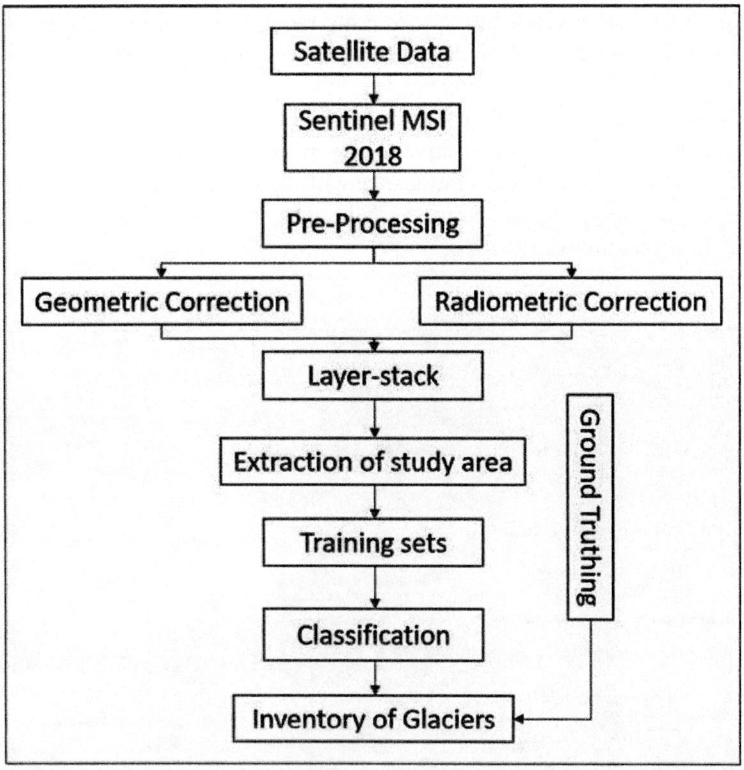

Fig. 2 Methodology flow chart

Table 1 Area of important glaciers

S. No	Glacier	Area (km^2)
1	Kafini	1.86
2	Pindari	4.29
3	Sundardhung	5.36
4	Baljuri Base	1.40
5	Others	13.25

confluence (30°10′36.55′ N latitude and 79°59′31.48). Kafini glacier is situated in the easternmost representative of the entire study area, Kafini Glacier extended from the upper portion of the Kumaun Himalayan region to the southeast part of Nanda Devi mountain. Kafini Glacier, the shifting blaze at an elevation of 3860 m on the lap of Nanda Kot (6861 m), is one of the hubs for the trekkers in the Kumaun Himalayan region that can be accessed during the summer and autumn seasons. The trek to Kafini Glacier snakes through some steep uphill sections laid with stones and expunges into the thick wood till manoeuvring to the gentle meadows.

In Fig. 3 the Kafini glacier is shown in the red highlighted mark. The slope of the glacier is 10° in the tongue and gradually increases to above 60° near the source. The aspect of the glacier is towards the west. The approximate length of the glacier is 500 m and covers an area of ~1 km^2.

Figure 4 represents the ground survey imagery of the Kafini glacier. A is a closer view of the glacier, and B is the farther view of the glacier. Many human-induced and climatic factors have resulted in the melting of snow, which covered the entire region in previous times.

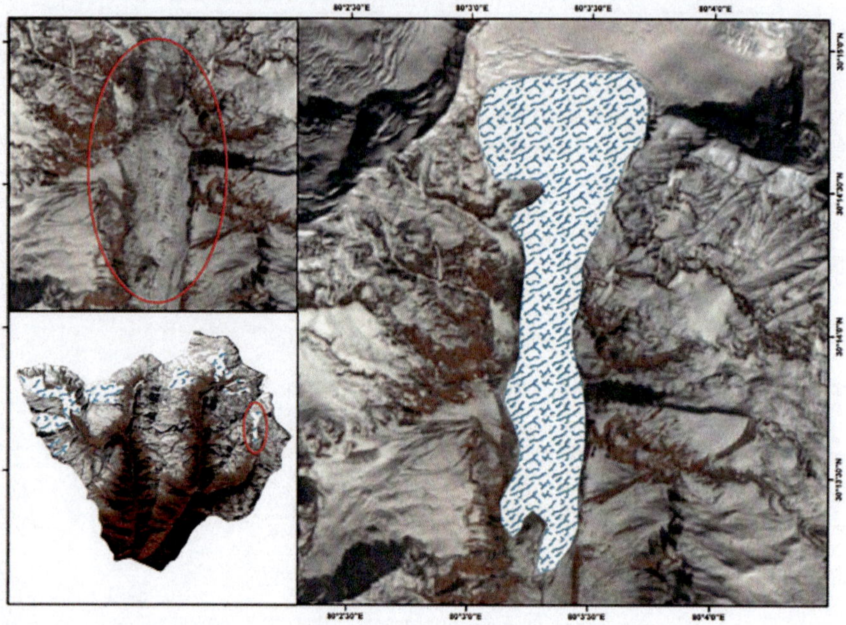

Fig. 3 Kafni glacier in the red marked area in 2018

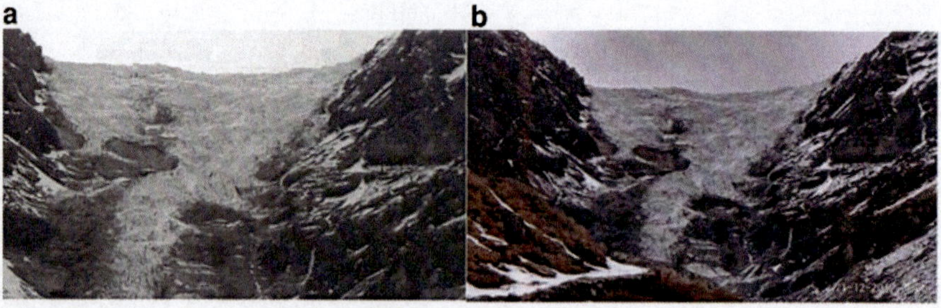

Fig. 4 Field survey imagery of Kafini glacier in 2017

4.2 Pindari Glacier

Pindari Glacier is the second important glacier in this study which is situated between the upper part of Kumaun Himalaya to the southeast of Nanda Devi and Nanda Kot. The source of the Pindar River, which is an important tributary of the Alaknanda River, is the Pindari glacier. The Pindar River travels around 118 km, nourishing the beautiful Pinder valley of 339.39 km^2 surrounding area before it confluence with the Alaknanda river at Karnprayag. Pindar River flows along a fault [28], who gave a general description of the glacier and augmented it with a few sketches, first visited the Pindari Glacier [29]. Chhayanguchha is the tributary glacier of Pindari Glacier.

Tiwari and Jangpangi resurveyed Pindari Glacier in 1958, and since 1906, reported the retreat of their snout by 1040 m [30]. The snout had shown a further retreat of approximately 200 m during their 1966 expedition. The division of the Chhanguch Branch, a tributary to the Pindari glacier, was also observed by them. The two glaciers have two separate ice caves as a consequence of this retreat. Pandey et al. [21] reported a significant magnitude of retreats in some of the selected glaciers, namely Pindari, Sundardhunga, Kafni, and Baljuri base camp glaciers. From 1972 to 2018, approximately 1719.95 m, 1751.21 m, 1057.01 m, and 810.78 m retreats, respectively, were observed from Pindari, Sundardhunga, Kafni, and Baljuri base camp glaciers. The most remarkable change was observed in the Pindari and Sundardhunga glaciers of over 1700 m [31]. The impact of climate change is the leading cause of net shrinkage, glacier retreat, and river snout transformation [32, 33].

Figure 5 is showing the location of the Pindari glacier in the study area, viz., the Pindar watershed (Kumaun Himalaya) (Fig. 1) extends between 30°6′15.429″ N 30°19′15.45″ N latitudes and 79°47′57.434″ E to 80°5′41.672″ E longitudes. The aspect of the glacier is towards the west.

Figure 6 is a ground survey photograph of the Pindari glacier, where A is showing the Pindari Glacier while B is showing Chhayanguchha Glacier. The picture shows the ground condition of the glacier. Both glaciers were the tributary glacier in the past decades. Pindari glacier goes back 600 m from the confluence and Chhayanguchha glacier is 500 m in length.

4.3 Sundardhung Glacier

Sundardhung Glacier is the third major glacier studied in this research, which confluence with the Pindar River at latitude 30°6′48.73″ and longitude 79°55′39.48″ E down part of the Khati village, the last valley village. Nine glaciers have been identified between Pindari and Sundardhung glaciers, such as Baljuri base camp, Maktoli Glacier, and some other unnamed glaciers (Figs. 7 and 8).

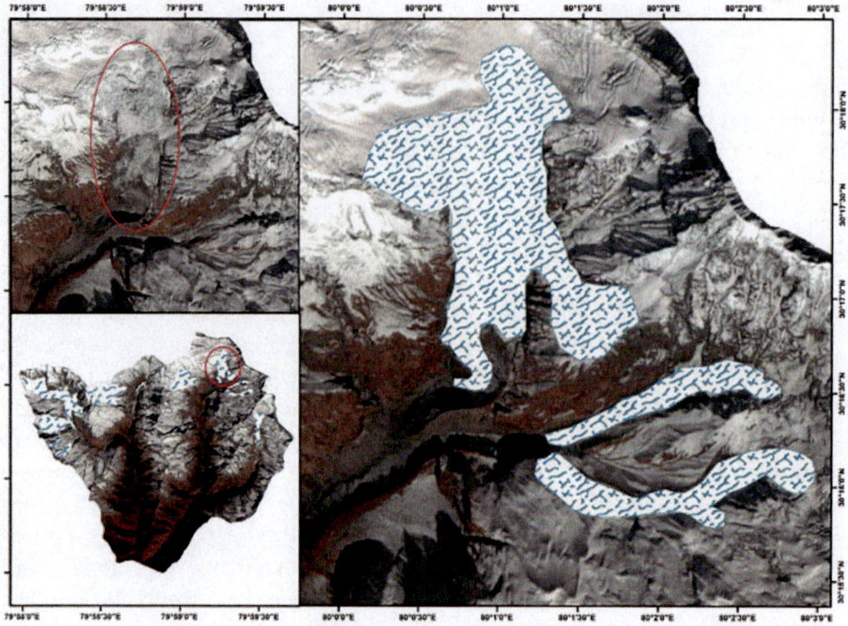

Fig. 5 Pindari (Chhayanguchha) glacier in 2018

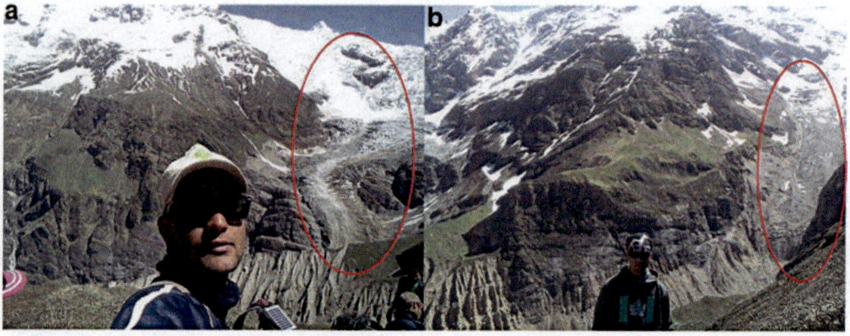

Fig. 6 Field photograph-Pindari glacier **a** and Chhayanghuchh glacier **b** in 2012

4.4 Baljuri Base Camp Glacier

Compared to other selected glaciers from the study, Baljuri base camp glacier is a relatively small glacier-formed at the base of Baljuri peak and the primary source of the Sundardhung river tributary river and it confluences with the Sundardhung river at latitude 30°13′30.32″ N and longitude 79°55′0.12″ E. Maktoli Glacier formed in the upper side of the Sundardhung glacier and contributed with Sundardhung glacier and is a significant source of water of Sundardhung river (Fig. 9).

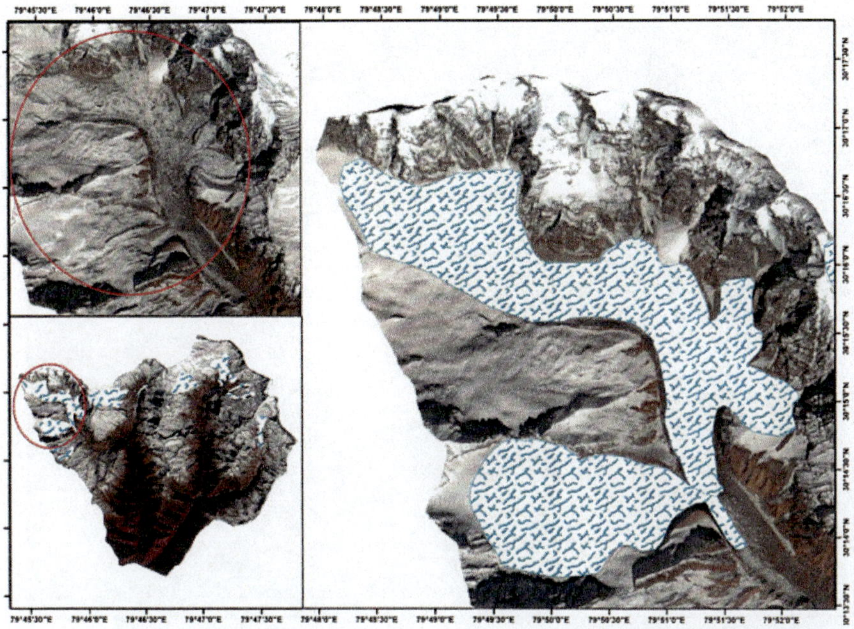

Fig. 7 Sundardhung glacier in 2018

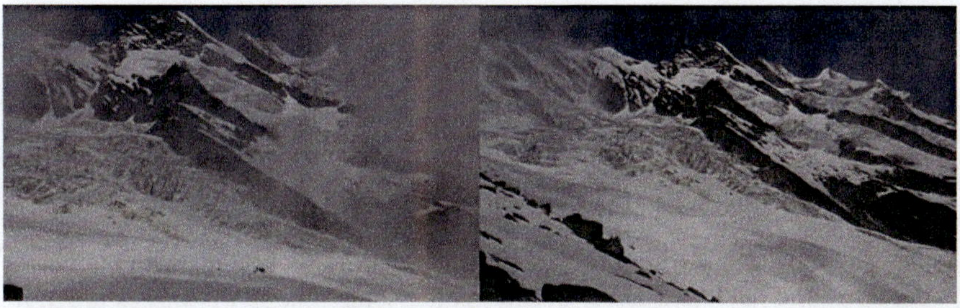

Fig. 8 Field photograph

4.5 Other Unnamed Glacier

Some unnamed glaciers also have been included during the fieldwork in this research. The location of these unnamed glaciers is described as following.

(A) In between Baljuri Base Camp Glacier and Sundardhung,
(B) the upper left side of Sundardhung and
(C) in between Pindari glacier and Baljuri Base camp.

All the unnamed glaciers are newly formed glaciers (Fig. 10).

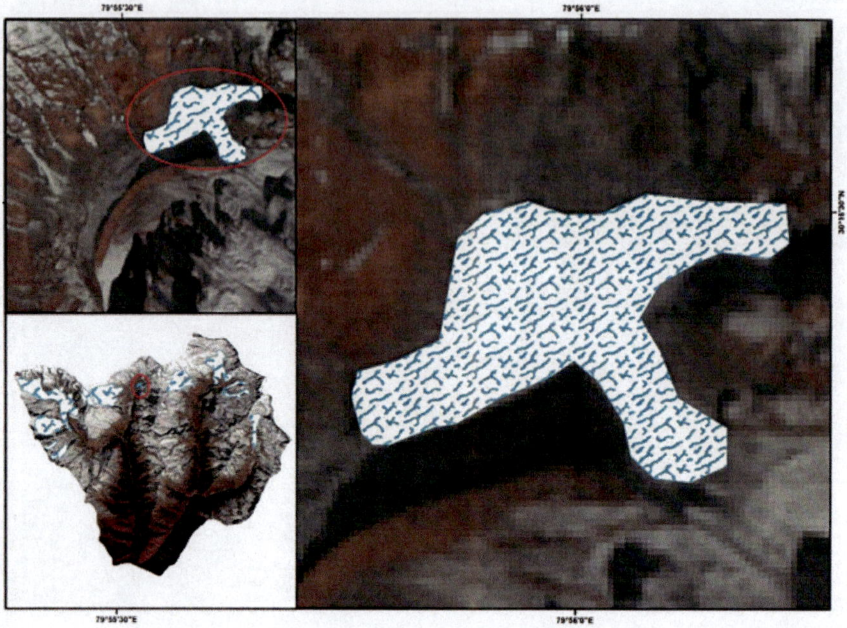

Fig. 9 Location of Baljuri base camp glacier

5 Discussion

Fastly increasing in the accessibility of the technology of remote sensing and geographic information system allows to the users a financially low-cost method of monitoring and mapping of glacier region in a large coverage of area. At present there are several freely satellite data available such as Landsat, Sentinel, NOAA, and Modis which makes growth in research of the field of cryosphere. With the worse condition for the field visit in the glacier region some time it will makes tough to study of the glaciers but with the application of remote sensing and Geographic information system it is now possible to take attention on glaciers study. With the wide range of electromagnetic spectrum range the better study of cryosphere is possible. The technology of remote sensing uses visible, microwave, thermal, hyperspectral and Lider technology for the study of a snow covered region.

In the last few decades, it was noticed by the scientific community that the rate of deglaciation is increasing very rapidly, which is very harmful to the overall earth's ecosystem. There are several causes for the increasing rate of deglaciation such as global warming, increasing in concertation of black carbon, and extensive use of fossil fuels, which causes the changing in the overall climate system and retreat of a glacier. Snow and glacier-stored water is an essential natural resource in the Himalaya region. They feed many North Indian rivers, and water supplies from this source are accessible in lean summer months. Therefore, for the optimum management of Himalayan

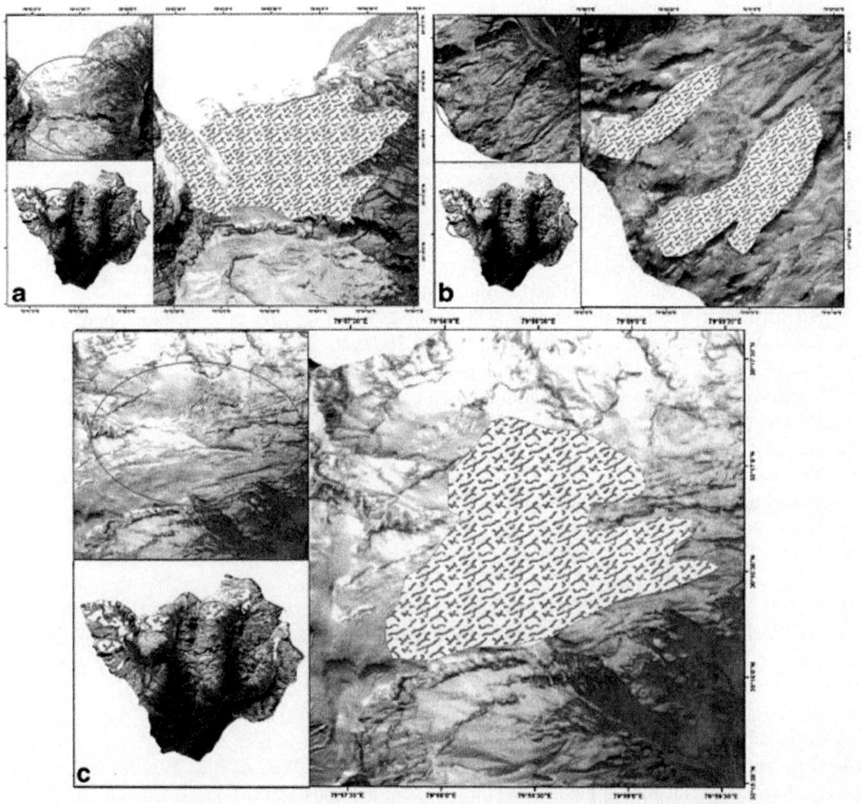

Fig. 10 Location map of some other unnamed glacier included in the research study

water resources, understanding the distribution and variation of snow and glaciers is essential. Monitoring of snow cover needs to understand different characteristics such as knowledge of reflectance characteristics, spatial resolution requirement, repeat data acquisition duration, sensor saturation, discrimination toward snow clouds and monitoring of snow cover under mountain shadow. These aspects and salient results of inventory on 1:250,000, 1:50,000 scales and glacial retreat estimates in the Pindar basin using ortho-images are discussed in this topic. The accuracy is better than 3% for debris-free glacier areas.

References

1. Wulder, M.A.; Masek, J.G.; Cohen, W.B.; Loveland, T.R.; Woodcock, C.E. Opening the archive: How free data has enabled the science and monitoring promise of Landsat. Remote Sens. Environ. 2012, 122: 2–10.

2. Vaughan, D.G.; Comiso, J.C.; Allison, I.; Carrasco, J.; Kaser, G.; Kwok, R.; Mote, P.; Murray, T.; Paul, F.; Ren, J.; et al. Observations: Cryosphere. In Climate Change 2013: The Physical Science Basis; Contribution of Working Group I to the Fifth Assessment Report of the IPCC; Cambridge University Press: Cambridge, UK, 2013: 317–378.
3. Paul, F., Winsvlod, S.H., Kaab, A., Nagler, T., and Schwaizer, G. Glacier Remote Sensing Using Sentinel-2. Part II: Mapping Glacier Extents and Surface Facies, and Comparison to Landsat 8. MDPI Remote sensing. 2016, pp. 02–10.
4. Dozier, J. Spectral signature of alpine snow cover from Landsat 5 TM. Remote Sens. Environ. 1989, 28:9–20.
5. Paul, F.; Kääb, A.; Maisch, M.; Kellenberger, T.; Haeberli, W. The new remote-sensing-derived Swiss glacier inventory: I. Methods. Ann. Glaciol. 2002, 34: 355–359.
6. Albert, T.H. Evaluation of remote sensing techniques for ice-area classification applied to the tropical Quelccaya Ice Cap, Peru. Polar Geogr. 2002, 26: 210–226.
7. Paul, F.; Kääb, A. Perspectives on the production of a glacier inventory from multispectral satellite data in Arctic Canada: Cumberland Peninsula, Baffin Island. Ann. Glaciol. 2005, 42: 59–66.
8. Paul, F.; Bolch, T.; Kääb, A.; Nagler, T.; Nuth, C.; Scharrer, K.; Shepherd, A.; Strozzi, T.; Ticconi, F.; Bhambri, R.; et al. The glaciers climate change initiative: Methods for creating glacier area, elevation change and velocity products. Remote Sens. Environ. 2015, 162: 408–426.
9. Bolch, T.; Menounos, B.; Wheate, R. Landsat-based glacier inventory of western Canada, 1985–2005. Remote Sens. Environ. 2010, 114: 127–137.
10. Frey, H.; Paul, F.; Strozzi, T. Compilation of a glacier inventory for the western Himalayas from satellite data: Methods, challenges and results. Remote Sens. Environ. 2012, 124: 832–843.
11. Rastner, P.; Bolch, T.; Mölg, N.; Machguth, H.; LeBris, R.; Paul, F. The first complete inventory of the local glaciers and ice caps on Greenland. Cryosphere 2012, 6: 1483–1495.
12. Paul, F.; Mölg, N. Hasty retreat of glaciers in northern Patagonia from 1985 to 2011. J. Glaciol. 2014, 60:1033–1043.
13. Guo, W.; Liu, S.; Xu, J.; Wu, L.; Shangguan, D.; Yao, X.; Wei, J.; Bao, W.; Yu, P.; Liu, Q.; et al. The secondChinese glacier inventory: Data, methods and results. J. Glaciol. 2015, 61: 357–372.
14. Paul, F.; Huggel, C.; Kääb, A. Combining satellite multispectral image data and a digital elevation model for mapping debris-covered glaciers. Remote Sens. Environ. 2004, 89: 510–518.
15. Shukla, A.; Arora, M.; Gupta, R. Synergistic approach for mapping debris-covered glaciers using optical-thermal remote sensing data with inputs from geomorphometric parameters. Remote Sens. Environ. 2010, 114: 1378–1387.
16. Racoviteanu, A.; Williams, M.W. Decision tree and texture analysis for mapping debris-covered glaciers in the Kangchenjunga area, Eastern Himalaya. Remote Sens. 2012, 4: 3078–3109.
17. Robson, B.A.; Nuth, C.; Dahl, S.O.; Hölbling, D.; Strozzi, T.; Nielsen, P.R. Automated classification of debris-covered glaciers combining optical, SAR and topographic data in an object-based environment. Remote Sens. Environ. 2015, 170: 372–387.
18. IPCC (2014) In: Core Writing Team, Pachauri RK, Meyer LA (eds) Climate change 2014: synthesis report. Contribution of working groups I, II and III to the fifth assessment report of the intergovernmental panel on climate change. IPCC, Geneva.
19. Rashid, I.; Majeed, U. Recent recession and potential future lake formation on Drang Drung glacier, Zanskar Himalaya, as assessed with earth observation data and glacier modelling. Environ. Earth Sci. 2018, 77(12): 1–13.
20. Mir, R.A.; Jain, S.K.; Lohani, A.K.; Saraf, A.K. Glacier recession and glacial lake outburst flood studies in Zanskar basin, western Himalaya. J. Hydrol. 2018, 564: 376–394.
21. Pandey, A., Parashar, D., Bhatt, N.C., Palni, S., Pundir, C., Yadav, A.S., Singh, A.P., Bhatt, P.K. Impact of climate on vegetation in Pindari watershed of Western Himalayas, Kumaun, India using spatiotemporal analysis: 1972–2018. Environ. Sci. Pollut. Res. 2022. https://doi.org/10.1007/s11356-022-19711-z

22. Azmat, M.; Liaqat, U.W.; Qamar, M.U.; Awan, U.K. Impacts of changing climate and snow cover on the flow regime of Jhelum River, Western Himalayas. Reg. Environ. Chang. 2017, 17: 813–823.
23. Pandey, A., Singh, G., Palni, S., Chandra, N., Rawat, J.S., Singh, A.P. Application of remote sensing in alpine grasslands cover mapping of western Himalaya, Uttarakhand, India. Environ. Monit. Assess. 2021, 193: 166. https://doi.org/10.1007/s10661-021-08956-9
24. Immerzeel, W.W.; van Beek, L.P.H.; Konz, M.; Shrestha, A.B.; Bierkens, M.F.P. Hydrological response to climate change in a glacierized catchment in the Himalayas. Clim. Change 2012, 110: 721–736.
25. Singh, A.P., Bhakar, P. Development of groundwater sustainability index: a case study of western arid region of Rajasthan, India. Environ. Dev. Sustain. 2021, 23: 1844–1868. https://doi.org/10.1007/s10668-020-00654-9
26. Rashid, I., Majeed, U., Aneaus, S., and Pelto, M. Linking the Recent Glacier Retreat and Depleting Streamflow Patterns with Land System Changes in Kashmir Himalaya, India, Water, 2020, 12(4): 1168.
27. Bali, R., Agarwal, K.K., Patil, S.K., Ali, S.N., Rastogi, S.K., Krishna, K. Record of Neotectonic Activity in the Pindari Glacier valley: study Based on Glacio geomorphic and AMS Fabric evidences. IIG Institutional Repository, 2011. http://localhost:8080/xmlui/handle/123456789/633
28. Strachey R. A description of the glaciers of the Pindari and Kuphinee Rivers in the Kumaon Himalayas. The Journal Asiatic Society Bengal, 1847, 16(2): 203.
29. Tewari, A.P. Recent changes in the position of the snout of the Pindari glacier (Kumaon Himalaya), Almora District, Uttar Pradesh, India. In The role of snow and ice in hydrology, Proceedings of the Banff symposia, 1972, 2: 1144–1149.
30. Tiwari, A.P. and Jangpangi, B.S.. The retreat of the snout of the Pindari Glaciers. In *Proceedings of the symposium of Obergurage. International Association of Hydrological Sciences, 1962:* 245–248.
31. Pandey, A., Parashar, D., Baruah, B., Palni, S., Yadav, A.S., Singh, A.P., Bhatt, P.K., Pundir, C. Assessment of snout analysis of Himalayan glaciers: Impact studies on Pindari, Kafni, Sundardhunga and Baljuri Base Camp Glaciers. Environ. Monit. Assess. 2022, 194: 338. https://doi.org/10.1007/s10661-022-09925-6
32. Bajracharya, S.R., Mool, P.K. and Shrestha, B.R. Global climate change and melting of Himalayan glaciers. *Melting glaciers and rising sea levels: Impacts and implications*, 2008: 28–46.
33. Dobhal, D.P., Gergan, J.T. and Thayyen, R.J. Recession and morphogeometrical changes of Dokriani glacier (1962–1995) Garhwal Himalaya, India. *CURRENT SCIENCE-BANGALORE*, 2004, *86*(5): 692–696.

Spatio-Temporal Change Detection Using Remote Sensing Application—A Case Study of District Bageshwar, Uttarakhand, India

Arvind Pandey, Deepanshu Parashar, Kanak Matiyani, Sarita Palni, J. S. Rawat, Ajit Pratap Singh, and Aditi Bisht

Abstract The study of changes over the ground has become an important component of contemporary natural resource management and environmental monitoring techniques. The current study depicts the spatiotemporal dynamics of LULC in the district of Bageshwar, Uttarakhand, India. Landsat satellite images from three distinct periods, including Landsat MSS 1976, Landsat ETM+ 1999, and Landsat 8 TIRS 2016, were obtained by the official webpage of Earth Explorer and used to assess spatial–temporal change analysis in the Bageshwar District from 1976 to 2016. The supervised classification approach and the maximum likelihood algorithm were used to group pixels in a meaningful way. Based on satellite imagery, the research area's categorization result was divided into seven separate classes, including forest, agricultural, waterbody, settlement, barren land/grassland, scrubland, and snow. The study results show that forest cover has faced the highest negative amount of change from 1976 to 2016, about (−)76.07 km^2 and on the other hand, barren land/grassland shows the highest positive change of land cover about (+)43.86 km^2. The LULC classification shows an overall accuracy of 76%, 82%, and 84% for the different years of 1976, 1999, and 2016 respectively. The present study helps understand the spatial–temporal pattern of LULC changes and highlights the significance of digital image change detection methods in determining the amount of LULC change and shifting in Bageshwar district Uttarakhand, India.

A. Pandey · D. Parashar · K. Matiyani · S. Palni
Department of Remote Sensing and GIS, Soban Singh Jeena University, Almora 263601, India

J. S. Rawat
Department of Geography, Soban Singh Jeena Campus-Almora, Kumaon University, Nainital 263001, India

A. P. Singh (✉)
Civil Engineering Department, Birla Institute of Technology and Science, Pilani 333031, India
e-mail: aps@pilani.bits-pilani.ac.in

A. Bisht
Department of Zoology and Environment Science, Gurukula Kangri (Deemed to be) University, Haridwar, Uttarakhand 249404, India

K. D. Bahukhandi et al. (eds.), *Environmental Pollution and Natural Resource Management*, Springer Proceedings in Earth and Environmental Sciences,
https://doi.org/10.1007/978-3-031-05335-1_16

Keywords Landsat time series · Land use/cover · Remote sensing · Maximum likelihood (ML) · Change detection · District Bageshwar

1 Introduction

Mapping of land features is critical for land management, ecological and climatic studies [1]. Changes in land cover provide information for appropriate planning and optimization of natural resource usage and management [2]. Shifting over the ground pattern with time serves as a warning system to keep an eye on actions that harm the environment while also highlighting the outcomes of environmental conservation efforts such as afforestation [3]. Remote sensing is one of the most effective methods for monitoring changes on Earth, providing valuable sources of information about different ecosystems, including terrestrial ecosystems, and ideal solutions for LULC mapping [4]. In order to make a decision, a contemporary nation, like a modern company, need enough information on many complicated linked elements of its activity. Land use is only one of these aspects. However, understanding land use and land cover is becoming increasingly important as the country plans to address the issues of haphazard, deteriorating environmental quality, destruction of essential wetlands, uncontrolled development, loss of prime agricultural lands, and loss of fish and wildlife habitat [5]. Data regarding Land use information is required to evaluate ecological processes and issues that must be comprehended to improve or maintain living conditions and standards at present levels. The range of land use and land cover data requirements is enormous. To implement any development plan, it is essential to know the present information about the status of land use and land cover [6].

Central, state, and municipal governments also require land use and land cover data for water resource assessment, planning for water supply, flood control, water waste treatment, etc. [7]. To better public land management, several Federal agencies require updated complete maps of existing public lands and the present and evolving uses of neighboring land properties. Land use data is also required by federal agencies to monitor the environmental degradation by the development of energy resources, to manage the conflicts between the ecosystem of wildlife resources and man-wildlife, to prepare a national-level overview of land use patterns and changes for the formulation of national policy, and to generate environmental impact statements and assess future impacts on environment condition. The physical state of the ground surface is described as land cover [8], i.e., forest, grassland, barren land, sandy land, etc., while on the other hand, the land use reflects human activities [9], i.e., residential area and industrial area, etc.

Land use influences land cover, and land cover shift over a time period directly influences land use. Changes in land cover due to land use may not always reflect a deterioration of the land. However, changing land-use patterns, induced by various socioeconomic reasons, result in the land cover shift that affects radiation and water budgets, trace gas emissions, biodiversity of a particular region, and other processes that contribute to climate and biosphere change [10]. Detection of change and shift over land is critical for better knowledge about landscape dynamics over a specified time period with sustainable management. Land use/cover changes are an increasing and continuous process primarily driven by natural occurrences and anthropogenic activities, resulting in changes that influence natural ecosystems [11, 12]. Identifying the pattern of landscape, transformations, and interactions between mankind activities and natural phenomena is not easy for effective management and decision making for land resources. Satellites, which are used to gather information about earth resources, are now extremely relevant and valuable for detecting change and shifting over LULC research [13, 14].

2 Study Area

The division of Kumaon is one of two regions and administrative divisions of the Indian state of Uttarakhand, which is located in the Central Himalaya. It covers an area of roughly 20,397 km^2 and is primarily hilly with a forest-dominated ecology. The study area, viz., district Bageshwar is one of the districts of the state of Uttarakhand situated in the part of Lesser Himalayan in India. It extends in between 29°40′ and 30°20′ N latitude and 79°25′ and 80°10′ E longitudes and encompasses an area of about 2231.31 km^2 (Fig. 1). The district's geological structure is highly complicated due to recurrent tectonic disturbances induced by distinct orogenic cycles. Valdiya

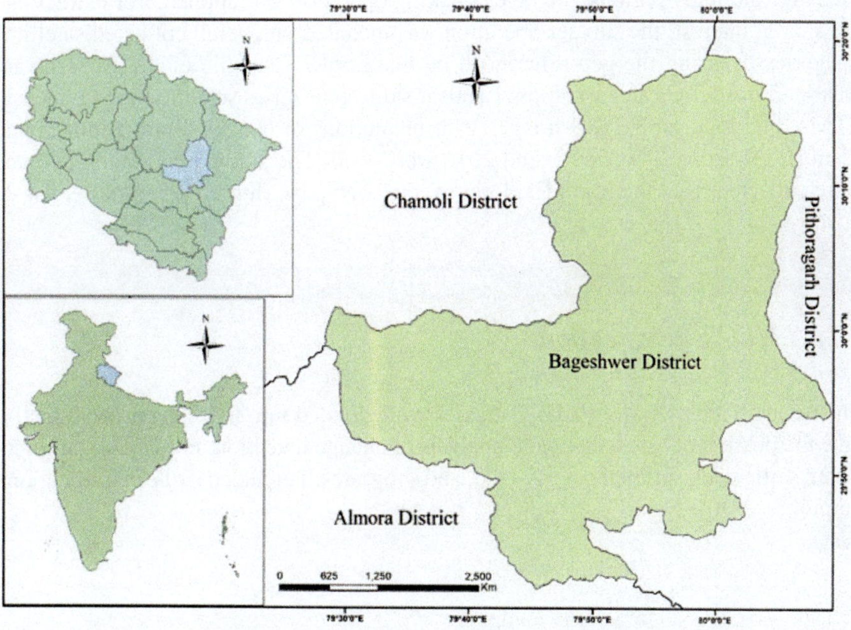

Fig. 1 Location map of the study area, viz., Bageshwar District, Uttarakhand (India)

(1980) conducted comprehensive geological and structural mapping in the region [15]. The total area of Bageshwar is 2231.31 km^2, in which the share of the rural area is approximately about 2225.50 km^2, and 5.81 km^2 is about the urban area. The district's total population is 259,898 people as per the census of India 2011 [16]. The study region in the Bageshwar district has a temperate to sub-humid climate. The northern portion of the area is sub-zero temperature practically all year, but the centre and parts of the southern area are fairly humid and warm. Except for the northern portion, which has a cool temperate temperature, the district has a tropical to sub-tropical and sub-humid climate. The coldest month is January, about ten °C of maximum mean temperature and two °C of minimum mean temperature. In the northern portion of the area, temperatures can dip below –6 °C in January and February. The hottest month is June, with an average maximum of 25 °C and the lowest 15 °C temperatures. The highest recorded temperature in the district was about 43 °C in May 2003, while the lowest recorded temperature was about 4 °C (January 2003).

3 Methodology

For conducting this study, ERDAS Imagine remote sensing application (Leica Geosystems, Atlanta, U.S.A.) is used to import satellite data sets and also used to process the satellite image to generate a false colour composite (FCC). To create FCCs for the study regions, the layer stacking operation was applied. For extracting the area of interest, the sub-set operation was included on overall collected satellite imageries by using the geo-referenced outline border shapefile of the Bageshwar District. For LULC categorization, Landsat's different series satellite MSS, Landsat ETM+, and Landsat 8 TIRS images with resolutions of 60, 30, 30 m (Multi), and 15 m (PAN) from 1976, 1999, and 2016 were used. The study area's satellite data were collected from the Earth Explorer official web page (http://earthexplorer.usgs.gov/).

3.1 LULC Classification

The research area's various LULC classes were divided into five categories to facilitate the process of assessment and analysis of change detection. Forest, agriculture, water, settlement, scrubland, grassland, and snow are all included in the classification output.

3.2 Image Classification

The method of supervised classification was performed using the algorithm of maximum likelihood and a band composition of Green, Red, and NIR to select training samples for the feature classes of water bodies and shrubs/grass, while the true composite of blue, green, and red bands was included for collecting the training samples of features classes of settlement, agriculture, and other classes. The training sample contributes to the map's production by identifying a region in the image on the basis of colour given to that category and the spectral homogeneity of the pixel in the designated area.

The two independently classified images were compared after classification to provide a change detection analysis. The matrix table of the "from-to" change class was generated for the post-classification assessment. Finally, this classification proved to be useful since it had the benefit of identifying the type and degree of shift and change that has occurred via pixel-level comparison.

The accuracy evaluation compares the classified imagery against testing samples or collected ground data by going on the field to measure how the classified image is accurate with respect to ground. This results in a matrix table that displays four various varieties of accuracies. The accuracy assessment is one of the initial parts of any image analysis part. The main objective of this part is to validate the classified image with respect to the available testing sets. There are several types of accuracy assessment methods available for validating the correctness of the classified images. In this present study, we adopted the most commonly used accuracy assessment method consisting of consumer, producer, and overall accuracy, generated by an error matrix.

3.3 Magnitude Assessment of Change

The magnitude assessment of change is a measure of how much the LULC size has grown or shrunk. A negative number indicates a reduction in the size of the LULC class, whereas a positive value indicates an increase in the size of the LULC class.

4 Results

The multi-temporal satellite imagery analysis findings are depicted in Figs. 2, 3 and 4, and data are given in Tables 2. Figure 3 depicts land use/cover status, categories and Fig. 4 illustrates the magnitude of change in the significant features classes of the forest, grassland, and settlement.

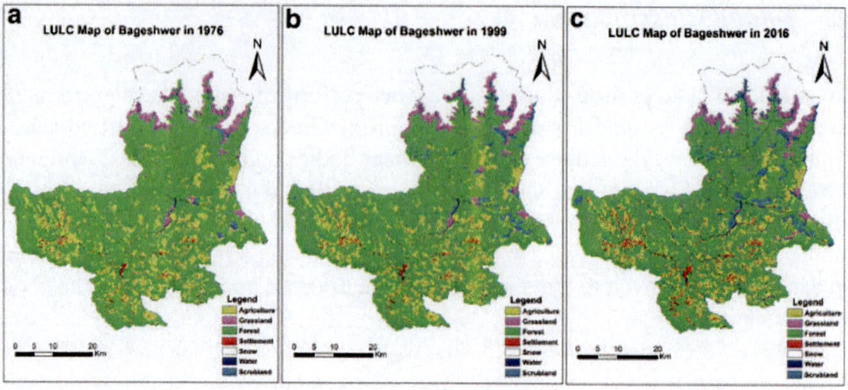

Fig. 2 Land use/cover status of the Bageshwar district in 1976 (**a**), 1999 (**b**) and 2016 (**c**)

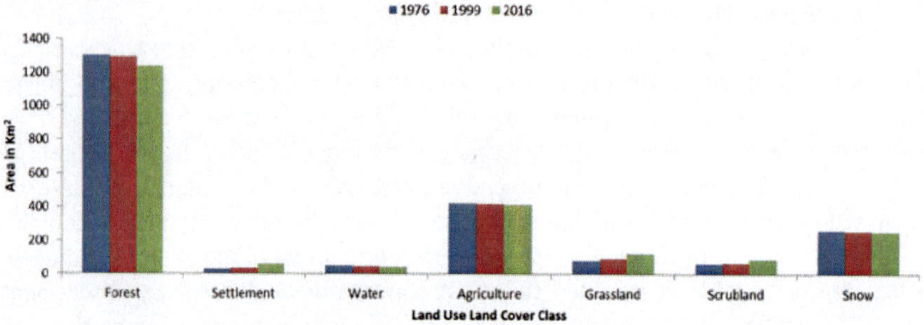

Fig. 3 Graphical representation of LULC statics from 1976 to 2016 of the Bageshwar District of Uttarakhand India

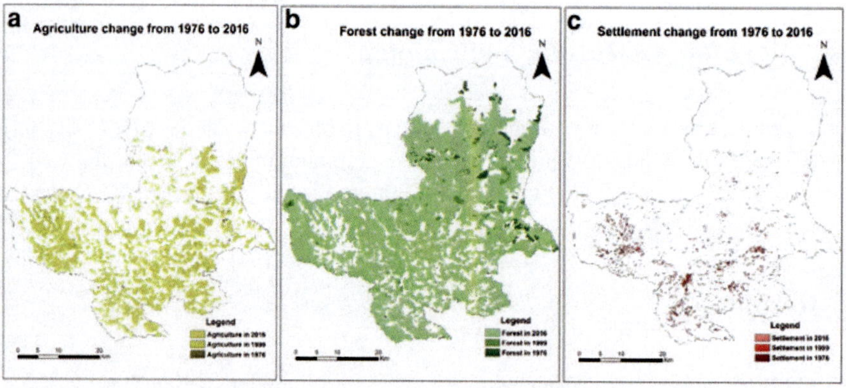

Fig. 4 Changes in pattern of agricultural land (**a**), forest area (**b**) and built up area (**c**) during 1976–2016 in Bageshwar district

4.1 Land Use/Cover Status

The operation of accuracy assessment for the obtained result of the land use/cover classification shows an overall accuracy of 76%, 82%, and 84% for the different years of 1976, 1999, and 2016 respectively (Table 1). The user, producer, and overall accuracies for the different years are also shown in Table 1. The statics of different years LULC is represented in Table 2. The classified result of 1976 shows the area of different features class viz; forest 1306.04 km^2, settlement 32.27 km^2, water 52.69 km^2, agriculture 425.11 km^2, grassland 82.43 km^2, scrubland 66.24 km^2, snow 266.53 km^2 (Figs. 2 and 3; Table 2). The distribution of feature classes in 1999 showed about 1296.26 km^2 for forest cover, which was decreased by 9.78 km^2. The area of feature classes settlement, grassland, and scrubland showed positive changes from 1976 to 1999 by 32.27–32.87 km^2, 82.43–95.06 km^2, and 66.24–70.04 km^2, respectively (Table 2). The feature classes of water, agriculture, and the snow showed

Table 1 Accuracy assessment of the LULC classification for Bageshwar district during the years of 1976, 1999, and 2016

LULC category	1976		1999		2016	
	UA	PA	UA	PA	UA	PA
Forest area	81	86	83	80	89	79
Agricultural land	65	71	74	83	84	81
Grassland/Barren land	77	63	80	83	82	80
Built up	75	79	81	84	86	81
Scrubland	80	87	86	79	76	82
Water bodies	76	80	84	76	81	86
Snow cover	76	69	84	80	89	77
Overall accuracy	76		82		84	

User accuracy is denoted by UA and producer accuracy is denoted by PA

Table 2 Land use land cover statics from 1976 to 2016 of the Bageshwar District of Uttarakhand India

LULC categories	1976		1999		2016	
	km^2	%	km^2	%	km^2	%
Forest area	1306.04	58.53	1296.26	58.09	1232.97	55.26
Agricultural land	425.11	19.05	421.43	18.89	416.41	18.66
Grassland/Barren land	82.43	3.69	95.06	4.26	126.29	5.66
Built up area	32.27	1.45	35.87	1.61	63.04	2.83
Scrubland	66.24	2.97	70.04	3.14	89.36	4
Water bodies	52.69	2.36	49.41	2.21	45.98	2.06
Snow cover	266.53	11.95	263.24	11.85	257.26	11.53

a decrease in the overall land cover from 1976 to 1999 by 52.69–49.41 km^2, 425.11–421.43 km^2, and 266.53–263.24 km^2 respectively.

Figure 3 shows the graphical representation of the change in the area of different feature classes from 1976 to 2016 of the study area. The obtained result from the year 2016 illustrated in Fig. 3 shows that the land use land cover area for the feature class of forest was 1232.97 km^2, 63.04 km^2 for settlement, 45.98 km^2 for a water body, 416.41 km^2 for agriculture, 126.29 km^2 for grassland, 89.36 km^2 for scrubland and 257.26 km^2 for snow cover (Table 2). The positive changes in the area coverage for the different features classes are noticed in the settlement, grassland, and snow cover, while on the other hand, the negative changes are noticed in the features class of forest cover, water, agriculture, and snow cover (Fig. 3).

5 Discussion

5.1 Overall Changes on LULC Pattern

The data in Table 3 and Figs. 4 and 5 show the positive and negative shifts and changes in the study area's ground surface.

Over the previous two decades, the study region's forest has shrunk from 1296.26 km^2 in 1999 to 1232.97 km^2 in 2016, accounting for 0.09% of the overall study area. The settlement grew from about 35.87 km^2 in 1999 to 63.04 km^2 in 2016, a 0.04% growth. The water body has a negative shift from 49.41 km^2 in 1999 to 45.98 km^2 in 2016, accounting for a 0.01% change. Agriculture practices are also decreased from 421.43 km^2 in 1999 to 416.41 km^2 in 2016, which accounts for 0.01% of the total study area. The grassland showed positive growth in the land cover of about 95.06 km^2 in 1999 to 126.29 km^2 in 2016, accounting for 0.05% of the total study area. The scrubland is also increased from 70.04 km^2 in 1999 to 89.36 km^2 in

Table 3 The overall loss and gain (in km^2) in LULC categories during different time periods in the Bageshwar District of Uttarakhand

LULC categories	1976–1999	1999–2016	1976–2016	Rate of change per year (1976–2016)
Forest area	(−)9.78	(−)63.29	(−)73.07	(−)2.09
Agricultural land	(−)3.68	(−)5.02	(−)8.7	(−)0.25
Grassland/Barren land areas	12.63	31.23	43.86	1.25
Built up area	3.6	27.17	30.77	0.88
Scrubland	3.8	19.32	23.12	0.66
Water bodies	(−)3.28	(−)3.43	(−)6.71	(−)0.19
Snow cover	(−)3.29	(−)5.98	(−)9.27	0.26

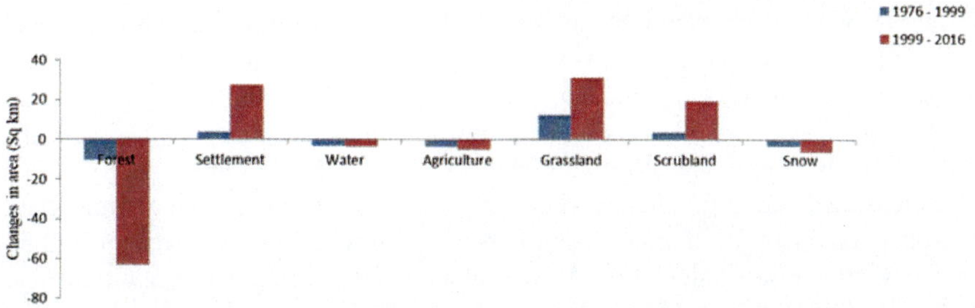

Fig. 5 Graphical representation of loss and gain in LULC cover categories during 1976–1999 and 1999–2016 of Bageshwar district of Uttarakhand in India

2016, which accounts for 0.03% of the total study area. The snow has decreased from 263.24 km^2 in 1999 to 257.26 km^2 in 2016, accounting for the same as scrubland 0.01% of the study site.

The study shows the change in classes where agriculture, forest, water, and snow area decreases from 1976 to 2016 and other classes settlement, grassland and scrubland are increase. Therefore, studying changes has become essential in sustainable development, especially when systematic approaches and techniques like remote sensing are available [17].

5.2 *Pattern of LULC Over the Time Period of 1976–1999*

Figures 4 and 5 show the magnitude of change from 1976 to 1999. The major positive changes occurred in the feature classes of settlement, grassland, and scrubland, which were about 3.6 km^2, 12.63 km^2, and 3.8 km^2, respectively (Table 3). On the other hand, the major negative changes were about 9.78, 3.28, 3.68, and 3.29 km^2 for the feature classes of the forest, waterbody, agricultural land, and snow cover. The greatest positive change in the area of feature class was noticed in the grassland (12.63 km^2), and forest cover showed the maximum negative changes over the study site, which was about 9.78 km^2 (Figs. 4 and 5; Table 3).

5.3 *Pattern of LULC Over the Time Period of 1999–2016*

During the time period between 1999 and 2016, the forest cover and grassland cover showed maximum changes of −63.29 km^2 and 31.23 km^2, respectively. The major positive changes were noticed in the features classes of settlement and scrubland, which were about 27.17 and 19.32 km^2. The negative LULCC were found in the feature classes of water body, agriculture, and snow cover near about 3.43 km^2, 5.02 km^2, and 5.98 km^2, respectively (Figs. 4 and 5; Table 3). All LULC classes except forest, water, agriculture, snow cover has shown positive growth in the land over time, the built-up feature class were increased from 1976 to 2016 by 30.77 km^2, about 49.86 km^2 increased grassland, and the features class of scrubland got increased by 23.12 km^2, between 1976 and 2016 while agriculture and water have witnessed a decrease in the overall land cover area about 8.7 km^2, and 6.71 km^2 respectively due to increasing demand for shelter within the study area. The second main important reason for this change is that several farmers in the study region have given up farming to pursue other minor enterprises. As a result, part of the abandoned agricultural land has been converted to shrub/grass.

6 Conclusion

The current study examined and tracked changes in the LULC pattern over the Bageshwar district of Kumaun Himalaya, India, using LANDSAT-MSS, LANDSAT-7 ETM+, and LANDSAT-8 TIRS from 1976 to 2016. The study is useful for analyzing the pattern of change in the study site over time and predicting the future pattern of the LULC pattern over the study site. The analysis indicates that the most important land feature class in the study region is forest cover. The study site's significant variations in LULC classes were seen in the characteristics of forest cover, built-up land, grassland, and scrubland, which were about 73.07 km^2, 30.77 km^2, 43.86 km^2, and 23.12 km^2, respectively. The current study uses the geospatial technique, which is currently one of the most prominent technologies for LULCC analysis, which is not feasible with other standard mapping methods. These technologies enable change detection in less time, at a lower cost, and with better precision. The study's research and conclusions have significant policy implications for sustainable land use/cover practices in the Kumaun Himalaya.

Acknowledgements The authors thank their parent organisations: Soban Singh Jeena University, Almora-Uttarakhand, Kumaon University, Nanital-Uttarakhand, Gurukula Kangri (Deemed to be) University, Haridwar-Uttarakhand, India for providing the necessary facilities to carry out this research work. Thanks are due to the Advanced Research Laboratory in Environmental Engineering and Faecal Sludge Management (ARLEE-FSM) of Civil Engineering Department, BITS Pilani, India for all support given by the research staff of the laboratory.

References

1. Munsi, Madhushree, et al. "A landscape approach for quantifying land-use and land-cover change (1976–2006) in middle Himalaya." Regional Environmental Change 10.2 (2010): 145–155.
2. Sekertekin, A., A. M. Marangoz, and H. Akcin. "Pixel-based classification analysis of land use land cover using Sentinel-2 and Landsat-8 data." *Int. Arch. Photogramm. Remote Sens. Spat. Inf. Sci* 42 (2017): 91–93.

3. Kala, Devkant, and S. C. Bagri. "Barriers to local community participation in tourism development: Evidence from mountainous state Uttarakhand, India." *Tourism: An International Interdisciplinary Journal* 66.3 (2018): 318–333.
4. Pandey, Arvind, et al. "Impact of climate on vegetation in Pindari watershed of Western Himalayas, Kumaun, India, using spatiotemporal analysis: 1972–2018." Environmental Science and Pollution Research (2022). https://doi.org/10.1007/s11356-022-19711-z
5. Mohammad, Nedal A., Sharifah Mastura SA, and Johari Mat Akhir. "Land use evaluation for Kuala Selangor, Malaysia using remote sensing and GIS technologies." *Geografia-Malaysian Journal of Society and Space* 3.1 (2017).
6. Anderson, James Richard. *A land use and land cover classification system for use with remote sensor data.* Vol. 964. US Government Printing Office, 1976.
7. Kumar, Kunal, R. K. Sahu, and Shankar Yadav. "Assessment of land use/land cover change using geo-informatics in catchment of Burhi Gandak River, Bihar." *Journal of Agricultural Engineering* 53.3 (2020): 377.
8. Turner, Billie L., and William B. Meyer. "Global land-use and land-cover change: an overview." *Changes in land use and land cover: a global perspective* 4.3 (1994).
9. Goldewijk, Kees Klein, and Navin Ramankutty. "Land cover change over the last three centuries due to human activities: The availability of new global data sets." *GeoJournal* 61.4 (2004): 335–344.
10. Riebsame, W. E., et al. "Integrated modeling of land use and cover change." *Bioscience* 44.5 (1994): 350–356.
11. Alonso-Pérez, Fernando, et al. "Land cover changes and impact of shrimp aquaculture on the landscape in the Ceuta coastal lagoon system, Sinaloa, Mexico." *Ocean & Coastal Management* 46.6–7 (2003): 583–600.
12. Pandey, Arvind, et al. "Assessment of snout analysis of Himalayan glaciers: Impact studies on Pindari, Kafni, Sundardhunga and Baljuri Base Camp Glaciers". *Environmental Monitoring and Assessment* 194.338 (2022). https://doi.org/10.1007/s10661-022-09925-6
13. Yuan, Fei, et al. "Land cover classification and change analysis of the Twin Cities (Minnesota) Metropolitan Area by multitemporal Landsat remote sensing." *Remote sensing of Environment* 98.2–3 (2005): 317–328.
14. Brondizio, Eduardo S., et al. "Land use change in the Amazon estuary: Patterns of Caboclo settlement and landscape management." *Human Ecology* 22.3 (1994): 249–278.
15. Valdiya, K. S. "The two intracrustal boundary thrusts of the Himalaya." *Tectonophysics* 66.4 (1980): 323–348.
16. INDIA, POMPI. "Census of India 2011 provisional population totals." *New Delhi: Office of the Registrar General and Census Commissioner* (2011).
17. Pandey, Arvind, et al. "Application of remote sensing in alpine grasslands cover mapping of western Himalaya, Uttarakhand, India." *Environmental Monitoring and Assessment* 193.4 (2021): 1–14. https://doi.org/10.1007/s10661-021-08956-9

A Review of Municipal Solid Waste: Its Generation, Composition, Impacts, Management and Challenges in Urban Areas with Special Focus on India

Kanchan Deoli Bahukhandi and Saswat Ollemman

Abstract Municipal Solid waste (MSW) is a rapidly growing issue in the world of today along with countless number of other issues as well in all grounds. The trend of urbanization, growth of industrialization, advancements in technology etc. seen in urban areas of the world are some of the perks that attract the population towards living an urban life for more income, better comfort and improved lifestyles. But the same has led to a number of issues over the years as well, including a very high generation rate of MSW (with varied compositions) due to certain and ever-growing urban lifestyles and also ever-increasing demands of the population to adjust to the modern way of living. Today, these generation rates have reached such high levels that proper and effective management has become a huge challenge for the urban areas of the world. The developing nations of the world like India are suffering quite a lot due to it compared to certain developed nations despite generating comparatively lesser amounts of MSW, reasons being both economical as well as various social grounds too. An effort to describe MSW along with its generation and composition figures in various parts of the world has been made through this paper keeping a primary focus towards India. Studies have also been made about the various impacts MSW could have towards people, other living beings and also the environment along with the means for its management. Different steps like: waste segregation, collection, transportation, recycling, treatment and disposal are involved in the process of MSW management and the same needs to be executed effectively in order to call the MSW management which has been or is being performed as a proper one. But doing so is easier said than done as various challenges are faced along the way towards MSW management of an area/city/country. Several factors may be responsible for it ranging from economic, political and even social, all of which have been discussed by means

K. D. Bahukhandi (✉)
Sustainability Cluster (HSE and Civil Engineering), University of Petroleum and Energy Studies (UPES), Bidholi, Dehradun, India
e-mail: kanchan@ddn.upes.ac.in

S. Ollemman
Department of HSE and Civil Engineering, University of Petroleum and Energy Studies (UPES), Bidholi, Dehradun, India

K. D. Bahukhandi et al. (eds.), *Environmental Pollution and Natural Resource Management*, Springer Proceedings in Earth and Environmental Sciences,
https://doi.org/10.1007/978-3-031-05335-1_17

of this review work. Once again, special attention has been given towards the nation of India while doing the same.

Keywords Municipal solid waste (MSW) · MSW management · Urbanization · Generation and composition

1 Introduction

The wastes which are generally solid in nature and generated as a result of certain deeds of human beings and also animals could be termed as Municipal Solid Wastes (MSW). They are usually regarded as futile and having no purpose in daily life [13]. Various terms like "trash" or "garbage" are seen to be in common use to denote Municipal Solid Waste which might include even several categories of waste products ranging from durable, non-durable, packaging and other household or even office refuses. MSW could also range from the refuse generated by various household jobs to certain industrial/manufacturing ventures. Some examples are: plastic cups/plates, wrappers, food wastes etc. while other types like industrial, constructional or toxic wastes do not fall under this category [1, 18]. The rise of industrialization has been a major reason for the increase in the solid waste amounts in developing countries like India [38]. Vast rise in industrialization has led to the movement of more and more of the population from rural to urban areas thus leading to the increase in the amounts of MSW which is produced and the same is expected in the future as well [8, 9, 12, 92, 22]. The amounts of solid wastes produced in China has increased due to rise in population over the decades and also due to its economic development as a result. The rise in the amounts of solid wastes produced hasn't been observed in any other nation such as in China [7, 32, 48].

According to Hoornweg and Bhada-Tata [49], urbanized areas of the world generate as much as over a billion tonnes of MSW annually with that figure predicted to rise up to over 2.2 billion by the year 2025 with no signs of lowering down anytime soon. Such kind of results are not just adversely affecting the lands and water bodies by causing pollution, but certain other aspects as well. Methane gas could be produced in higher fractions from solid waste products with is a Green House Gas thus resulting in air pollution and a contributor towards global warming as well. Along with that, un-managed or poorly managed MSW could lead to issues like floods in the cities, adverse health effects such as dengue, diarrhoea, respiratory issues etc. According to [10, 74, 77, 119, 125], the issues related to management of MSW in developing countries are enhanced by factors like extreme rates of generation which is a resultant of sharp increases in population and the preferred practice of urban dwelling by people in today's world. Along with that, lack of education and poor economic conditions leading to insufficient financial aid towards waste management also act as contributing factors for the same.

The generated solid wastes in households could vary up to nine different types with food/vegetable wastes being the most common and highest generated category and that a number of issues could be faced in proper management of the same by the municipal corporations who bear the responsibility of MSW management in urban cities. This is a common issue in highly populated developing countries [104]. The biggest issue related to MSW management in developing nations is lack of proper means for treatment and discarding of the gigantic amounts of solid wastes that are produced in their urban areas. Along with that, lack of ample facilities for collection and means of transportation of the wastes further adds to the already existing issues. As a result of all, the management and discarding of the wastes is done in various unscientific or rather absurd manners which in turn affects the health and well-being of residents of the city and also its surrounding environment [44, 59, 62, 87]. The above-mentioned issues are the prime reasons for the designing of a proper waste management system in developing nations to satisfy both economy and environment being an extremely challenging job [3].

2 Review of Literature

Sharholy et al. [96], have highlighted how generation of municipal solid waste in developing countries like India is rapidly increasing stating various reasons behind it. It has also been claimed that such wastes can adversely affect human health if not managed effectively and doing so by various irrational means like dumping in open grounds surrounding public areas. It has also been claimed that as much as 90% wastes are being managed by such manners adversely affecting human health. Various treatment and disposal methods of MSW have been discussed while also throwing light on waste management rules existent in India. Suggestions on spreading awareness among general public encouraging participation on MSW management via NGOs have also been made. Shekdar [98], has primarily focus his study about waste management in Asian countries arguing that the various approaches made towards management of MSW should be adaptable with the surroundings or environment in which they are performed as different areas with different conditions require different management techniques. Various situations have been analysed throughout the study to know the reasons for high waste generation rates in Asian countries which are quite complex and varying in terms of culture. Finally, discussions on various integrated approach methods have been discussed which could be followed by all Asian countries in order to achieve sustainable solid waste management while also pointing towards the importance of "sustainable development".

Imam et al. [54], focussed on the study in management of solid waste in the city of Abuja, Nigeria. Rates of waste generation, its increase and the problems arising due to it in the entire nation of Nigeria with prime focus on Abuja have been discussed along with the means of managing wastes that are followed. Finally, it has been argued that improvements are required in the fields of waste management also suggesting that developing certain management techniques like waste composting plants (small-scale) could also turn out to be a means of employment and a source of income for the public. Zhang et al. [126], made a review study on the management of MSW in the largest population of the world China having put forward generation rates and composition of wastes that are seen in China. Along with that, the various methods for management of MSW in the nation and issues in the management processes have also been discussed. At last, certain ways or methods have also been studied and put forward about potential opportunities of future improvemens in the management of MSW in China.

Minghua et al. [70], have also made a study on solid waste management methods followed in China with prime focus on the "Pudong New Area". The different solid waste management techniques followed in Pudong have been discussed in a very detailed manner with special attention towards the collection, transportation and treatment techniques of the generated waste. It was brought into light that a geographic information system (GIS) was also in place in the waste management system of Pudong. This provided easier access of routes for trucks and helped reduce fuel consumption. A prominent focus upon the separation of source of the wastes to reduce the requirement for its disposal and also the need to utilize more and more of recycled items have been recommended. Wilson et al. [124], analysed the management of waste in 20 different cities from varied continents all around the world and made a comparative study by means of integrated and sustainable waste management (ISWM). Comparative study about the area, population and its growth rate etc. of the 20 cities along with population size, GNI per capita etc. of the respective countries to which they belonged was done thus recognizing the respective income levels of the different cities (categorized as: High, Upper-middle, Lower-middle and Low). Per capita waste generation and also its composition were calculated based on the income levels. Certain major indicators or dimensions were made use to carry out the comparative study of waste management and hence concluding that a certain type of management system doesn't work for all and that the prime strength is the usage of diverse methods.

Tanskanen [106], discussed the concept of integrated municipal solid waste management (MSWM) by usage of computer-based model designed for the Helsinki Metropolitan Area (HMA) with the goal of designing strategies for waste separation to achieve the intended recovery rate of municipal solid waste that was in place for the country of Finland. The developed HMA model, consisting of different stages (six to be precise) and certain formulas to calculate various components, proved to be fruitful in achieving the intended goals, thus qualifying to be an ideal tool for the integrated municipal solid waste management planning strategies. Demirer et al. (2005), performed a case study about the life cycle assessment of municipal solid waste management in the Turkish capital of Ankara by means of integrated waste management (IWM) model. Certain scenarios were under consideration in the course of study and the various types of wastes produced as well as the amounts of waste managed in different scenarios were compared. This helped attain the most suitable system for MSW in the city of Ankara in terms of its effect towards the environment.

3 MSW Generation and Composition: A Global Scenario

According to Guerrero et al. [43], the management of solid wastes in urban areas is a great challenge for officials responsible for its management in city areas of developing countries due to various reasons such as: increment in waste generation, lower capital available with the authorities and complexity in the processes involved in waste management, improper knowledge and understanding of factors which can affect the waste management process etc. A study depicting the GDP and solid wastes produced from certain cities of various countries along with its area of origin was presented and is shown in Table 1.

In 2015, the total population of the earth was predicted to be about 7.3 billion out of which the total strength that resided in urban areas was calculated to be around 49% [112]. Due to this, the exact amounts of MSW produced globally is very complicated to be known precisely. According to World Bank data [61], in the year 2016 the amount of MSW that was produced globally ranged to accounted for around 2 billion tonnes. Most of that generated waste (about 33%) was managed in an unsuitable manner which affected both people and environment. Around 115 tonnes of organic waste are produced by just 37 megacities in the world every year. Urban areas of the world play vital roles in both the development as well as production of maximum amounts of MSW all over the world. An estimate of total waste generation performed by World Bank is shown in Figs. 1a–b and 2.

In Canada, the households were reported to generate around 12.9 million tonnes of MSW in the year of 2008, out of which about 8.5 million tonnes were sent for disposal while the remaining 4.4 million tonnes were recycled. Due to this, the disposal rate of MSW reduced by 4% between the years of 2006–2008 [72]. The total amount of MSW produced in USA amounted to about 238 million tonnes accounting for about 2.02 kg per capita/day in the year 2015. Out of this, about 28% of the waste produced was organic forms of waste [89]. According to a US-EPA report (EPA 2021), the total MSW generation in USA increased to 292.4 million tonnes in 2018 or the per capita produce increasing to 2.2 kg/day. 69 million tonnes of waste out of this were sent for recycle process. Like other areas of the world, the MSW generated in USA came in varied compositions of different products or items. The MSW which was generated with varied compositions in the year 2018 has been shown in Fig. 3. The European Union has 27 member nations generating wastes of different forms and in different quantities based on varied factors like total consumption by its population and also the economic conditions of the different nations. It could also depend upon the ways in which waste collection and management is done. Based on such factors the MSW produced vary in amounts such as 280 kg per capita in the nation of Romania while countries like Denmark producing up to 844 kg per capita in the year 2019 [35]. Figure 4 shows the total MSW generated in different nations of the European Union as well as comparisons between the difference in amounts produced in 2005 and 2019.

The total generated waste in Australia between 2016 and 2017 amounted to be around 67 million tonnes out of which 54 million tonnes accounted for core waste.

Table 1 Waste generation rate and area of origin in various cities with the numbers indicating different areas as follows: 1: household, 2: offices, schools, 3: construction, 4: health care, 5: agriculture, 6: industry and 7: shops

Continent	Country	GDP (In US Dollars)	Year of study	City	Waste origin arriving at the official disposal site	Waste generation rate (kg/capita/day)
Africa	Ethiopia	344	2009	Adis Ababa	1,2,4,6,7	0.32
	Kenya	738	2009	Nakuru	1,2,3,4,5,6,7	0.50
	Malawi	326	2009	Lilongwe	1	0.50
	South Africa	5786	2009	Pretoria	1,2,3,4,7	0.65
	South Africa	5786	2009	Langeberg	1,3,4,5,6,7	0.65
	South Africa	5786	2009	Emfuleni	1,3,6	0.60
	Tanzania	509	2010	Dar es Salam	1,2,4,5,6,7	0.50
	Zambia	985	2010	Lusaka	1,2,3,4,6,7	0.37
Asia	Bangladesh	551	2007, 2008, 2009	Gazipur	1,4	0.25
	Bhutan	1805	2010	Thimphu	1,2,3,7	0.54
	China	3744	2010	Beijing	1,3,4,7	0.80
	India	9232	2010	Doddaballapur	1,2,3,6,7	0.28
	Indonesia	2349	2009, 2010	Banda Aceh	1,4	0.90
	Indonesia	2349	2009, 2010	Ambon	1,4	0.90
	Indonesia	2349	2010	Jogjakarta	1,2,5,7	0.90
	Nepal	364	2007	Kathmandu	1,2,6,7	0.35
	Pakistan	495	1995	Lahore	1,2,6,7	0.84
	Philippines	1995	2009	Quezon City	1,2,3,4,7	0.67
	Sri Lanka	2068	2010	Balangoda	1,2,3,4,6,7	0.83
	Sri Lanka	2068	2010	Hambantota	1,2,3,4,7	0.81
	Thailand	4043	2009, 2010	Bangkok	1,2,3,4,6,7	1.10
	Turkey	8215	2010	Kutahya	1,2,4,6,7	0.60
	Turkey	8215	2010	Bitlis	1,2,3,4,5,6,7	0.90
	Turkey	8215	2010	Amasya	1,2,4,7	1.20
Central & South America	Costa Rica	4084	1985, 1995	Cartago	1,2,3,4,5,7	0.7–0.8
	Costa Rica	6386	2011	San Jose	1,2,3,4,6,7	1.10
	Costa Rica	3370	1991	Talamanca	1,7	0.30
	Costa Rica	4084	1992, 1995	Tarcoles	1,7	0.30–0.50
	Costa Rica	5529	2001	Tuis	1,7	0.30

(continued)

Table 1 (continued)

Continent	Country	GDP (In US Dollars)	Year of study	City	Waste origin arriving at the official disposal site	Waste generation rate (kg/capita/day)
	Ecuador	1771	1995	Pillaro	1,7	0.50
	Ecuador	1771	1995	El Carmen de los Colorados	1,7	0.50
	Nicaragua	1069	2008, 2009, 2010	Managua	1,2,3,4,5,6,7	0.48
	Nicaragua	1069	2009, 2010	Masaya	1,2,4,7	0.40
	Peru	4447	2008, 2009, 2010	Canete	1,2,3,4,5,6,7	0.47
	Suriname	5888	2008, 2009	Paramaribo	1,7	0.47
	Suriname	5888	2008	Asidonhopo	–	0.28

Source Guerrero et al. [43]

Out of that total amount of core waste, about 13.8 of municipal solid waste was produced from different areas in different parts of the country. This amounted for around 560 kg per capita of waste which was generated (Grant et al. 2018). A sharp increase of 145% in waste generation rate has been observed in Australia between the years of 1997 to 2012 which was much higher than the rate of population increase [6].

4 Waste Generation in India

The urban dwelling population of India is in the range of over 370 million which accounts for over 31% of the total population of the country that was estimated to be about 1.2 billion [17]. With a population strength accounting for 18% of the total population of the entire world, India is predicted to overcome China in becoming the most populated nation in the world by 2022 and by 2050, the figures predicted to be as high as 1.6 billion [81]. India being second largest populated nation of the world and with an ever-increasing population has seen an increase in the rates of MSW generation in recent years. An estimated generation of over 147,000 metric tonnes (MT) of MSW was calculated from over 84,000 wards in January of 2020 by India. The figures for MSW generation by India are predicted to increase for in future as: 276,342 tonnes per day (in 2021); 450,132 tonnes per day (in 2031); 1,195,000 tonnes per day (in 2050). An increase of 1.3% has been observed in the per capita waste

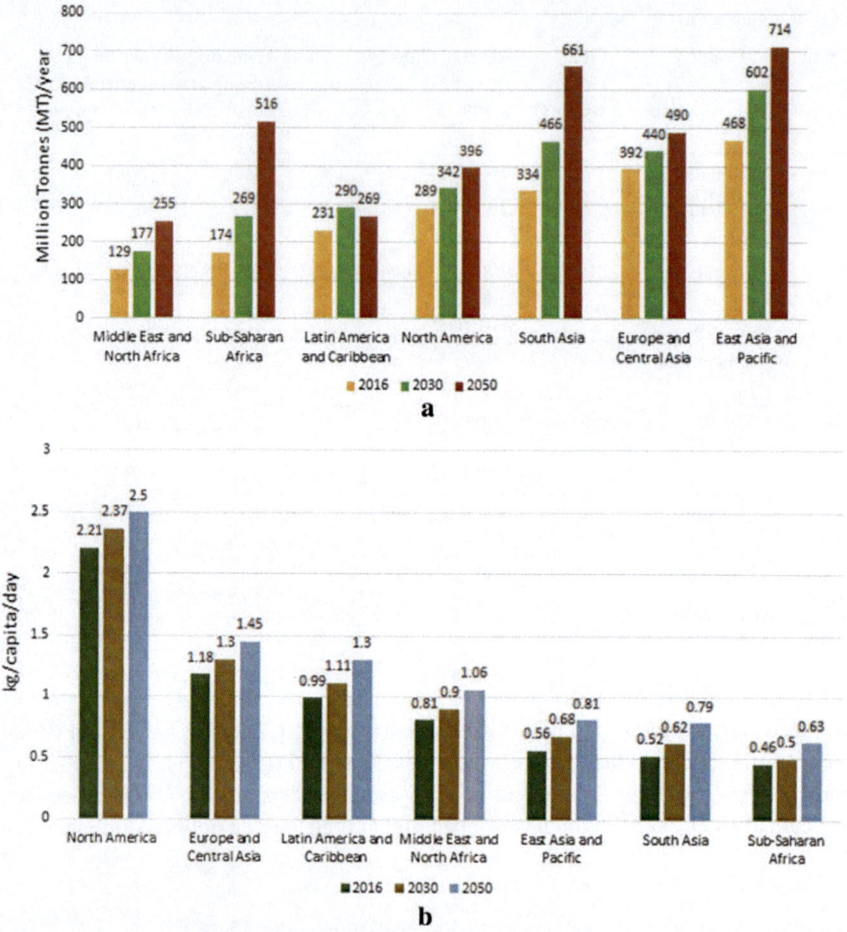

Fig. 1 **a** Global waste generation figures. (*Source* Kaza et al.[61]), **b** Global Waste Generation per capita figures. (*Source* Kaza et al. [61])

generation ate in India per annum [100]. Waste Generation in India was estimated as 100,000 MT for the year 2000 as per the Ministry of Urban Development. Survey conducted by the Central Pollution Control Board over the period of 1999 to 2016 in various parts of India has shown the amounts of waste which was generated by various cities/urban areas of the country, some of which has been provided in Table 2 [20] (Fig. 5).

The MSW that is generated from a particular city, state or even area can be linked with its population. In India, the Northern portion has been found to contribute the most (around 30%) whereas the Eastern portion has been found to contribute the least (around 17%) of MSW generated in the entire country with figures of its components varying with different classes of cities [5].

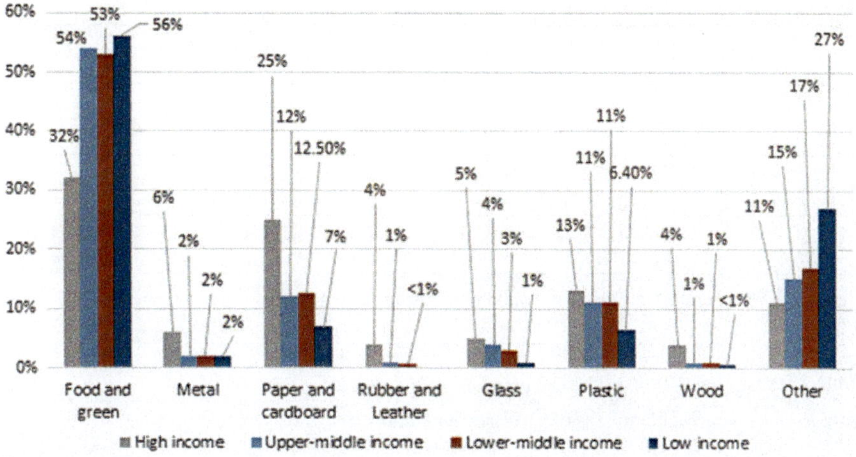

Fig. 2 Compositions of Waste in different income categories. (*Source* Kaza et al. [61])

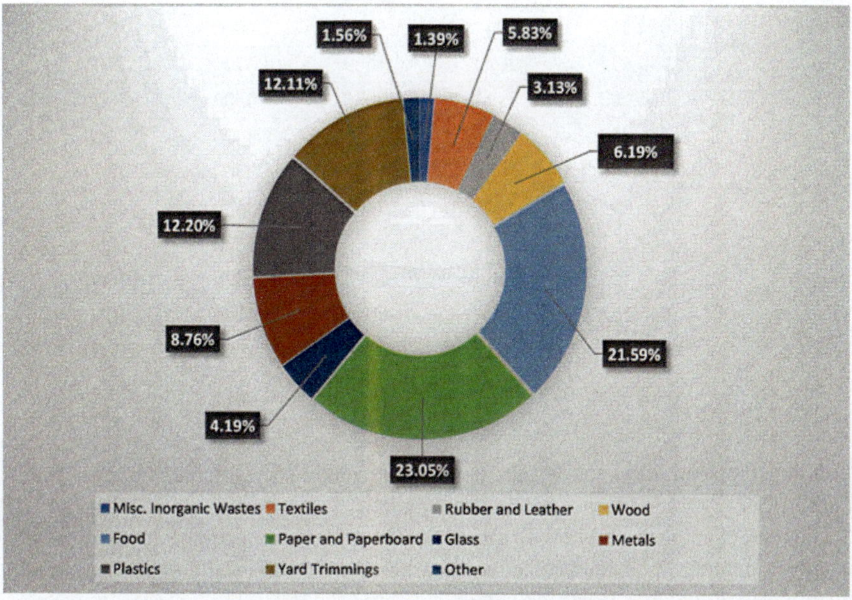

Fig. 3 MSW Generated in 2018 by USA with varied compositions. (*Source of data* EPA 2021)

The solid wastes containing organic fragments that could be decomposed via microorganisms can be termed as compostable wastes [94]. A recycling process is one where previously utilized material (waste products) could be processed into newer products considering that the life cycle of the material does not get aborted

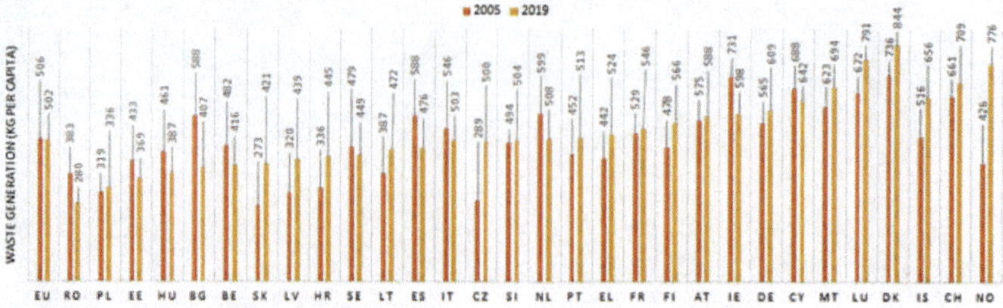

Fig. 4 MSW generated in European Union Nations. (*Source* Eurostat [35])

Table 2 Waste generation in 11 most populated cities of India

City	Population (as of 2011)	Waste generation			
		1999–00 (TPD)	2004–05 (TPD)	2010–11 (TPD)	2015–16 (TPD)
Mumbai	12,442,373	5355	5320	6500	11,000
Delhi	11,034,555	400	5922	6800	8700
Bangalore	8,443,675	200	1669	3700	3700
Chennai	7,088,000	3124	3036	4500	5000
Hyderabad	6,731,790	1566	2187	4200	4000
Ahmedabad	5,577,940	1683	1302	2300	2500
Kolkata	4,496,694	3692	2653	3670	4000
Surat	4,467,797	900	1000	1200	1680
Pune	3,124,458	700	1175	1300	1600
Jaipur	3,046,163	580	904	310	1000
Lucknow	2,817,105	1010	475	1200	1200

Source of data CPCB [20]

upon consumption and such wastes could be termed as recyclable wastes (Ramos et al. 2013) (Figs. 6, 7, 8 and 9).

The total generation of MSW in the city of Pune is estimated to be about 1600 MT per day which could be categorized into certain types but broadly as biodegradable and non-biodegradable waste as shown in Table 4 [101]. The average figures of MSW generation of the country are around 450 gms/per capita/day compared to over 1 kg/per capita/day generated by many developed nations. The waste generated by India also has the capability of being a potential source of (electrical) energy with potential up to 500 megawatts (MW) of electricity generation which is predicted to keep increasing in the future [110] (Fig. 10).

Fig. 5 Top 10 most waste generating states of India [here, MT = Metric tonnes]. (*Source of data*: [100])

Table 3 MSW components with compositions in different classes of cities

City or Region	MSW (TPD)	Compostables (%)	Recyclables (%)	Inerts (%)	Moisture (%)	Cal. Value (MJ/kg)	Cal. Value (kcal/kg)
Metros	51,402	50.89	16.28	32.82	46	6.4	1,523
Other cities	2,723	51.91	19.23	28.86	49	8.7	2,084
East India	380	50.41	21.44	28.15	46	9.8	2,341
North India	6,835	52.38	16.78	30.85	49	6.8	1,623
South India	2,343	53.41	17.02	29.57	51	7.6	1,827
West India	380	50.41	21.44	28.15	46	9.8	2,341
Overall Urban India	130,000	51.3	17.48	31.21	47	7.3	1,751

Source Annepu [5]

5 Impact of Municipal Solid Waste Generated

So far, details on the enormous amounts of solid wastes that are generated or were generated at certain periods of time has been observed. Such amounts of solid waste

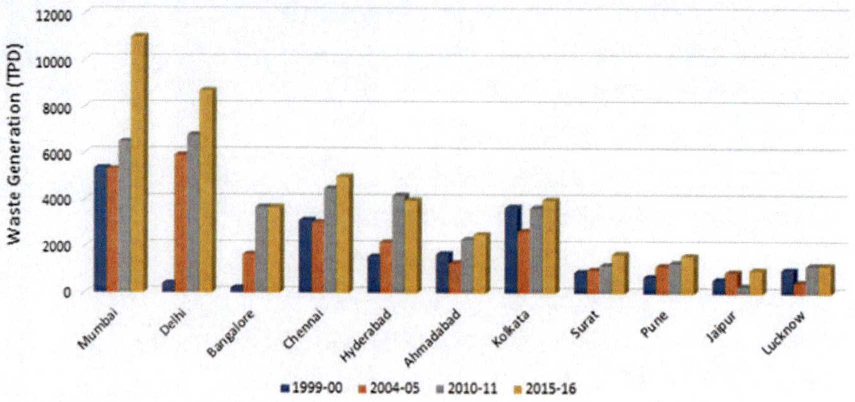

Fig. 6 Comparison of total waste generated over different time periods in the top 11 most populated cities in India (as per Table 2)

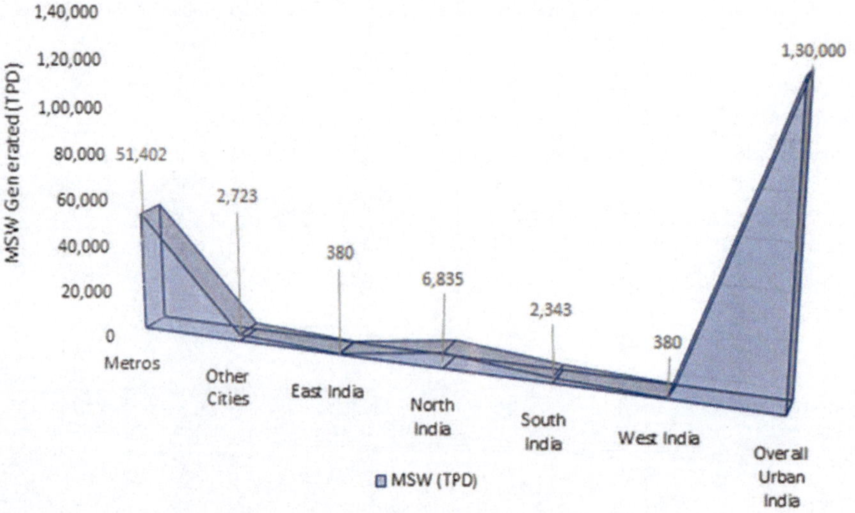

Fig. 7 Comparison of MSW generated in different classes of cities (as per Table 3)

could bring about adverse effects to the public, the surroundings as well as atmosphere if not fruitfully and efficiently managed. Effects of MSW could be varied and may depend mostly on the amounts generated and also how it is managed or what techniques are being utilized for its management. The study put forward by Vergara and Tchobanoglous [118] lists the major impacts that solid waste could have as follows:

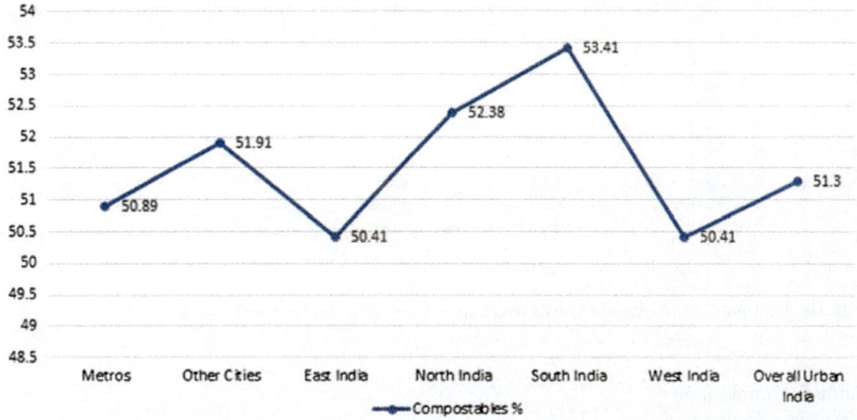

Fig. 8 Comparison of % compostables in different classes of cities (as per Table 3)

Fig. 9 Comparison of % recyclables in different classes of cities (as per Table 3)

Table 4 Types of MSW generated in Pune city

Biodegradable waste (720 MT/D)	Non-biodegradable waste (880 MT/D)
Organic waste-400 MT	Inorganic waste-560 MT
Household waste-320 MT	Inert waste–320 MT

Source Soni et al. [101]

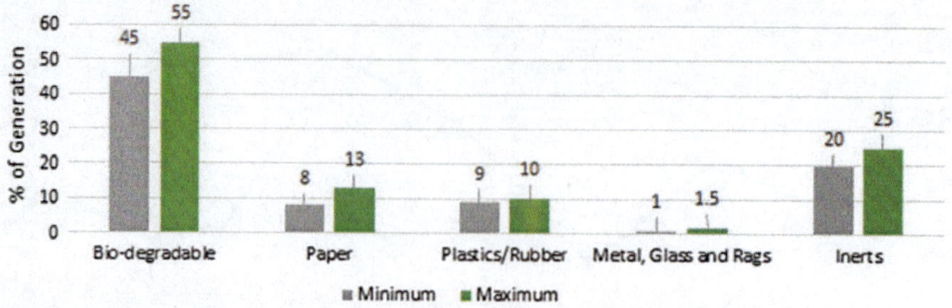

Fig. 10 Comparison of generated MSW compositions in India (as per Table 5)

Table 5 Composition of MSW generated in India

Type of waste generated	% of Generation
Bio-degradable	45–55
Paper	8–13
Plastics/Rubber	9–10
Metal, glass and rags	1–1.5
Inerts	20–25

Source of data Tripathy [110]

5.1 *Pollutant Emission to the Environment*

Pollutants of varying sizes, properties, types and proportions could get released into the environment and adversely affect its various components like air, water bodies or land leading to their pollution if the generated MSW is not effectively managed. Sometimes certain management techniques could also result in release of pollutants. According to Cheng et al. [23], the burning of MSW in the open could be hazardous and release gaseous pollutants like CO, CO_2, SO_2 etc. into the atmosphere. According to Bahukhandi and Aaron [8], generation of MSW in very high rates along with its poor management has led to pollution of underground water which is a major source for drinking purposes for large sections of population in the country of India. An increase in amounts of TDS, SO_4, NO_3, total hardness etc. of water beyond permissible limits has been an outcome for the same reasons.

5.2 *Climate Change and Rise in Average Temperature*

Improperly managed solid waste could be a source of Green House Gases (GHGs) which could get released into the air and could be a prime factor leading up to possible climate change although not at a rapid pace [15]. Climate change could lead to issues like increment in the average temperatures, rise in sea level, global warming, glacier melting, harsh weather conditions etc. thus affecting humans as well as the nature

[53]. Decrease in GHG emissions, which can aid in avoiding the above conditions, can be achieved either by reduction in the generation of waste or by practicing better management methods such as recycling [25].

5.3 Adverse Effect on Public Health

Improperly managed waste could affect human/public health in different ways. The prime reasons behind this include attraction of disease spreading vectors due to its mass accumulation, blockage of drainage systems serving as habitat for those vectors etc. [52]. Sometimes even waste management techniques could end up affecting the health of the public, such as birth defects observed in people with work or habitat near landfills, respiratory issues observed in people living or inhaling the air in the proximity of composting zones etc. [39]. It has been observed that the economically weaker sections of the society are affected more by wastes as they reside in close proximities to the waste dumping areas or disposal grounds. Along with them, waste workers also remain at high risk of getting affected by the ill-effects of solid waste generated [111]. The unethical export activity of certain toxic wastes to another country or region also puts their population at greater risks to face its ill-effects [24, 95, 122].

5.4 Adverse Effect on Ecological Health

The management and disposal of waste acts as a kind of alteration in the use of a particular land or zone which in turn could remould the livelihood/dwellings of certain species belonging to that zone. Exposure to hazardous wastes cause ill-effects on plants, animals and also water bodies like oceans upon direct release to the environment [118]. MSW like plastics get released into the ocean bodies in the range of "millions of tons" which has resulted in the ratio between plastic and other marine wastes to be around 6:1 [42], Moore et al. [71]. Plastic wastes have showcased its effects on about 267 kinds of species who get affected by either its consumption or entanglement by it, thus leading to health issues and fatal conditions of the species [31, 42]. Plastics can also serve as disease carrying vectors within the food chain as it has the ability for absorption of organic pollutants and harmful toxins [66].

6 Management of Municipal Solid Waste

The management of MSW has been a big issue for almost all parts of the world, be it developed or in a developing stage. Improper handling or management of solid wastes has proved to be one of the chief sources of causing adverse effects on human health while also affecting the environment by polluting the air, water bodies and also bringing about climatic changes by causing overall rise of average Earth temperature thus leading towards global warming as discussed previously. This makes proper and effective management of MSW a necessity in the world of today. According to Rajput et al. [83], the figures of collection of generated wastes in many major cities of the world being performed effectively is as low as 25–55% with about 60% of the of the world's nations facing major issue related to environmental due to inefficient management and disposal of waste.

The practice of control over the generation of waste along with its storage, collection, transportation and discarding can be termed as solid waste management. Meanwhile reducing the overall amount of waste generation by virtue of re-using or recycling of goods could be termed as waste minimization [108]. Pollution of the environment is a possibility if any of those factors aren't performed effectively [65]. Ensuring proper health conditions of the general public, especially the poorer sections of the society is the primary focus of Municipal Solid Waste Management (MSWM). Along with that, maintenance of proper environmental conditions, economic stability, overall sustainability, efficient productivity as well as generation of employment are also some of major goals which are aims by MSWM and could only be achieved by the virtue of practicing proper and effective systems for MSWM [93]. History has been evidence of waste management being an engineering function. Advancement in technology in the modern society has led to amounts of solid waste generation reach big figures thus making MSWM the need of the hour [107].

There are quite a lot of differences observed between the developing and developed countries of the world when it comes to waste management. This difference could also be observed in regards to management in urban or rural areas or whether the area of generation is residential or an industrial one [28]. A waste management process is hugely influenced by the composition of waste that is being generated, which also differs largely according to income levels of the population. MSW such as paper, glass, plastics etc. are generated in more amounts among the communities with higher income as usage of packaged items are more common in those areas [102]. According to Joshi and Ahmed [57], waste management practice in India has always faced various issues with urban local bodies not taking necessary steps towards its proper implementation and also due to inefficiencies observed in other grounds like collection and processing of wastes as well. In order for a successful MSWM programme to take place, the following practices with solid wastes are compulsory to be performed in an efficient manner:

- Segregation
- Collection
- Recycle/Reuse
- Disposal

6.1 Segregation of MSW

Waste segregation could be described as source separation and is generally the method of separating items which could prove to be of some utility from the MSW which has been generated. This method could prove beneficial in many ways such as reducing the overall amounts of waste which would require further management thus saving time and energy during the waste management process. Segregation also helps keep the environment clean thereby also aiding towards well-being of the public health conditions [68]. The cost as well as complications in waste management could be reduced to a greater extent by the segregation process [37]. Segregation also acts as an important agent for the recycling process. Waste management is improved by a large scale if various items amidst the generated MSW could be treated as resources with economic values and be re-utilized for other activities while ensuring that sustainability is being maintained in the process [115].

Inappropriate, low scale or even absence in the practice of source segregation process is a prime reason for the various issues which are observed in many developing Asian countries like India, Bangladesh, Sri Lanka, Malaysia etc. during the practice of MSW management. Another issue is the loss of ability of a lot of generated wastes to be re-utilized the reason being it getting mixed with certain other hazardous wastes [46]. While carrying out segregation in countries like India, no scientific method is technique is generally observed at any level. Also, the conditions under which it takes place has been observed to be unhygienic for human health and the overall process hasn't proven to be very much fruitful due to it not being performed by experts and thus only items with visible economic value in the market being segregated [60].

6.2 Collection and Transportation of MSW

Waste collection which could be the first step towards management is performed via diverse means by different nations considering various factors with health concerns of public, economic issues, political conditions etc. being some of them. Generated waste could present adverse effects towards health and nature on the basis of the way it has been stored, handled, collection or managed [121]. The process of waste collection could end up being the largest cut out of the total expenditure bill of the entire MSW management process. This is usually the case in city/urban dwellings with to the large concentrations of the generated waste. Quite a few nations in Asia still struggle for possession of effective and efficient waste collection systems [120]. The methods of waste collection observed in European nations could be termed as perfect due to the diverse methods being utilized as well the effectiveness in collection of MSW in fractions [34]. In certain huge and developed cities of Latin America and the Caribbean regions waste collection methods observed have also been satisfactory even though there exist various cities that struggle to do the same [113].

According to Olukanni et al. [78], various challenges are encountered during waste collection process in various nations of the African continent owing to certain factors like insufficient capital for proper techniques to be practised, low knowledge of the general population about threats posed by untreated waste etc. The major waste collection methods described in the study are:

- Mode of Operation based waste collection: The major systems discussed under this category include:
  - Haul container system
  - Stationary container system
- Type of Waste based waste collection: Here, the systems of collection described are:
  - Collection of unseparated waste
  - Collection of separated waste

According to Hui et al. [51], a fruitful way to attain "Integrated Solid Waste Management (ISWM)" is the source separation of the MSW which is collected, but is not at practice in Chongqing, China. The major collection mechanisms of MSW in practice at Chongqing are said to be the following:

- Wastes by residential areas: Household wastes are collected in all towns/villages in dumping vessels which is transferred to waste collection stations. The Chongqing Municipal Administration Commission's (CMAC) department of Environmental Sanitary Protection takes on the responsibility to further transport it to waste treating zones.
- Wastes by institutions: The particular institution generating the waste is held responsible its transfer to treatment zones for which private parties could be hired by them as well.
- Wastes by Commercial areas: Here the job is mainly carried out by the management office of that commercial area who transfers the generated waste to treatment sites.
- Wastes by cleaners of public places: In this case well, CMAC is held responsible for the waste generated to be transferred to treating locations.
- Wastes by construction sites: The organization whose construction activity is in progress is given the responsibility of transferring non-recyclable wastes to treatment zones generally without involvement of a third party.

According to Jin et al. [55], the responsibility for the collection of generated MSW in Macao lies on the shoulders of the Government who in turn had appointed privately owned waste service company (WSC) for the same purpose from the year 1992. The activities like collection of wastes generated in households, various institutions, industries, commercial areas etc. and its transportation is carried out by the WSC. A total of 380 appointed staff members, 25 waste collection trucks, 4 high pressure water jet vehicles and 6 sweeping vehicles are under the possession of WSC who utilizing the same collects about 500 tpd of solid waste. The Port Authority of the SAR Government of Macao are the ones responsible for removal of refuse in water bodies of Macao.

According to Nandan et al. [73], the responsibility of collection of generated waste in India is to be taken up by the Government's Municipality Department. Collection of waste from different households has to be performed by the Municipal Authority by virtue of:

- Collection of MSW in Community containers
- Household to household collection
- Waste collection in regular intervals
- Provision of collecting vehicles with ringing bells to alert the public

The collected waste is then transferred into treatment zones in time intervals of generally two times in a week.

Community vessels constructed out of different materials ranging from concrete, metals etc. act as collection means for wastes which is generated from different households in a city/town/area. Along with that, wastes generated from industries, commercial areas, roadside garbage etc. also makes its way into these community waste collecting containers/vessels with the exception of any organization/body who may make payments to municipal authorities for transferring the generated waste directly from their premises to treatment zones [64].

Gonzàlez-Torre et al. [41], have discussed about the various types of container systems that could be utilized for collection of MSW in urban cities. Some of them are:

- *Household Containers:* Comprises of plastic vessels received by families for collecting wastes. Each household may receive a single container to store all wastes or multiple colour coded ones to store wastes of different compositions accordingly which later helps in easier to sort up the waste.
- *Neighbourhood Containers:* A common vessel/storage container for a group of people to collect all their generated waste together. Generally seen in flats/apartments.
- *Zone Containers:* Larger sized collection containers placed in particular areas outside one's premises to collect waste generated by a number of societies or neighbourhoods of people and could be easily collected again by waste collecting vehicles who transfers it to appropriate locations.

6.3 Recycling of MSW

The process by virtue of which goods discarded off as refuse or waste is remodelled or re-utilized for other purposes is termed as recycling. Examples include: paper recycling to avoid cutting down of trees, reduction in mining jobs by re-utilization of refused metals etc. [80]. Recycling could be achieved through different means, either by reusing a refused item for the same or some different purpose and also by re-modelling the wastes generated into other products, like composting [47]. In recent times, a "throwaway mentality" has been instilled in the minds of the society with the advancements in the fields of industrialization and progress in economic grounds as well. Recycling process has also become complicated and got limited to just industrial wastes [11]. The increased usage of plastics in the world of today has become a big issue due to it being a chief source of non-biodegradable waste. The cities with very high plastic (single-use) waste generation figures face problems like floods as a result of drainage systems getting choked and pollution of marine bodies [4].

Carrying out recycling process by virtue of mechanical means may be reasonable for some certain wastes such as bioplastic polymers like PLA but it could become less efficient due to limited supply of such polymers in required amounts [27]. The barrier properties of various biopolymers may require to be intensified for usage in certain areas such as packaging of food items, multilayer lamination etc [69]. Countries like India are coming up in the face of the world as to having one of the most prominent markets in the sector of waste recycling but still it is not being performed up to satisfiable levels [14]. In the Indian city of Pondicherry, a lot of the recycling activities are performed by waste/rag-pickers who separate useful goods like glass, metals, plastics etc. from the refuse collected from community containers and makes an earning out of it [82].

6.4 Treatment and Disposal of MSW

The disposal process could also be termed as the final step towards the completion of an MSW management process. It could be the biggest decisive factor of how successful or efficient the solid waste management performance of a country, city or even area has been which could also have an influence on the public health, environmental conditions etc. of the particular place. According to Hamer [45], waste disposal has an inevitable relation with pollution. Since waste is referred as unnecessary remnants with adversely affecting characteristics and on the other hand, pollution can be termed as the introduction of objects that could adversely affect the environment, hence disposal of waste is seemed to always have some kind of pollution associated with it. According to Williams [123], effective waste treatment and proper disposal became a necessity with the rise of urbanization and increasing communities of people in city areas, which has led to higher waste generation rates and quantities which turned into a serious issue with passing time.

The disposal of solid wastes into open grounds by virtue of irrational practices is very commonly seen in India without any proper management methods being followed which as a reuslt has displayed many adverse effects on both public and environment [20]. Methane gas in the range of around 48% is emitted from the waste dumping grounds of just about seven megacities of India due to ineffective waste management techniques that are practiced [67, 86]. Therefore, it has become the need of the hour to practice some scientific and efficient waste management techniques so that adverse effects as such could be avoided for the present as well as future generations. Some of the commonly followed methods for achieving proper waste management are as follows.

6.5 Landfilling

Landfilling can be simply defined as the deposition of the generated waste on earth (land). It is known by different terms that are used to denote this practice in different parts of the world such as "sanitary landfill" in USA, "controlled tipping" in the United Kingdom but "dumps" in the whole world as common [29]. It could also be described as the waste disposal by means of stuffing it (under pressure) in selected sites or area of land. This is a commonly practiced waste management method in many nations of the world owing to the convenience it offers from a financial point of view. For this reason, landfilling practices for MSWM is very common in developing nations [36, 40, 58, 117], made a study on landfilling practices in many different parts of the world for MSWM and the acquired data has been provided in Table 6.

Landfilling is a common practice for MSWM in many urban areas of India but is also facing issues at the same time due to unavailability of large areas of land required for it to be performed in many major cities [96]. As many as 59 sites for landfill practices have been developed in India with plans of further 1305 more such areas in sight for development in the near future. Many landfilling zones have been either constructed or the pre-existing ones were increased in size in many major states in India such as: Delhi, Maharashtra, Andhra Pradesh, Madhya Pradesh, Punjab etc [21].

Although landfilling process for MSWM seems a very efficient method especially for developing nations, improper practice of the same could lead to adverse effects on both people and environment. According to Vallero et al. [116], there exists a fine line between landfilling and dumping with the latter being a practice of depositing the solid waste without providing a separation from the underlying rock beds and also where excavations made for landfills reach beyond groundwater levels which

Table 6 Disposal of municipal waste by different countries by landfilling method

Country	% MSW managed by landfilling (%)
USA	52.6
Brazil	59.1
Saudi Arabia	85
Malaysia	94.5
China	79
Venezuela	32
Mexico	65
Thailand	27

Table 7 Collection, treatment and landfilling amounts in some high solid waste producing states of India

State	Solid waste generated (TPD)	Waste collected (TPD)	Waste treated (TPD)	Waste landfilled (TPD)
Maharashtra	23,844.551	23,675.7	12,623.33	11,052.37
Uttar Pradesh	17,377.3	17,329.4	4615	0
West Bengal	14,613.3	13,064.63	916	334
Tamil Nadu	13,968	13,968	7196	5654
Karnataka	11,958	10,011	4515	–
Telangana	8497	8360	5747	869
Madhya Pradesh	8000	7500	6100	1400
Rajasthan	6625.56	6475.39	780.18	780.18
Punjab	4634.48	4574.93	917.56	3657.37
Kerala	3903.023	742.23	437.74	–

Source of Data CPCB [19]

results in pollution of the groundwater and subsequently causing ill-effects to the health of people who utilize it. Thus, the concept of "sanitary landfilling" is needed to be practiced for efficient management. According to [16, 88], improper landfilling of MSW in India could rate as high as 90% in total. When groundwater zones are invaded by the waste from landfills, leachate is formed as the wastes come into contact with water. They are generally comprised of both organic and inorganic constituents along with certain metallic components as well forming a complex mixture and having potential to adversely affect the public health and environment as well. According to Hudgins [50], performing landfilling on certain types of wastes like bioplastic polymers along with certain organic waste as well could result in methane gas formation under specific conditions which could affect the environment being a greenhouse gas.

But with the practice of correct methods and application of modern technology, such drawbacks could actually be turned into an advantage by converting the produced "landfill gases" into sources of energy. According to Joshi and Ahmed [57], landfill grounds could act as a potential source of energy as methane emitted from those grounds constitute for about 13% of the global methane. Siddiqui and Khan [99], provided data through their study of various amounts of potential energy that could be generated from landfill zones of various cities in India, which is as follows:

Delhi—8.4 Megawatts
Mumbai—5.6 Megawatts
Ahmadabad—1.3 Megawatts
Pune—0.7 Megawatts.

6.6 Incineration

Incineration is a technique to achieve management of MSW via means of thermal treatment. There exists other process like pyrolysis as well under this category [57]. Thermal treatments methods like incineration are used in very limited countries with a reduction in waste volume and volume of leachate formed being some of the major intensions for its applications [91]. Complete elimination of solid waste is not attained by incineration but rather a reduction of it in terms of both weight and volume, thus aiding in the landfilling process. This is a common practice in Europe and helps greatly in two areas of MSWM: i reduction in weight and volume of MSW; ii use of the method as a potential source of energy such as heat by burning of the wastes. Hence it is feasible in areas where huge land areas aren't available for landfilling [85].

One of the major setbacks offered by waste incineration in the country of Japan is the high amounts of residue which is produced as a result and which ends up requiring separate treatment. Such residue is also signified as "air pollution control residue (AR)" and requires certain treatment before landfilling processes could be performed [33]. In the nation of India, incineration of MSW is not often a suitable waste management technique owing to the large amounts of organic natured wastes that is generated in the country. Also, setbacks towards the method are put forward by the high values of moisture content in the generated waste with 800 to 1100 kcal/kg being a common range of its calorific value (Jalan and Srivastava 1995), [103]. Certain high-capacity incineration plants in the country like a one developed in the Tirampur region of Delhi in 1987 resulted in failure due to unavailability of right amounts of MSW with properties suitable for performing incineration [97].

6.7 Biological Treatment

As the name suggests, this is a waste treatment mechanism which is performed by the aid of certain living organisms. This method is generally suited to the treatment of wastes containing greater fractions of organic constituents. According to Hamer [45], for the treatment of biodegradable wastes, biological treatment has long been a suitable method. This method generally is performed by two means namely: Aerobic and anaerobic treatment. Composting of waste is a very widely known and practiced waste treatment method all over the world. According to Bharadwaj et al. [13], composting is a famous treatment technique seen in the UK where annual waste composting figures reach up to 2 MT/year in more than 300 zones for composting activities which have been developed. Composting is more widely observed in developing countries like India where generation of organic wastes is much higher than in developed countries.

6.8 Aerobic Composting

This is the organic solid waste composting method which is performed in the presence of air (oxygen) with under warm conditions. The end product of the process performed is known as compost and is a commodity which is very rich in nutrients (Bhide and Shekdar, 1998). This is a very widely practiced method in India with composting zones for MSW having been setup in many major urban cities of the country like Delhi, Mumbai, Indore, Bengaluru etc. including composting plants of treating capacities reaching up to 300 tonnes per day in many of those cities [97]. One of the most commonly practiced aerobic composting methods is "vermicomposting", where treatment of MSW is carried out with the aid of earthworms that consume the organic fractions from the solid waste and decomposed to form a nutrient rich product via actions of microorganisms. Major vermicomposting plants are also situated in many major urban areas of India [57].

6.9 Anaerobic Composting

Anaerobic composting which is also referred to as "anaerobic digestion" is the treatment/decomposition of organic wastes without the presence of air (oxygen). The magnitudes of required energy levels in this type of composting are generally lesser compared to aerobic composting [26]. This type of composting method is usually carried out in 2 stages namely: (i) acid formation stage where proteins, lipids or carbohydrates are transformed into certain fatty acids by action of microorganisms and (ii) conversion of the end products of previous stage into methane and CO_2 by the action of special type of bacteria [109]. Another term which can be used for this method is known as "bio methanation" and in India, it is a process often supported well by the Government in the treatment of agricultural and industrial wastes as well along with MSW. Developmental plans for a number of bio methanation plants are underway for many major urban cities of India like Delhi, Bengaluru etc. [2] (Fig. 11).

7 Challenges Faced in MSW Management

Throughout the study, we have come across the waste generation and composition data of various regions of the world and also the needs and techniques for its effective management. But that job is easier said than done. Development of a fully effective technique for MSW management for a particular nation or urban city is very difficult owing to many different reasons. This is an issue faced more often by the developing nations of the world than the developed ones. The reasons behind it might vary from nation to nation with some common points probably shared by them as well.

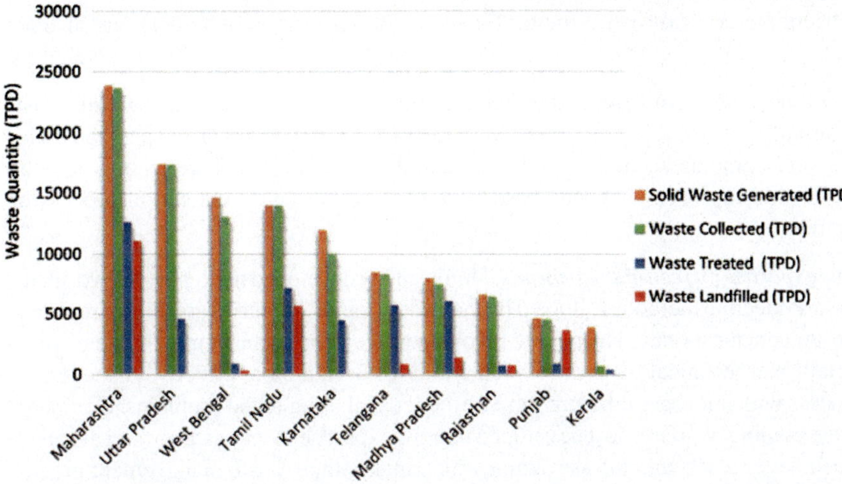

Fig. 11 Comparison of generation, collection, treatment and landfilling quantities of solid waste by some highest solid waste generating Indian States (as per Table 7)

Sulemana et al. [105], through their study have highlighted certain challenges which are commonly encountered by develop nations of the world in the process of waste collection. Some of the issue described are as follows:

Financial Issues: Developing countries often do not receive the required financial assistance from their central or state governments due to which an effective and efficient waste management process cannot be carried out. Governments show unwillingness in providing required funds to carry out the process successful due to which proper waste collecting vehicles like collection trucks could not be bought in required numbers. Because of reasons such as these, collection of MSW do not reach satisfactory figures and most of it gets openly dumped in unscientific manners.

Lacking in willingness and political support: Participation of the public is equally important along with the Government for a proper waste management process to take place. In most developing countries, it has been observed the mentality of people and their responsibility shown towards MSW management is very disappointing. Public awareness towards the importance of proper waste collection is very poor. On top of that, lack of political or governmental support is also very often seen in developing countries due to which an effective waste management process never becomes a reality.

Unsatisfactory schedule for collectors and routing of vehicles: This is perhaps the biggest roadblock in the path towards a proper waste collection system. Waste pickers/collectors aren't provided with proper schedules to perform the collection activities which results in non-uniformity in the process. Along with that the poor quality of waste collection components such as community containers which makes collection process further difficult. Finally, required number of vehicles are also not available in hand for which waste collection process goes further away from success.

Improper routes towards collection grounds: In many areas of the world, waste dumping areas are very commonly located in very compact or narrow passages. This is a very common scene in developing countries due to which, collection vehicles aren't able to properly reach those sites making the waste collection process an inefficient one.

Low expertise on technical grounds: Unlike many developed nations of the world, the waste collection methods followed by many developing countries are quite simple and most mechanical ones. The people involved in the waste management systems often lack proper technical expertise which could result in a more effective management process with improper infrastructure further contributing to the problems. Compared to developing nations, most developed nations expertise in technicality and also have much better assets and infrastructure which makes their waste management process much better and effective.

Kumar and Agrawal [63], have discussed various issues which faced while MSW management process is practiced in India, which are as follows:

- *Issues of public health and environment:* Methods like landfilling are very commonly utilized in India which results in high rates of leachate formations, high CH_4 emissions into the atmosphere and other factors like pollution of groundwater. The environment gets greatly affected by this which also results in the ill-effects to the public health who are very much dependent upon the environment for their life and activities.
- *Lack of financial aid and proper infrastructure:* Like many other developing nations, the Government of India often fails to provide adequate financial support which is required for an effective waste management process to take place. Although many steps have been taken throughout the years to bring improvement in this factor.
- *Necessary policies not effectively implemented:* Certain national policies are necessary to be formulated and followed for an effective waste management. But very often, a sense of negligence is observed among the people of India when it comes to waste management due to which importance towards certain is not given in required sense.
- *Challenges faced in various MSWM processes:* MSWM consists of many processes as discussed previously ranging from segregation to proper disposal. For the waste management to take place, each of those processes have to be performed effectively. But this could not happen in India due to various reasons like lack of capital and sound infrastructure along with certain social issues as well.
- *Social issues:* Along with financial and infrastructural, MSW management in India faces certain social issues as well du to which it is not as effective as desired. One such issue is lack of proper education and understanding about the importance of MSW management and how it can impact the health of individual and environment as well. Lack of space is also an issue in many urban areas of India due to which

proper disposal could not take place. Along with that, the issue of division of jobs according to classes to people has always existed in India. The ones working in management of wastes are often considered to be at lower levels than those at the upper sections of the society. Hence, the upper classed society often neglects the importance of MSW management considering it a job lower for their standards. Hence, waste management process suffers as a whole.

8 Conclusion and Recommendation

On completion of the study, it could be concluded that MSW generation has become a serious issue all over the world with its generation rates with varied compositions increasing with time and rapid urbanization in the modern day. There is quite a difference observed in the compositions of MSW generated by developed countries compared to developing countries with the former generating more of paper based solid wastes. The adverse effects of MSW could be seen in large extent over the developing nations of the world due to financial grounds and also other social and communal issues. The country of India is a good example of that. Due to this, the practice of MSW management could not be carried out in the urban areas of those nations either due to lack of proper and required funding towards it by the Government, lack of required infrastructure and also lack of a culture where everyone takes active participation towards waste management to mitigate its possible hazards without depending completely on the Government and also without letting any social barriers interfere the same. A few suggestions to overcome various issues and have an efficient MSW management process in order to keep the people and environment free from ill-effects are as follows:

(i) People, especially from the developing countries lacking proper MSW management systems must be provided proper education on the same and its importance and effects if not performed effectively. Waste Management should be made a compulsory course in schools/colleges regardless of whatever line of study a student pursues.

(ii) State as well as central Governments of all nations must make their utmost efforts so that the required funds could be provided for MSW management to be carried out without failure.

(iii) It should be the responsibility of every individual of the country to make an effort towards the reduction in the generation rates of MSW in the first place. More the generation and composition, more complicated and expensive the management process becomes. Hence, unnecessary desires and demands need to be subsidised so and the concept of "re-use" should be promoted and practiced.

(iv) It is also necessary that the correct management techniques for MSW are followed by a certain nation, city, town or area so that the overall process turns out to be effective. For e.g.: In India the generation of organic waste is observed to be in greater amounts than many other developed nations of the world. On the contrary, there is shortage of free land areas for practicing landfilling. So, the practice of certain biological treatments of waste such as vermicomposting should be practiced and promoted rather than landfilling. This would not only make waste management better but also act as an employment opportunity and potential source of income.

(v) Effective treatment of waste is needed to be done before disposal. Direct disposal of collected waste in open dumping grounds or landfilled is not termed as MSW management. Steps like segregation, incineration etc. need to be carried out before disposal. Otherwise, the disposed waste could lead to pollution of environment and ill-effects on human health

(vi) Recycling process of generated wastes should be given priority all over the world. The 4R policy of: Reduce, Re-use, Recycle and Recover should be given priority. This way, the total volume of waste needed to be disposed decreases which protects the environment and the ones depending on it. On the other hand, this could also become a potential source of income for people.

(vii) Use of plastics should be reduced as it constitutes for highest amounts in MSW generation in all parts of the world, be it developed or developing nations and is non-bio-degradable.

(viii) A focus should be given towards "waste to energy" practices in India, especially for landfilled wastes as methane is extracted in very large amounts and could be converted to biogas making the generated MSW into a potential source of non-conventional energy.

References

1. Akhtar, M., Hannan, M. A., Basri, H. & Scavino, E. (2015), "Solid Waste Generation and Collection Efficiencies: Issues and Challenges", *Jurnal Teknologi*, Vol. 75(11), pp. 41–49, https://doi.org/10.11113/jt.v75.5331.
2. Ambulkar, A.R & Shekdar, A.V. (2004), "Prospects of biomethanation technology in the Indian context: A pragmatic approach", *Resources, Conservation and Recycling*, Vol. 40(2), pp. 111–128
3. Amoah, S. & Kosoe, E. (2014), "Solid Waste Management in Urban Areas of Ghana: Issues and Experiences from Wa", *Journal of Environment Pollution and Human Health*, Vol. 2, 110–117https://doi.org/10.12691/jephh-2-5-3
4. Anderson, J. C., Park, B. J., & Palace, V. P. (2016), "Microplastics in aquatic environments: Implications for Canadian ecosystems", *Environmental pollution (Barking, Essex: 1987)*, 218, 269–280, ISSN 0269-7491.https://doi.org/10.1016/j.envpol.2016.06.074
5. Annepu, R.K. (2012), "Sustainable Solid Waste Management in India", Submitted to: *Department of Earth and Environmental Engineering, Fu Foundation School of Engineering and Applied Science*, Columbia University, New York
6. Asefi H., Zhang Y., Lim, S., Maghrebi, M., Shahparvari, S. (2020), "A multi-criteria decision support framework for municipal solid waste landfill siting: a case study of New South Wales (Australia)", *Environmental Monitoring and Assessment*, 192: 682https://doi.org/10.1007/s10661-020-08565-y

7. B. Xue, X. Chen, H. Lu, C. Lu and M. Yang, "Analysis of Transition Process from Waste Management towards Resource Management System", *4th International Conference on Wireless Communications, Networking and Mobile Computing*, 2008, pp. 1–4, https://doi.org/10.1109/WiCom.2008.2737.
8. Bahukhandi, D. K. & Aaron, A. (2017), "Impact of Improper Disposal of Municipal Solid Waste on Ground Water Quality in and Around the Solid Waste Dumping Site of Visakhapatnam, Andrapradesh, India", *Journal of Global Ecology and Environment*, Vol. 5(3), pp. 133–143
9. Bahukhandi, Kanchan Deoli, Saini Ritesh, Ahuja Neelu Jyoti, Thakur Divya, Siddiqui Nihal Anwar. 2021. Assessment of the Impact of the Municipal Solid Waste on hydrochemistry of the Surface and Groundwater Quality of Dehradun District of Uttarakhand: Seasonal Variation in Water Quality." *Himalayan Geology* 42 (1).
10. Bahukhandi KD & Ghosh D (2012) Solid Waste Management. *International Journal of Sustainable Development and Green Economics (IJSDGE)* Vol. 1(1)
11. Bai, R. & Sutanto, M. (2002), "The practice and challenges of solid waste management in Singapore", *Waste Management*, Vol. 22(5), pp. 557–567
12. Bartarya SK, Bahukhandi DK (2012) Impact Assessment of Urbanization and Industrialization on Surface and Groundwater Quality of Dehradun district of Uttarakhand. *Global Journal of Engineering, Design and Technology* Vol. 1(1), pp. 11–22. ISSN 2319 - 7293, Global Institute of Research & Education, GJET
13. Bharadwaj, A., Yadav, D. & Varshney, S. (2015), "Non-Biodegradable Waste – Its Impact & Safe Disposal", *International Journal of Advanced Technology in Engineering and Science*, Vol. No. 3(1), pp: 184–191.
14. Bhattacharya, R.R.N.S., Chandrasekhar, K., Deepthi, M.V., Roy, P., Khan, A., (2018), "Challenges and opportunities: plastic waste management in India", *The Energy and Resources Institute*, Vol. 24(2), Available on: https://www.teriin.org/sites/default/files/2018-06/plastic-waste-management_0.pdf.
15. Bogner, J., Pipatti, R., Hashimoto, S., Diaz, C., Mareckova, K., Diaz, L., Kjeldsen, P., Monni, S., Faaij, A.P.C., Gao, Q., Zhang, T., Ahmed, M., Sutamihardja, R.T.M. & Gregory, R. (2008), "Mitigation of Global Greenhouse Gas Emissions from Waste: Conclusions and Strategies from the Intergovernmental Panel on Climate Change (IPCC) Fourth Assessment Report. Working Group III (Mitigation)", *Waste management & research: the journal of the International Solid Wastes and Public Cleansing Association*, ISWA, 26(1), 11–32https://doi.org/10.1177/0734242X07088433
16. Budi, S.; Suliasih, B.A.; Othman, M.S.; Heng, L.Z.; Surif, S. (2016), "Toxicity identification evaluation of landfill leachate using fish, prawn and seed plant", *Waste Management*, 55, 231–237
17. Census (2011), "Provisional Population Totals", *Office of the Registrar General and Census Commissioner*, Ministry of Home Affairs, India, Accessed from: https://censusindia.gov.in/2011-prov-results/data_files/india/pov_popu_total_presentation_2011.pdf
18. Center for Sustainable Systems, University of Michigan (2020), "Municipal Solid Waste Factsheet." Pub. No. CSS04–15.
19. Central Pollution Control Board (CPCB), "Annual Report For the year 2018–19 On Implementation of Solid Waste Management Rules (As per provision 24(4) of SWM Rules,16)", *Ministry of Environment, Forest and Climate Change, Govt. of India*, December, 2019.
20. Central Pollution Control Board (CPCB), "Solid Waste Generation in 46 Metrocities", India, 2018, Accessed from: https://cpcb.nic.in/uploads/MSW/trend_46_cities_list.pdf
21. Central Pollution Control Board (CPCB), "Status Report on Municipal Solid Waste Management", *Ministry of Environment and Forests*, Delhi, 2013.
22. Central Pollution Control Board (CPCB), Management of Municipal Solid Waste, *Ministry of Environment and Forests, New Delhi*, India, 2004.
23. Cheng K., Hao W., Wang Y., Yi P., Zhang J. & Ji W. (2020), "Understanding the emission pattern and source contribution of hazardous air pollutants from open burning of municipal solid waste in China", *Environmental Pollution (Barking, Essex: 1987)*, Aug;263(Pt A):114417. DOI: https://doi.org/10.1016/j.envpol.2020.114417. PMID: 32220777.

24. Chi, X., Streicher-Porte, M., Wang, M. Y., & Reuter, M. A. (2011), "Informal electronic waste recycling: a sector review with special focus on China", *Waste management (New York, N.Y.)*, Vol. 31(4), pp. 731–742. https://doi.org/10.1016/j.wasman.2010.11.006
25. Christensen, Thomas H. (2011), "Solid waste technology & management", *Chichester, West Sussex, U.K.: Wiley*
26. Chynoweth, D.P., Bosch, G., Earle, J.F.K., Legrand, R. & Liu, K. (1991), "A novel process for anaerobic composting of municipal solid waste", *Applied Biochemistry and Biotechnology,* Vol. 28, pp. 421–432. https://doi.org/10.1007/BF02922622
27. Claesen C. (2005), "Hycail—more than the other PLA producer", Presented at *RSC Symposium sustainable plastics: biodegradability vs recycling.*
28. Cointreau, J, S. (1982), "Environmental management of urban solid wastes in developing countries: a project guide", First Edition, World Bank, Washington DC, Vol. no. 5.
29. Crawford, J.F. & Smith, P.G., "Landfill Technology", First Edition, *Butterworth-Heinemann*, December, 1984.
30. Deoli Kanchan Bahukhandi & Ghosh Digantika "Solid waste Management" 2012, International Journal of Sustainable Development and Green Economics (IJSDGE), ISSN.No.2315–4721, Vol-1 Iss-1.
31. Derraik J. G. (2002), "The pollution of the marine environment by plastic debris: a review" *Marine pollution bulletin,* Vol. 44(9), pp. 842–852.https://doi.org/10.1016/s0025-326x(02)00220-5106
32. Dong, S., Tong, K. & Yuping, W. (2001), "Municipal solid waste management in China: Using commercial management to solve a growing problem", *Utilities Policy,* Vol. 10, pp. 7–11.https://doi.org/10.1016/S0957-1787(02)00011-5
33. Ecke, H., Sakanakura, H., Matsuto, T., Tanaka, N. & Lagerkvist, A. (2000), "State-of-the-art treatment processes for municipal solid waste incineration residues in Japan", *Waste Management and Research.* Vol. 18(1), pp. 41–51
34. European Research and Consulting Limited (2002), "Costs for Municipal Waste Management in the EU", Available on: http://ec.europa.eu/environment/waste/studies/pdf/e ucowaste.pdf.
35. Eurostat (2020), "Municipal waste statistics", Accessed from: https://ec.europa.eu/eurostat/statistics-explained/index.php?title=Municipal_waste_statistics
36. Feng, S.J., Chen, Z.W., Chen, H.X., Zheng, Q.T. & Liu, R. (2018), "Slope stability of landfills considering leachate recirculation using vertical wells", *Engineering Geology*, Vol. 241, pp. 76–85.https://doi.org/10.1016/j.enggeo.2018.05.013
37. Fish, Jim D.; Massey, Charles D. & Ward, Stephen J. (1990), "Waste segregation procedures and benefits", Article, Albuquerque, New Mexico, Available on: https://digital.library.unt.edu/ark:/67531/metadc1192664/, University of North Texas Libraries, UNT Digital Library, https://digital.library.unt.edu.
38. Gautam Priti, Bahukhandi Kanchan Deoli, 2018. Sustainable options for E-Waste Management, International Journal of Engineering, Science and Mathematics Vol.7 Issue 4, April 2018, ISSN: 2320–0294 Impact Factor: 6.765, http://www.ijesm.co.in, (UGC Approved), https://www.ugc.ac.in/journallist/4305_Journals.pdf
39. Giusti, L. (2009), "A review of waste management practices and their impact on human health", *Waste Management*, Vol. 29(8), pp. 2227–2239
40. Gonzalez-Valencia, R., Magana-Rodriguez, F., Cristóbal, J., & Thalasso, F. (2016), "Hotspot detection and spatial distribution of methane emissions from landfills by a surface probe method", *Waste management (New York, N.Y.)*, 55, 299–305, https://doi.org/10.1016/j.wasman.2016.03.004.
41. González-Torre, P. L., Adenso-Díaz, B., & Ruiz-Torres, A. (2003), "Some comparative factors regarding recycling collection systems in regions of the USA and Europe", *Journal of environmental management*, Vol. 69(2), pp. 129–138.https://doi.org/10.1016/s0301-4797(03)00109-9
42. Gregory M. R. (2009), "Environmental implications of plastic debris in marine settings--entanglement, ingestion, smothering, hangers-on, hitch-hiking and alien invasions", *Philosophical transactions of the Royal Society of London. Series B, Biological sciences*, 364(1526), pp. 2013–2025.https://doi.org/10.1098/rstb.2008.0265

43. Guerrero, L.A., Maas, G. & Hogland, W. (2013), "Solid waste management challenges for cities in developing countries", *Waste Management*, Vol. 33(1), pp. 220–232.https://doi.org/10.1016/j.wasman.2012.09.008
44. Gupta, S., Mohan, K, Prasad, R. & Kansal, A. (1998), "Solid waste management in India: Options and opportunities", *Resources Conservation and Recycling*, 24, pp:137–154.
45. Hamer G. (2003), "Solid waste treatment and disposal: effects on public health and environmental safety", *Biotechnology advances*, Vol. 22(1-2), pp. 71–79.https://doi.org/10.1016/j.biotechadv.2003.08.007
46. Hargreaves, J.C., Adl, M.S. & Warman, P.R. (2008), "A review of the use of composted municipal solid waste in agriculture", *Agriculture, Ecosystems & Environment*, Volume 123, Issues 1–3, Pages 1–14.https://doi.org/10.1016/j.agee.2007.07.004
47. Heimlich, J., Hughes, K. & Christy, A. (2007), "Integrated Solid Waste Management", *The Ohio State University Extension.*
48. Hoornweg, D., Lam, P., Chaudhry, M. (2005), "Waste Management in China: Issues and Recommendations May 2005", *Urban Development working Papers No. 9,* Washington, D.C.: World Bank Group, Available on: https://documents1.worldbank.org/curated/en/237151468025135801/pdf/332100CHA0Waste1Management01PUBLC1.pdf
49. Hoornweg, Daniel & Bhada-Tata, Perinaz. (2012), "What a waste: a global review of solid waste management", *Urban Development Series Knowledge Papers*, Knowledge papers no. 15, World Bank, Washington, DC.
50. Hudgins, M. (1999), "Aerobic landfill studies from the USA", *Paper presented at the 1st Int. Conf. Solid.*
51. Hui, Y., Li'ao, W., Fenwei, S. & Gang, H. (2006), "Urban solid waste management in Chongqing: Challenges and opportunities", *Waste Management*, Vol. 26(9), pp. 1052–1062
52. Hunt, C. (1996), "Child waste pickers in India: The occupation and its health risks", *Environment and Urbanization*, Vol. 8(2), pp. 111–18
53. IPCC (2007), "Climate Change 2007: Impacts, adaptation and vulnerability", Contribution of Working Group II to the Fourth Assessment Report of the Intergovernmental Panel on Climate Change, Edited by: ML Parry, O.F. Canziani, J.P. Palutikof, PJ van der Linden & CE Hanson, *Cambridge University Press*, Cambridge, UK, 976pp.
54. Imam, A., Mohammed, B., Wilson, D.C., Cheeseman C.R. (2008), "Solid waste management in Abuja, Nigeria", *Waste Management*, Volume 28(2), pp. 468–472, ISSN 0956-053X.https://doi.org/10.1016/j.wasman.2007.01.006
55. Jin, J., Wang, Z. & Ran, S. (2006), "Solid waste management in Macao: Practices and challenges", *Waste management (New York, N.Y.)*, Vol. 26(9), pp. 1045–1051, doi: https://doi.org/10.1016/j.wasman.2005.08.006.
56. Joardar, S. D. (2000), "Urban residential solid waste management in India: Issues related to institutional arrangements", *Public Works Management and Policy*, Vol. 4(4), pp. 319–330.https://doi.org/10.1177/1087724X0044006
57. Joshi, R. & Ahmed, S. (2016), "Status and challenges of municipal solid waste management in India: A review", *Cogent Environmental Science*, Vol. 2(1), https://doi.org/10.1080/23311843.2016.1139434.
58. Jovanov, D., Vujić, B., & Vujić, G. (2018), "Optimization of the monitoring of landfill gas and leachate in closed methanogenic landfills", *Journal of environmental management*, 216, 32–40.https://doi.org/10.1016/j.jenvman.2017.08.039
59. Kansal, Arun. (2002), "Solid Waste management Strategies for India", *Indian Journal of Environment Protection*, 22(4): 444–448
60. Kaushal, R., Varghese, G. & Chabukdhara, M. (2012), "Municipal Solid Waste Management in India-Current State and Future Challenges: A Review", *International Journal of Engineering Science and Technology*, 4(04), pp. 1173–1189
61. Kaza, Silpa; Yao, Lisa C.; Bhada-Tata, Perinaz; Van Woerden, Frank. (2018), "What a Waste 2.0: A Global Snapshot of Solid Waste Management to 2050", *Urban Development*; Washington, DC: World Bank, http://hdl.handle.net/10986/30317

62. Krishna, V. & Chaurasia, S. (2017), "Aspects of Municipal Solid Waste Management in Allahabad City: A Questioner Survey of the Citizens", *IOSR Journal of Environmental Science, Toxicology and Food Technology*, 11, 11–16.https://doi.org/10.9790/2402-1102011116
63. Kumar, A. & Agrawal, A. (2020), "Recent trends in solid waste management status, challenges, and potential for the future Indian cities – A review", *Current Research in Environmental Sustainability*, Vol. 2, 100011, https://doi.org/10.1016/j.crsust.2020.100011.
64. Kumar, S., Bhattacharyya, J. K., Vaidya, A. N., Chakrabarti, T., Devotta, S., & Akolkar, A. B. (2009), "Assessment of the status of municipal solid waste management in metro cities, state capitals, class I cities, and class II towns in India: an insight", *Waste management (New York, N.Y.)*, Vol. 29(2), pp. 883–895. https://doi.org/10.1016/j.wasman.2008.04.011
65. Kyessi, A. & Mwakalinga, V. "GIS application in coordinating solid waste collection: the case of Sinza Neighbourhood in Kinondoni municipality, Dar es Salaam city, Tanzania", *Municipality and Natural Resources Management*, 2009.
66. Law, K. L., Morét-Ferguson, S., Maximenko, N. A., Proskurowski, G., Peacock, E. E., Hafner, J., & Reddy, C. M. (2010), "Plastic accumulation in the North Atlantic subtropical gyre", *Science (New York, N.Y.)*, Vol. 329(5996), pp. 1185–1188, https://doi.org/10.1126/science.1192321.
67. Lv, W., Wang, Z., Cao, H., Sun, Y., Zhang, Y., & Sun, Z. H. (2017), "A critical review and analysis on the recycling of spent lithium-ion batteries", *ACS Sustainable Chemistry & Engineering*, Vol. *6*(2), pp. 1504–1521
68. Medina, M. (2002), "Globalization, Development, and Municipal Solid Waste Management in Third World Cities", El Colegio de la Frontera Norte [College of the northern border], Tijuana, Mexico, *Human Settlement Development*, Volume: 3.
69. Miller, R., "The Landscape for Biopolymers in Packaging", *Miller-Klein Associates*, March 2005.
70. Minghua, Z., Xiumin, F., Rovetta, A., Qichang, H., Vicentini, F., Bingkai, L., Giusti, A., & Yi, L. (2009), "Municipal solid waste management in Pudong New Area, China", *Waste management (New York, N.Y.)*, Vol. 29(3), pp. 1227–1233, https://doi.org/10.1016/j.wasman.2008.07.016
71. Moore, C. J., Moore, S. L., Leecaster, M. K., & Weisberg, S. B. (2001), "A comparison of plastic and plankton in the north Pacific central gyre", *Marine pollution bulletin*, Vol. 42(12), pp. 1297–1300.https://doi.org/10.1016/s0025-326x(01)00114-x
72. Mustapha, Imam (2013), "Composting by households in Canada", *Environment Accounts and Statistics Division*, Volume 7(1), ISSN 1913–4320.
73. Nandan, A., Yadav, B. P., Baksi, S. & Bose, D. (2017), "Recent Scenario of Solid Waste Management in India", *World Scientific News*, 66, 56–74
74. Naninja, W. "Optimizing Transportation Cost of Solid Waste A Case Study in The Sunyani Municipality", Submitted to: *The College of Science, Department of Mathematics, Kwame Nkrumah University of Science and Technology*, September, 2013
75. National Waste Report, 2018, *Blue Environment Pty Ltd,* Prepared for: Department of the Environment and Energy, Available on: https://www.environment.gov.au/system/files/resources/7381c1de-31d0-429b-912c-91a6dbc83af7/files/national-waste-report-2018.pdf.
76. Neelu Ahuja, Deoli Kanchan Bahukhandi (2012) 'Expert systems for solid waste management: A review' has been published in an international journal 'International Review on Computers and Software [Praise Worthy Prize publishing] Print ISSN: 1828–6003 Cd-Rom ISSN: 1828–6011 with an impact factor of 6.14., Vol. 7 N. 4, 1608–1613 (SCI. Scopus Index)
77. Oduro-Kwarteng, S. (2011), "Private Sector Involvement in Urban Solid Waste Collection", Submitted to: *The Erasmus University, Rotterdam*, Available on: http://www.limpezapublica.com.br/textos/finalthesis.pdf.
78. Olukanni, D.O., Adeleke, J.O. & Aremu, D.D. (2016), "A Review of Local Factors Affecting Solid Waste Collection in Nigeria", Pollution, Vol. 2(3), pp. 339–356, ISSN: 2383–4501.
79. Ozeler, D & Yetis, Ulku & Demirer, Goksel (2006), "Life cycle assesment of municipal solid waste management methods: Ankara case study", *Environment International*, 32, 405–411.https://doi.org/10.1016/j.envint.2005.10.002

80. Pappu, A., Saxena, M. & Asolekar, S. R. (2007), "Solid wastes generation in India and their recycling potential in building materials", *Journal of Building and Environment,* Vol. 42(6), pp. 2311–2320
81. Parishwad, O.G., Shukla, P. & Mitkari, M.D., "Sustainable Municipal Solid Waste Management and Economics of Informal Sector, for Inclusion in India", *Waste Management & Resource Utilisation,* 2016.
82. Pattnaik, S., & Reddy, M. (2010), "Assessment of Municipal Solid Waste management in Puducherry (Pondicherry), India", *Resources Conservation and Recycling, 54,* 512–520
83. Rajput, R., Prasad, G. & Chopra, A. (2009), "Scenario of solid waste management in present Indian context", *Caspian Journal of Environmental Sciences,* Vol. 7(1), pp. 45–53
84. Ramos, T., Gomes, M. & Barbosa-Povoa, A. (2014), "Economic and environmental concerns in planning recyclable waste collection systems", *Transportation Research Part E: Logistics and Transportation Review,* 62, pp.34–54, https://doi.org/10.1016/j.tre.2013.12.002.
85. Rarotra, S., Sahu, S., Kumar, P., Kim, K. H., Tsang, Y. F., Kumar, V., Kumar, P., Srinivasan, M., Veksha, A., & Lisak, G. (2020), "Progress and Challenges on Battery Waste Management: A Critical Review", ChemistrySelect, Vol. 5(20), pp. 6182–6193.
86. Rarotra, S., Sahu, S., Kumar, P., Kim, K. H., Tsang, Y. F., Kumar, V., Kumar, P., Srinivasan, M., Veksha, A., & Lisak, G. (2020), "Progress and Challenges on Battery Waste Management: A Critical Review", *ChemistrySelect,* Vol. *5*(20), pp. 6182–6193
87. Rathi, S. (2006), "Alternative approaches for better municipal solid waste management in Mumbai, India", *Waste Management,* Volume 26, Issue 10, Pages 1192–1200, ISSN 0956-053Xhttps://doi.org/10.1016/j.wasman.2005.09.006
88. Ratna, M.V., Kumar, G.V. & Dileep, G. (2021), "The Effects of leachate from Municipal Solid Waste Landfill Dump Sites on Ground Water Contamination", *Levant Journal,* Vol. 20(7), pp. 194–207
89. Ricci-Jürgensen, M., Gilbert, J. & Ramola, A. (2020), "Global Assessment of Municipal Organic Waste Production and Recycling", *International Solid Waste Association (ISWA),* Available on: https://www.altereko.it/wp-content/uploads/2020/03/Report-1-Global-Assessment-of-Municipal-Organic-Waste.pdf
90. Ritesh Saini, Nihal Anwar Siddiqui, Kanchan Deoli Bahukhandi, Neelu J Ahuja (2018) : "Study of effects of Municipal solid waste on water quality of Dehradun City" Journal of Environmental Science, Computer Science and Engineering & Technology , An International Peer Review E-3 Journal of Sciences and Technology, JECET; June 2018- August 2018; Sec. A; Vol.7. No.3, 409–416, E-ISSN: 2278–179X, PP 409–416, [DOI: https://doi.org/10.24214/jecet.A.7.3.40916]
91. Sabbas, T., Polettini, A., Pomi, R., Astrup, T., Hjelmar, O., Mostbauer, P., Cappai, G., Magel, G., Salhofer, S., Speiser, C., Heuss-Assbichler, S., Klein, R., Lechner, P., & pHOENIX Working Group on Management of MSWI Residues (2003), "Management of municipal solid waste incineration residues", *Waste management (New York, N.Y.),* Vol. 23(1), pp. 61–88, https://doi.org/10.1016/S0956-053X(02)00161-7
92. Saini Ritesh, Ahuja Neelu, Bahukhandi Kanchan Deoli (2016) "Fuzzy logic based advisory for handling landfill operational problems for early warning and emergency response planning" International Journal of ChemTech Research CODEN (USA): IJCRGG, ISSN: 0974–4290, ISSN(Online):2455–9555 Vol.9, No.08, August, .282–297 pp 282–297, 2016, Pub : Sphinx Knowledge House http://www.sphinxsai.com/2016/ch_vol9_no8/1/(282-297)V9N8CT.pdf,
93. Schübeler, P., Wehrle, K. & Jurg, C. (1996), "Conceptual Framework for Municipal Solid Waste Management in Low-Income Countries", 1st Edition, *St. Gallen: SKAT, Swiss Centre for Development Cooperation in Technology and Management,* Urban Management Programme working paper series: 9.
94. Seo, S.; Asce, M.; Aramaki, T.; Hwang, Y.; Hanaki, K. (2004), "Environmental Impact of Solid Waste Treatment Methods in Korea", *Journal of Environmental Engineering,* Vol. 130(1), pp. 81–89

95. Sepulveda A, Schluep M, Renaud FG, Streicher M, Kuehr R, et al. (2010), "A review of the environmental fate and effects of hazardous substances released from electrical and electronic equipments during recycling: Examples from China and India", *Environmental Impact Assessment Review*, Vol. 30(1), pp. 28–41.https://doi.org/10.1016/j.eiar.2009.04.001
96. Sharholy, M., Ahmad, K., Mahmood, G., & Trivedi, R. C. (2008), "Municipal solid waste management in Indian cities - A review", *Waste management (New York, N.Y.)*, Vol. 28(2), pp. 459–467, https://doi.org/10.1016/j.wasman.2007.02.008
97. Sharholy, M., Ahmad, K., Vaishya, R. C., & Gupta, R. D. (2007), "Municipal solid waste characteristics and management in Allahabad, India", *Waste management (New York, N.Y.)*, Vol. 27(4), pp. 490–496, https://doi.org/10.1016/j.wasman.2006.03.001.
98. Shekdar, Ashok V. (2009), "Sustainable solid waste management: An integrated approach for Asian countries", *Waste Management*, Vol. 29(4), pp. 1438–1448.https://doi.org/10.1016/j.wasman.2008.08.025
99. Siddiqui, F.Z. & Khan, M.E. (2011), "Landfill gas recovery and its utilization in India: Current status, potential prospects and policy implications", *Journal of Chemical and Pharmaceutical Research*, Vol. 3(5), pp.174–183
100. Singh, S. "Solid Waste Management in Urban India: Imperatives for Improvement," *ORF Occasional Paper No. 283*, November 2020, Observer Research Foundation.
101. Soni, A.; Patil, D.; Argade, K. (2016), "Municipal Solid Waste Management", *Procedia Environmental Sciences, 35*, pp. 119–126. https://doi.org/10.1016/j.proenv.2016.07.057.
102. Sridevi, V., Musalaiah, M., Ch, M.V.V., Lakshmi, A., Kesavarao, L., (2012), "A Review on Integrated Solid Waste Management", *International Journal of Engineering Science & Advanced Technology*, Vol. 2(5), pp. 1491–1499
103. Sudhir, V., Muraleedharan. V.R. & Srinivasan, G. (1996), "Integrated solid waste management in Urban India: A critical operational research framework", *Socio-Economic Planning Sciences*, Vol. 30(3), pp. 163–181.https://doi.org/10.1016/0038-0121(96)00012-2
104. Sujauddin, M., Huda, S. M., & Hoque, A.T.M.R. (2008), "Household solid waste characteristics and management in Chittagong, Bangladesh", *Waste management (New York, N.Y.)*, Vol. 28(9), pp. 1688–1695, https://doi.org/10.1016/j.wasman.2007.06.013.
105. Sulemana, A., Donkor, E.A., Forkuo, E.K. & Oduro-Kwarteng, S. (2018), "Optimal Routing of Solid Waste Collection Trucks: A Review of Methods", *Journal of Engineering*, 28(11), https://doi.org/10.1155/2018/4586376.
106. Tanskanen, Juha-Heikki (2000), "Strategic planning of municipal solid waste management", *Resources, Conservation and Recycling*, 30, 111–133. https://doi.org/10.1016/S0921-3449(00)00056-2
107. Tchobanaglous, G. & Kreith, F. (2002), "Handbook of Solid Waste Management", Second Edition, McGRAW-HILL, Available on: https://sanitarac.pro/wp-content/uploads/2017/07/Solid-Waste-Management.pdf.
108. Tchobanaglous, G., Theisen, H. & Eliassen, R. (1977), "Solid Wastes: Engineering principles and management issues", *McGraw-Hill*, New York
109. Toerien, D.F. & Hattingh, W.H.J. (1969), "Anaerobic digestion I. The microbiology of anaerobic digestion", *Water Research*, Vol. 3(6), pp. 385–416.https://doi.org/10.1016/0043-1354(69)90002-5
110. Tripathy, U., "A 21st Century Vision on Waste to Energy in India: A Win-Win Strategy for Energy Security and Swachh Bharat Mission (Clean India Mission)", *Background Paper for Plenary Session 7 of the Programme*, 8th Regional 3R Forum in Asia and the Pacific, 9–12 April 2018, Indore, Madhya Pradesh, India, Available on: https://www.uncrd.or.jp/content/documents/6717FINAL-Background%20paper_Tripathy.pdf.
111. UN Habitat (2010), "Solid Waste Management in the World's Cities", London: *United Nations Human Settlements Programme*
112. United Nations (2015), "The World Population Prospects: 2015 Revision", *United Nations Department of Economic and Social Affairs,* Accessed from: https://www.un.org/en/development/desa/publications/world-population-prospects-2015-revision.html.

113. United Nations Environmental Programme (UNEP), "Solid Waste Management (Volume II: Regional Overviews and Information Sources)" *United Nations Publications*, 2005, Available on: https://wedocs.unep.org/20.500.11822/30734.
114. United States Environmental Protection Agency (EPA). 2021, "National Overview: Facts and Figures on Materials, Wastes and Recycling", Accessed from: https://www.epa.gov/facts-and-figures-about-materials-waste-and-recycling/national-overview-facts-and-figures-materials.
115. Upadhyay, V., Jethoo, A.S. & Poonia, M. P. (2012), "Solid Waste Collection and Segregation: A Case Study of MNIT Campus, Jaipur", *International Journal of Engineering and Innovative Technology (IJEIT),* Vol. 1(3), pp. 144–149
116. Vallero, D.A., Shulman, V. et al., "Waste", Second Edition, *Academic Press*, March, 2019.
117. Vaverková, M.D. (2019), "Landfill Impacts on the Environment-Review", *Geosciences*. Vol. 9(431), pp. 2–16
118. Vergara, Sintana & Tchobanoglous, George. (2012), "Municipal Solid Waste and the Environment: A Global Perspective", *Annual Review of Environment and Resources*, 37: 277–309. https://doi.org/10.1146/annurev-environ-050511-122532
119. Vidanaarachchi, C.K, Yuen, S.T.S. & Pilapitiya, S. (2006), "Municipal solid waste management in the Southern Province of Sri Lanka: Problems, issues and challenges", *Waste Management*, Vol. 26(8), pp: 920-930, https://doi.org/10.1016/j.wasman.2005.09.013.
120. Visvanathan, C. (2006), "Solid Waste Management in Asian Perspectives", Environmental Management Tools, 131–142, Available on: http://www.sacep.org/pdf/Male-Declaration-Publications/2006-Male-Declaration-EMT-Manual-Final.pdf#page=147.
121. Visvanathan, C., Norbu, T., Stegmann, R., Körner, I., Allgaier, G., Ritzkowski, M., Kuschke, C., Lehmann, K., Cossu, R., Gadia, R., Awang, M., Abdul Aziz, A., Hassan, M.N., Theng, L.C., Chee, T.M. & Tränkler, J. (2006), "Teaching and Training Modules for Higher Education in the Waste Management Sector", TUHH, *Hamburg University of Technology, Institute of Waste Resource Management, Germany.*
122. Williams, E., Kahhat, R., Allenby, B., Kavazanjian, E., Kim, J., & Xu, M. (2008), "Environmental, social, and economic implications of global reuse and recycling of personal computers", *Environmental science & technology*, Vol. 42(17), pp. 6446–6454. https://doi.org/10.1021/es702255z.
123. Williams, P.T. (2005), "Waste Treatment and Disposal", Second Edition, *John Wiley & Sons Ltd.*, West Sussex, England.
124. Wilson, D.C., Rodic, L., Scheinberg, A., Velis, C.A. & Alabaster, G. (2012), "Comparative analysis of solid waste management in 20 cities", *Waste Management & Research*, Vol. 30(3), pp. 237-254, https://doi.org/10.1177/0734242X12437569
125. Xue, W., Cao, K. & Li, W. (2015), "Municipal solid waste collection optimization in Singapore", *Applied Geography*, Vol. 62, pp: 182-190, https://doi.org/10.1016/j.apgeog.2015.04.002.
126. Zhang, D. Q., Tan, S. K., & Gersberg, R. M. (2010), "Municipal solid waste management in China: status, problems and challenges", *Journal of environmental management*, Vol. 91(8), pp. 1623–1633. https://doi.org/10.1016/j.jenvman.2010.03.012.

Environmental Impact of Emerging Electro World with Special Focus on Electrical-Vehicles and Its Impact on Global Environment

P. T. Hemand Chandran, Sarath Ajithkumar, and Kanchan Deoli Bahukhandi

Abstract With the introduction of the electric vehicle, plenty of lifecycle evaluations calculating CO_2 emissions have sprung up. Electric vehicle emissions, including battery and charging, are compared to those of conventional vehicles. According to the most recent evidence, an average electric car is already close to three times better than an equivalent conventional car today. As new evidence accumulates on: the longer lifetime of batteries thanks to innovation, the ramping up of battery reuse, repurpose, and recycling, and the accelerating uptake of renewables, the life cycle assessment emissions of electric cars are bound to decrease even further than what is presented in this case. The major disadvantage that's are been connected with the EV are the environmental impact of battery production for electric vehicles (EVs) and the mining of rare earth metals for the car production. The purpose of this paper is to investigate the environmental impact of electric vehicles. And how the batteries used in electric vehicles can pose a risk to humans.

Keywords Global warming · Electric vehicle · Global environment

1 Introduction

During the past 10-year the number of vehicles on the road have increases significantly. And it is the major contributor of greenhouse gases [1]. Recent trend in the automotive industry is the introduction of fully electric vehicles. By replacing internal combustion engine with electric vehicle will not resolve the problem. Energy needed to charge the battery pack is received by burning fossil fuel in many countries.

Scotsman Robert Anderson is credited with inventing the first electric car sometime between 1832 and 1839. But it was in 2008 when the tesla introduced its roadster

P. T. Hemand Chandran · S. Ajithkumar
HSE, College of Engineering, UPES, Uttarakhand, India

K. D. Bahukhandi (✉)
Sustainability Cluster (HSE and Civil Engineering), University of Petroleum and Energy Studies (UPES), Bidholi, Dehradun, India
e-mail: kanchan@ddn.upes.ac.in

K. D. Bahukhandi et al. (eds.), *Environmental Pollution and Natural Resource Management*, Springer Proceedings in Earth and Environmental Sciences,
https://doi.org/10.1007/978-3-031-05335-1_18

model people thought about electric deeply [2]. Electric cars have the reputation as a possible solution to many modern environmental and economic issues. People around the world also started realizing the importance of environmental changes and realize the fact that environmental change is real. To growth an alternative for the CO_2 emission from the gasoline vehicle. Zero tailpipe emission will reduce the amount of CO_2 produced.

The environment suffers from industrialized nations as the amount of greenhouse gas emissions reach new heights every year. The electric vehicle market has been rapidly expanding since 2008 [3]. Electric vehicles have also grown in popularity because of the increasing concern for the environmental impact that greenhouse gas emissions cause This increase in emissions is due to the rising population and increased production and consumption of energy and transportation the amount of greenhouse gas emissions increased by 41% internationally The emissions of fully electric vehicles (FEVs), when taking the energy production emissions into account, will be analysed to prove that FEVs will make a significant reduction in greenhouse gas emissions [4].

The major concern about electric vehicle Is the energy required in making the vehicle and energy spent on making and running fully electric vehicle. Only the emission is zero, carbon dioxide production is happening in other steps of process. Which we have to overcome with newer technologies.

Electric vehicles have a significant advantage in terms of energy efficiency. They also produce zero emissions at the time of use, reducing greenhouse gas emissions from the transportation sector significantly [5].

The electric car manufacturing is booming because the technology is widely in use and cost of production has significantly come down. The long charging durations of the batteries, the vehicle's relatively low range, and the high initial cost are all downsides of electric cars.

The environment suffers from industrialized nations as the amount of greenhouse gas emissions reach new heights every year. The electric vehicle market has been rapidly expanding since 2008. Electric vehicles have also grown in popularity because of the increasing concern for the environmental impact that greenhouse gas emissions because This increase in emissions is due to the rising population and increased production and consumption of energy [6]. In transportation the amount of greenhouse gas emissions increased by 41% internationally the emissions of fully electric vehicles (FEVs), when taking the energy production emissions into account, will be analysed to prove that FEVs will make a significant reduction in greenhouse gas emissions.

When we compare Cabon footprint of the electric and gasoline driven car, in the initial stages of the of the production electric car the carbon footprint is significantly larger as there is requirement of excavation for finding minerals for making battery and resources for powering the battery pack of the car [7].

Fully electric does produce carbon dioxide but it is only 50% of what gasoline produce. Driving an electric car instead of a gasoline car could save carbon emission of about 4096 pounds a year. The major concern about electric vehicle Is the energy required in making the vehicle and energy spent on making and running fully electric vehicle. Only the emission is zero, carbon dioxide production is happening in other steps of process. Which we have to overcome with newer technologies.

As the electric vehicle market grows and technology improves to accommodate bidirectional charging, EVs may become significant grid resources. Operators of electric vehicles will profit from lower energy and maintenance costs [8]. Benefits to society as a whole, frequently through the elimination of negative effects, are referred to as societal benefits. Environmental or health repercussions are examples of externalities.

Compared to conventional internal combustion engines, electric vehicles have much fewer moving parts. The drive train's battery, motor, and electronics don't need to be serviced on a regular basis. Oil changes are no longer necessary, and there are no additional fluids to replace except brake fluid. The wear on the brakes of an electric vehicle is greatly decreased due to regenerative braking, hence they require less maintenance than brakes on a conventional automobile [9].

Driving an EV saves roughly 344 gallons of gas each year, or 3440 gallons throughout the vehicle's lifetime. Widespread use of EV comes at a cost to the economy Job losses in the oil business, at petrol stations, and possibly in the auto maintenance and mechanic industry will occur. In the auto industry, direct jobs will be created in production, research and development, and battery manufacturing. Installation and maintenance of electric vehicle supply equipment will create indirect jobs (EVSE) [10].

1.1 Effect on Environment

There are many reasons to promote sustainable energy sources because the amount of greenhouse gas emission in most of the cities are already above the allowable level. This rise in greenhouse gases is resulting in premature death due to asthma, cardiovascular disease, lung cancer is the few.th global cost of air pollution is about 3 trillion dollars per year [11]. To achieve this goal, transportation must be electrified, electricity must be generated without carbon emissions, agricultural operations must be electrified, buildings must be heated with solar energy and electricity, and carbon emissions from construction, mining, and industrial production must be reduced. One significant advantage of all of this is improved air quality due to lower emissions.

The electrification of transportation is one of the global developments that has begun. In 2017, approximately 2 million electric vehicles were in use. Electric vehicles account for more than 25% of new car sales in Norway. There are 30 cities in California where the percentage of new car sales that are plug-in vehicles ranges from 6 to 18% [12].

For electric power companies, the last ten years have been a period of transition, with large increases in power generated from wind and solar energy. Brown et al. have written a book about the major shift to wind and solar energy [13]. Natural gas, wind, and solar energy are displacing coal-fired power plants. From 2008 to 2015, Obama claims that the cost of wind-generated electricity has decreased by

41%, while the cost of roof-top solar photovoltaic power has decreased by 54%. According to current sources, the amount of new wind capacity added in 2015 was first, while the amount of new solar capacity added was second. One of the causes for the decrease in greenhouse gas emissions is the expansion of wind and solar generating capacity [14].

The extensive deployment of a fleet primarily made up of electric vehicles would pose several issues to the grid's electricity producing infrastructure. Charging a large number of cars during peak hours could result in a major rise in electricity demand, resulting in grid overload.

1.2 Electricity Production in India

As per current study India is generating about 384,116 MW of energy each year. And is increasing each year due to increase in population, urbanization, and digitalization the few. When the electric market in India is fully fledged demand of electricity will increase exponentially.

To compensate for this increase in electricity demand, an additional 4 GWh of electricity per year will be required. The amount of carbon emissions from the electrical grid is predicted to increase slightly by 0.47 megatons CO_2 per year as power generation rises. Given that the amount of carbon dioxide produced by light internal combustion cars is currently 3.6 megatons per year [15] (Table 1).

Our country mainly produces electricity from thermal energy derived from coal which is a fossil fuel which will emit greenhouse gases that are harmful for the environment. Countries will be benefited from using electric vehicle only if they start to transform into renewable source of electricity production otherwise only the tailpipe emission remains zero and other pollution will remain the same (Fig. 1).

2 Cost Analysis

Here we are considering two vehicle of the same brand tata Nexon EV verses tata Nexon petrol versions, and analysing different cost that it has to be paid during the life of the car.

Every government is promoting the use of electric vehicle to make the environment cleaner. Because of the same reason they are proving incentives and wave off when we buy an electric vehicle. Each government have different policy (Fig. 2).

Table 2 shows the amount that we have to take into account when we buy the car. In the first look itself we might consider petrol vehicle as our first choice but when we analyse it in long span, we will come to the conclusion that EV is better with few drawbacks (Fig. 3; Table 3).

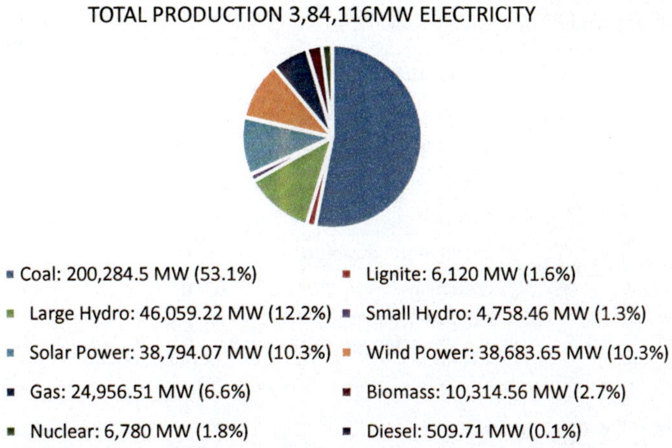

Fig. 1 Total power generation around the world

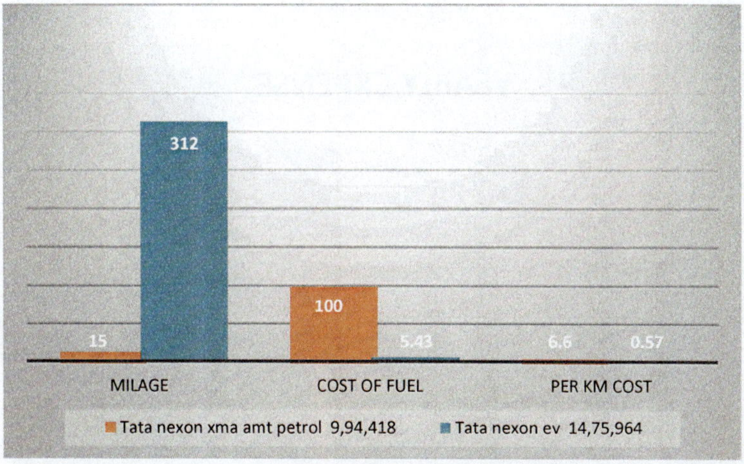

Fig. 2 Cost to run vehicle

Both cars need regular maintenance and fuel to run considering both these. EV is cheaper as it is easier to maintain, contains fewer moving parts as compared to internal combustion engine.

2.1 Cost Spend on Life of Car

Here we consider there is no problem to the parts of the vehicle. Price of fuel and electricity is considered to be constant during the life of the vehicle (Table 4).

Table 1 Power generated by different fuel sources

Fuel	MW	% of total
Total thermal	234,058	60.90
Coal	202,005	52.60
Lignite	6620	1.7
Gas	24,924	6.50
Diesel	510	0.10
Hydro (renewable)	46,322	12.10
Nuclear	6780	1.80
RES* (MNRE)	96,956	25.20
Total	384,116	

Table 2 Nexon EV verses nexon petrol

Model	On road price	Milage	Cost of fuel	Per km cost
Tata nexon xma amt petrol	994,418	15	100	6.6
Tata nexon ev	1,475,964	312	5.43	0.57

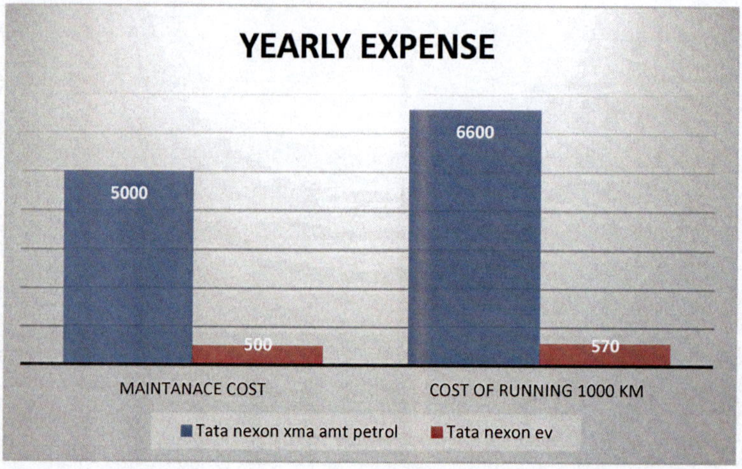

Fig. 3 Yearly expense of car

Table 3 Maintenance cost comparison

Model	Maintanace cost	Cost of running 1000 km
Tata nexon xma amt petrol	5000	6600
Tata nexon ev	500	570

Table 4 Amount that should be spent on insurance

Years	Nexon ev	Nexon petrol
1	42,777	20,231
2	18,406	8918
3	18,406	8918
4	9440	4261
5	9440	4261
6	9440	4261
7	9440	4261
8	9440	4261

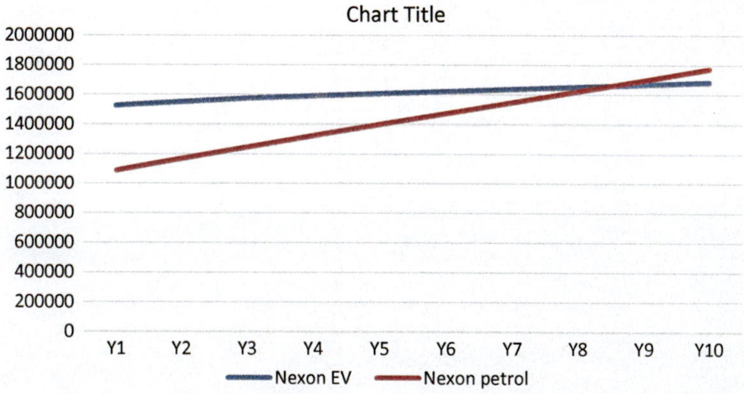

Fig. 4 Total distance covered in each year

Taking into account of the money spent on fuel, insurance, and considering both vehicles travel about 10,000 km per year the amount that the user will have to spent on each year are (Fig. 4; Table 5).

When we consider the amount that we have to spent on vehicle for 10 years it comes to the conclusion that electric vehicle is cheaper than internal combustion engine.

2.2 *What Will Happen at the End of Life?*

At the end of cars life, it will be demolished at will be recycled and used its elements for other purposes. In the case of internal combustion engine most of its part is being demolished and no much environmental damage is produced with it. But in the case of electric car, it's not the same. Battery is the component that will be affect the environment the most. A proper method to recycle battery should be in place to take care of the battery in these electric vehicles [16].

Table 5 Total distance covered in each year

Model	Y1	Y2	Y3	Y4	Y5	Y6	Y7	Y8	Y9	Y10
Nexon EV	1,524,941	1,549,547	1,574,153	1,589,793	1,605,433	1,621,073	1,636,713	1,652,353	1,667,993	1,683,633
Nexon petrol	1,085,649	1,165,567	1,245,485	1,320,746	1,396,007	1,471,268	1,546,529	1,621,790	1,697,051	1,772,312

In country like India where most of the recyclable materials thrown off to oceans and landfills. Electric car boom will be a concern unless we make necessary arrangements to recycling battery.

3 Working

An electric vehicle is one that is propelled by electric motors and draws power from an onboard electric source. It is more durable and has a simpler mechanical design than a gasoline vehicle. Because it does not emit emissions like an internal combustion engine, it has a higher fuel economy than gasoline.

Major components of motor are

- Motor
- Motor controller
- Battery

The primary purpose of a motor is to transform electrical energy into mechanical energy. There are only two primary elements to it. There are rotors and stators. The rotor is a revolving component that contains a permanent magnet, whereas the stator is a stationary component that houses the stator winding. The stator's construction is similar to that of an induction motor. It's made of steel lamination that's been axially sliced to allow for winding.

In most of the electric vehicle, Brushless DC motor is used which is better than brushed DC motor, permanent magnet DC motor. Because brushless DC motors do not have a brush or commutator, electric vehicles that use them must have a motor controller to manage the motor's numerous attributes and ensure that appropriate current and voltage are given to the motor. When the accelerating pedal is pressed, the linked variable resistor type controller sends a signal to the motor controller, which adjusts the speed according to our requirements.

A battery's primary function is to store electric energy. Lithium-ion batteries were employed in the majority of hybrid and electric vehicles. For hybrid electric vehicles, plug-in hybrid electric vehicles, and all types of electric vehicles, this storage system is frequently required. The majority of automotive manufacturers now employ rechargeable batteries as a result of advancements in technology [17] (Fig. 5).

4 Types of Electric Vehicle

1. Battery EV:—A battery electric vehicle is made up of an electric motor that is powered by a battery. In this sort of vehicle, the movement is powered by an electric motor. It doesn't emit anything. BEVs provide more torque to the wheels and smoother acceleration in traffic than internal combustion engines. While the

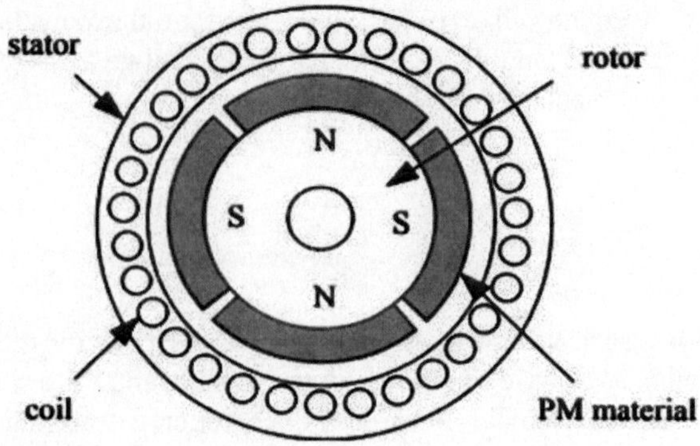

Fig. 5 Schematic diagram of electric motor in a car

motor is running, it makes no noise. However, there are several drawbacks, such as high production costs, a limited top speed, and a longer recharge time.

2. Hybrid EV:—A hybrid electric car is one that combines an electric motor with a modern internal combustion engine. This vehicle is compatible with both electric and gasoline engines. Switch to Fuel mode when the electric motor stops working. Low-speed mechanical electric motors are employed in HEVs for city traffic. There are no emissions when a car runs on electric power.
3. Range extended EV:—The majority of range-extended electric vehicles are designed to run on batteries, but they also contain a gasoline generator to replenish the batteries when the battery charge is low [18].

5 Types of Batteries Used in Electrical Vehicles

5.1 Lithium Ion

Li-ion batteries can be characterized as energy storage systems that rely on insertion reactions from both electrodes where lithium ions act as the charge carrier. Negative electrodes in Li-ion batteries are primarily made of carbon (e.g., graphite) or lithium titanate ($Li_4Ti_5O_{12}$), with certain unique materials, such as Li metal and Li (Si) alloys. The electrolyte utilised varies depending on the electrode materials employed, however it is commonly made up of a mixture of lithium salts and an organic solvent to facilitate ion transport. A separating membrane is used to allow lithium ions to pass between the electrodes while preventing an internal short circuit [19] (Fig. 6).

The electrolyte must offer the highest possible lithium-ion transport under use conditions.

The batteries must operate in the general environment, likely to extend from, example -30 °C for a vehicle that has been parked for a period of time in extreme cold, to +60 °C for a battery that has heated as a consequence of the combination of environmental conditions and heat generated by charging. The separator must likewise offer the highest possible lithium ion conduction under the same operational

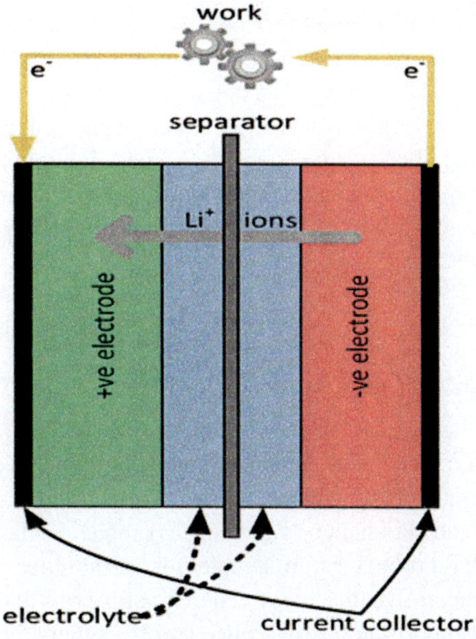

Fig. 6 Energy cycle in electric car

conditions and must offer the ability for a rapid thermal shutdown if significant overheating occurs to prevent a thermal runaway process [20].

In terms of rechargeable batteries, lithium-ion batteries have one of the greatest columbic efficiencies (CE) ratings. CE is the ratio of total charge extracted from the battery to total charge placed into the battery over a full cycle, and it describes the charge efficiency by which electrons are moved in batteries. The columbic efficiency of Li-ion batteries improves with cycling. To prove this, Panasonic, E-one Moil, Sony, LG and Samsung Li-ion batteries in 18,650 cell formats where cycled. Some cells began with a columbic efficiency of 99.1% and improved to 99.5% with 15 cycles. Some started at 99.5% and reached 99.9% with 30 cycles. The consistency on repeat tests was high, reflecting in Li-ion being a very stable battery system [12].

5.2 *Nickel–Metal Hydride*

The hydrogen-absorbing negative electrode replaces the cadmium-based electrode in nickel–metal hydride batteries, which are simply an extension of the established sealed nickel–cadmium battery technology. While this substitution enhances the battery's electrical capacity for a given weight and volume and removes the cadmium, which raises toxicity concerns, the nickel–metal hydride battery's rest of the components are quite similar to the nickel–cadmium product. Many application characteristics do not differ significantly between the two types of batteries. The nickel–metal hydride battery chemistry is a combination of the sealed nickel–cadmium battery's proven positive electrode chemistry and the energy storage properties of metal alloys developed for advanced hydrogen energy storage concepts [21].

The positive and negative electrodes are separated by separators and form the basic components. After injecting electrolyte, the sandwiched electrodes are wrapped together and inserted into a metallic container. nickel–metal hydride batteries are typically sealed designs with electrically insulated metallic casings and tops. The cover serves as the battery's negative terminal, while the top serves as the positive terminal.

5.3 Lead Acid Battery

The lead acid battery's major components are the container and plates. The plates transform chemical energy into electrical energy, which is stored in the container. The lead acid battery's container is made of glass, lead lined wood, ebonite, bituminous hard rubber, ceramic materials, or moulded plastics and is seated at the top to prevent electrolyte discharge. There are four ribs at the bottom of the container, two of which support the positive plate and the others the negative plate. The prism supports the plates while also protecting them from a short circuit [10]. The material used to make the battery containers must be resistant to sulphuric acid, not distort or porous, and not have pollutants that harm the electrolyte. The lead-acid cell plate comes in a variety of shapes and sizes, but they all have a grid constructed of lead and the active material. The grid is necessary for carrying electric current and evenly dispersing it across the active material. The active material will loosen and fall out if the current is not dispersed evenly. Thin sheets of non-conducting material, such as chemically treated Leadwood, porous rubbers, or glass fibre mats, are placed between the positive and negative to insulate them from each other. On one side, separators are grooved vertically and smooth on the other. The positive and negative terminals of a battery are connected. The positive terminal, which has a top diameter of 17.5 mm, is slightly larger than the negative terminal, which has a diameter of 16 mm.

When sulphuric acid dissolves, its molecules split into two types of ions: positive hydrogen ions (2H+) and sulphate negative ions (SO_4-) that can freely move around. If the two electrodes are immersed in solutions and linked to a DC supply, positively charged hydrogen ions will migrate towards the electrodes and attach to the supply's negative terminal. The negatively charged SO_4- ions gravitated toward the electrodes attached to the supply main's positive end (i.e., anode) [16]. Each hydrogen ion takes one electron from the cathode, and each sulphates ion absorbs two negative ions from the anodes, forming sulphuric and hydrogen acid when they combine with water. The oxygen produced by the above equation reacts with lead oxide to produce lead peroxide. As a result, although the lead cathode remains as lead during charging, the lead anode is transformed to lead peroxide.

6 Safety Issues in Battery

6.1 Over Charging

Li-ion cells have a very narrow range of operating voltage and temperature. The safe voltage charging occurs in three steps: constant current charge (bulk charge), constant voltage charge (saturation charge) and trickle charge. A Li-ion cell is considered fully charged when its voltage reaches the nominal value and the charging current drops to about 3% of the rated value. After the battery's designed capacity has been attained, surplus energy is pumped into the battery in the form of trickle current. Overcharge cannot be absorbed by lithium-ion batteries. After attaining full charge, forcing even a tiny continuous current causes the cell voltage to rise. The electrolyte and solvents breakdown and generate combustible fumes when the cell voltage hits roughly. Pressurization of these gases result in battery swelling along with high cell temperature, which, in worst condition, may lead to explosion and fire. The pressurisation of these gases causes battery swelling and a rise in cell temperature, which, in the worst-case scenario, might end in an explosion and fire.

6.2 Over Discharge

This is also known as current-collector dissolution. This phenomenon known as over-discharge happens when a cell is discharged below the designated lower voltage limit. This occurs in battery packs that use series–parallel cell combinations to achieve high voltage and current. Because it is difficult to have all of the cells in a pack at the same charge level, the cells with lower capacity are required to release the same amount of energy as the other cells while discharging. When a Li-ion cell is overcharged, the copper current collector at the anode dissolves. This copper can be deposited on other components, resulting in an internal short. This causes capacity to deteriorate or fail over time.

6.3 Thermal Runaway

Cycle life of the battery reduces significantly at temperature below 15 °C and slightly above 50 °C. However, beyond 60 °C cell temperature, fast reduction in life besides the danger of battery going in to thermal runaway is very high.

6.4 Gas Evalaution

Gas generation will grow under abnormal conditions such as overcharge, over-discharge, thermal runaway, and so on, potentially leading to pressurisation and explosion. The breakdown of the electrolyte and its reactions with the positive/negative electrodes are the primary causes of gas generation in Li-ion batteries. The amount of gas produced will vary depending on the state of the battery. A safety valve is supplied in some high-power Li-ion battery systems to expel the gas [22].

7 Life Span of Battery

The health and lifespan of Li-ion batteries are affected by a number of factors. The loss of either capacity or power is a symptom of Li-ion battery performance degradation. Capacity is lost when active materials are changed into inactive phases as a result of parasitic chemical events, though the problem is complex and difficult to model using basic methodologies. When coming to power, parasitic processes occur, the battery materials are converted to other compounds that act as transport barriers, raising the cell's internal impedance and, as a result, lowering the operational voltage at each discharge rate.

7.1 Temperature

One of the most important elements affecting the performance and lifespan of a Li-ion battery is the temperature of the cells, especially during charging and discharging. Overheating the batteries can cause a thermal runaway, with temperatures reaching 500 °C. Even a single cell's thermal runaway can cause a chain reaction with other cells, potentially resulting in fire and death.

All Li-ion batteries have a battery management system (BMS) that regulates and controls all aspects of the batteries, including charging, discharging, cell equalisation and monitoring, as well as controlling the overall temperature of the system, to improve the battery's lifespan and solve the greater safety issue [21]. A BMS in an electric vehicle can perform data logging, report to a Supervisory Control Module, and improve battery performance and optimise vehicle operation, such as protecting from safety hazards such as fire and shock, maintaining an optimal operating environment, state of charge, depth of discharge, charge/discharge power, and battery cell balancing for improved battery life and efficiency.

7.2 *Charge and Discharge Rate*

A common charging convention known as constant current constant voltage is recommended by most manufacturers (CC-CV). When charging, the system maintains a steady current until the battery reaches the maximum charging voltage, then decreases the current to keep the charging voltage constant to avoid the cell being overcharged. The charger evaluates the battery condition while it is being charged and adjusts the charge current as needed. An ultra-fast charger may charge at up to 10 ℃ at first, then drop to 1 ℃ or even lower when the battery reaches its typical working voltage. This extra current increases the stress on the battery during the initial charging time, which raises the battery temperature. The goal is to keep the electrodes from becoming lithium plated while maintaining a normal working temperature. As a result, only use ultra-fast chargers when absolutely essential [23].

7.3 *Ways by Which We can Increase the Life Span of Battery*

It is also feasible to improve the electrode design and boost battery capacity by examining the design and/or size of the materials used in the battery's manufacture. It is possible to reduce the heat created during operation by lowering the internal resistance of the electrodes, hence increasing the cell capacity and lifespan. It is feasible to optimise transport distances or route ion travel to increase overall conductivity and thermal performance by altering the material as it is manufactured. This can be accomplished by coating, doping, or depositing electrodes as they are manufactured. These modifications have improved the material's strength while also increasing its thermal efficiency and total capacity [16].

8 Impact on Human

One study estimates that the potential human toxicity effects of the production phase to be between 2.2 and 3.3 times greater for BEVs than ICEVs. Cobalt, graphite, and lithium are all used in the production of lithium-ion batteries.

8.1 *Cobalt*

Food and beverages are the most common sources of cobalt exposure for the general public. Cobalt is sometimes present in cement and other things that we use on a daily basis. When cobalt enters the environment through dust, water, or soil, it generates a slew of negative consequences including contamination. Estimates of cobalt exposure imply that more than 99 percent of it comes from the food we eat every day. Cobalt can induce dermatitis, rhinitis, and asthma in people who are sensitive to it. Some of the health effect of cobalt is cancer, respiratory effects cardiac effects, haematological effects, cardiovascular effects, renal effects and hepatic effects. Adsorption, bio sorption of metal by various fungal species, bio sorption by bacteria, bio sorption by yeast, and phytoremediation are all physical methods for removing cobalt. Chemical processes include chemical precipitation, coagulation, and flocculation, as well as electrochemical chemical treatments, ion exchange, membrane filtration, metal precipitation, and electro dialysis [24].

8.2 Lithium

Many lithium reactions can result in fire or explosion. In a fire, emits unpleasant or hazardous fumes (or gases). The chemical is harmful to the eyes, skin, and respiratory system. When consumed, it is corrosive. Lung oedema may occur if the chemical is inhaled. Lung oedema symptoms often do not appear until a few hours later, and they are exacerbated by physical exertion. As a result, rest and medical observation are required [6].

8.3 Lead

Lead serves no biological purpose in the human body. It builds up in the body and affects nearly every organ system. Lead poisoning can have long-term and devastating effects on people of all ages, but it is especially hazardous to children. Lead serves no biological purpose in the human body. It builds up in the body and affects nearly every organ system. Lead poisoning can have long-term and devastating effects on people of all ages, but it is especially hazardous to children. Lead-acid batteries are used in automobiles, to store energy generated by solar panels and wind turbines, and to provide backup power. The growing usage of renewable energy sources, which necessitates the need for storage batteries, as well as the rising need for motor vehicles as countries progress economically, means that lead-acid battery demand will continue to rise.

Lead can be emitted in the form of fumes, particles, and dust at any point during the battery recycling process, from draining and dismantling the batteries to smelting and purifying the lead. Recycling can pollute the environment, exposing workers and the surrounding community to dangerous levels of contamination. Recycling workers who do not wash and change their clothes before leaving work may transport lead pollution home with them, potentially exposing family members. Recycling workers who do not wash and change their clothes before leaving work may transport lead pollution home with them, potentially exposing family members [8].

9 Conclusion

As the E vehicle industry is expanding at a exponential rate, there are many advantages attracting people to get a hold of these new generational automobile. Of all the points discussed in the above paper, two of them which need utmost attention are the recycling of the battery which are used in the vehicle and the mining of rare earth metals for the manufacturing of automobile. With current raw material prices and battery composition, all recycling methods have been proved to be cost-effective in large volumes. When opposed to the extraction of virgin materials, recycling has a lesser environmental impact.

References

1. "running a car cost much more than people think." .
2. V. Rizza *et al.*, "Effects of deployment of electric vehicles on air quality in the urban area of Turin (Italy)," *J. Environ. Manage.*, vol. 297, p. 113416, Nov. 2021, doi: https://doi.org/10.1016/J.JENVMAN.2021.113416.
3. M. A. Andor, A. Gerster, K. T. Gillingham, and M. Horvath, "Running a car costs much more than people think — stalling the uptake of green travel," *Nature*, vol. 580, no. 7804, pp. 453–455, Apr. 2020, doi: https://doi.org/10.1038/D41586-020-01118-W.
4. W. Habla, V. Huwe, and M. Kesternich, "Electric and conventional vehicle usage in private and car sharing fleets in Germany," *Transp. Res. Part D Transp. Environ.*, vol. 93, p. 102729, Apr. 2021, doi: https://doi.org/10.1016/J.TRD.2021.102729.
5. F. R. Kalhammer, A. Kozawa, C. B. Moyer, and B. B. Owens, "Electric vehicle batteries," *Electrochem. Soc. Interface*, vol. 5, no. 1, pp. 32–35, 37, 1996, doi: https://doi.org/10.1149/2.f05961if.
6. L. E. Erickson, "Reducing greenhouse gas emissions and improving air quality: Two global challenges," *Environ. Prog. Sustain. Energy*, vol. 36, no. 4, pp. 982–988, Jul. 2017, doi: https://doi.org/10.1002/EP.12665.
7. I. Malmgren, "EVS29 International Battery, Hybrid and Fuel Cell Electric Vehicle Symposium Quantifying the Societal Benefits of Electric Vehicles," *World Electr. Veh. J.*, vol. 8, pp. 996–1007, 2016.
8. M. Baumann, M. Salzinger, S. Remppis, B. Schober, M. Held, and R. Graf, "Reducing the environmental impacts of electric vehicles and electricity supply: How hourly defined life cycle assessment and smart charging can contribute," *World Electr. Veh. J.*, vol. 10, no. 1, 2019, doi: https://doi.org/10.3390/wevj10010013.
9. M. Giansoldati, L. Rotaris, M. Scorrano, and R. Danielis, "Does electric car knowledge influence car choice? Evidence from a hybrid choice model," *Res. Transp. Econ.*, vol. 80, no. February, p. 100826, 2020, doi: https://doi.org/10.1016/j.retrec.2020.100826.
10. G. Milev, A. Hastings, and A. Al-Habaibeh, "The environmental and financial implications of expanding the use of electric cars - A Case study of Scotland," *Energy Built Environ.*, vol. 2, no. 2, pp. 204–213, 2021, doi: https://doi.org/10.1016/j.enbenv.2020.07.005.
11. M. Giansoldati, A. Monte, and M. Scorrano, "Barriers to the adoption of electric cars: Evidence from an Italian survey," *Energy Policy*, vol. 146, no. September, p. 111812, 2020, doi: https://doi.org/10.1016/j.enpol.2020.111812.
12. S. Bobeth and I. Kastner, "Buying an electric car: A rational choice or a norm-directed behavior?," *Transp. Res. Part F Traffic Psychol. Behav.*, vol. 73, pp. 236–258, 2020, doi: https://doi.org/10.1016/j.trf.2020.06.009.

13. L. Yao *et al.*, "A Review of Lithium-Ion Battery State of Health Estimation and Prediction Methods," 2021.
14. Z. H. Zhu, Z. Y. Gao, J. F. Zheng, and H. M. Du, "Charging station location problem of plug-in electric vehicles," *J. Transp. Geogr.*, vol. 52, pp. 11–22, Apr. 2016.
15. A. Schmid, "An Analysis of the Environmental Impact of Electric Vehicles Recommended Citation," *Missouri S&T's Peer to Peer*, vol. 1, no. 2, pp. 1–9, 2017, [Online]. Available: http://scholarsmine.mst.edu/peer2peerhttp://scholarsmine.mst.edu/peer2peer/vol1/iss2/2.
16. P. A. Nelson, W. H. Webster, and H. Shimotake, "Batteries for Electric Vehicles.," *New Mater. New Process.*, vol. 2, no. November, pp. 362–364, 1983, doi: https://doi.org/10.1017/cbo9781316090978.
17. D. A. J. Rand and P. T. Moseley, "Secondary Batteries - Lead- Acid Systems | Overview," *Encycl. Electrochem. Power Sources*, pp. 550–575, 2009, doi: https://doi.org/10.1016/B978-044452745-5.00126-X.
18. C. Mahmoudi, A. Flah, and L. Sbita, "An overview of electric Vehicle concept and power management strategies," *2014 Int. Conf. Electr. Sci. Technol. Maghreb, Cist. 2014*, no. November, 2014, doi: https://doi.org/10.1109/CISTEM.2014.7077026.
19. R. K. Lattanzio and C. E. Clark, "Environmental Effects of Battery Electric and Internal Combustion Engine Vehicles," 2020, [Online]. Available: https://crsreports.congress.gov/product/pdf/R/R46420.
20. R. Sarode and A. Eadke, "Electric Vehicle (Construction and Working principle)," *Int. Res. J. Eng. Technol.*, vol. 7, no. 10, pp. 140–142, 2020.
21. "Electric vehicle life cycle analysis and raw material availability," no. October, pp. 1–11, 2017.
22. F. R. Kalhammer, H. Kamath, M. Duvall, M. Alexander, and B. Jungers, "Plug-in hybrid electric vehicles: Promise, issues and prospects," *24th Int. Batter. Hybrid Fuel Cell Electr. Veh. Symp. Exhib. 2009, EVS 24*, vol. 3, no. February, pp. 2134–2144, 2009.
23. M. R. Palacín and A. De Guibert, "Batteries: Why do batteries fail?," *Science (80-.).*, vol. 351, no. 6273, 2016, doi: https://doi.org/10.1126/science.1253292.
24. L. Leyssens, B. Vinck, C. Van Der Straeten, F. Wuyts, and L. Maes, "Cobalt toxicity in humans—A review of the potential sources and systemic health effects," *Toxicology*, vol. 387, pp. 43–56, 2017, doi: https://doi.org/10.1016/j.tox.2017.05.015.

Flash Flood in Himalayan Region of Uttrakhand (A Case Study of Kedarnath Flood 2013 and Rishi Ganga Flash Flood, Reini Village 2021)

Nitesh Sharma, Kanchan Deoli Bahukhandi, and Siddiqui Nihal Anwar

Abstract The climate change significantly affect the glacial of Himalayan region. The flash floods are one of the most perilous climate-related disastrous events in the Himalayan region. Such floods grow under six hours after the rainfall that prompted risky circumstances for individuals and cause loss of life, property and environment. Just about seven years after such flash floods attacked the Kedarnath valley in Uttarakhand, around 5000 individuals were killed, and the scientists have cautioned that conditions were creating for a comparative misfortune going to happen in the district again. The cloudburst situation on every next day had occurred mainly in the Uttarakhand caused decimating floods and landslides turning into the nation's most exceedingly awful natural disaster event since the tsunami in the year of 2004, which was about 375% more than the benchmark rainfall during an ordinary rainstorm. Aside from the social, political, and affordable misfortunes, such natural disasters adjust the prior landscape of Kedarnath region. This paper provided the satellite data for the year 2007, 2013, and 2019. The information and data are compared with before the event of the flash flood and after the flash flood and its impact caused alongside. The comparative analysis have also been carried out by using the GIS and Remote Sensing techniques by comparing before and after flash flood event.

Keywords Flash flood · Himalayan region · Geographic Information System · Natural disaster · Remote Sensing · Risk assessment

1 Introduction

The weak idea of worldwide tourism is one of the significant worries for possible management. Disaster management is a significant part of any tourism destination

N. Sharma
HSE, School of Engineering, UPES, Dehradun, India

K. D. Bahukhandi (✉) · S. N. Anwar
Sustainability Cluster (HSE and Civil Engineering), School of Engineering, UPES, Dehradun, India
e-mail: kanchan@ddn.upes.ac.in

K. D. Bahukhandi et al. (eds.), *Environmental Pollution and Natural Resource Management*, Springer Proceedings in Earth and Environmental Sciences,
https://doi.org/10.1007/978-3-031-05335-1_19

(particularly despite an emergency). The particular possibilities, for example, war, crime waves, epidemic, and natural disasters impacts affect any community, region, state, or country. Any potential objective is presented to at least one of the above dangers, which can scrutinize the security of residents, travelers, and can hamper the market view of that particular destination [1].

The 2013 calamity in Uttarakhand is viewed as India's most noticeably awful natural disaster since the December 2004 caused by Indian Ocean tidal wave, these areas are in danger from the staggering impacts of future tidal wave because of the presence of a structural intelligent plate [2], absence of a tsunami warning system in the Indian Ocean, and absence of set up correspondence network giving opportune data to that region [3]. Weighty, constant downpours have made uncommon harm to life and property where heavy waterways from the Himalaya cleared away streets, scaffolds, houses, and structures in the whirling water, whirling is main driver of harm like exhaustion, scraped spot of toward the back harsh cylinder bearing and oil spillage at the fixing and makes different issues emerge, for example, propeller shaft, harsh cylinder orientation and harsh cylinder fixing gadget [2]. As per government authorities, in excess of 1000 people are normal dead with in excess of 6000 absent and many thousands have been uprooted. The heavy rainfall somewhere in the range of 15 and 17 June 2013 flooded the zone causing unreasonable gulley disintegrations and silt statements on its way. It is proposed that during the evening of 16 June 2013, because of perpetual precipitation, a huge volume of water conveying an immense measure of residue, and trash from cold moraines also, encompassing zones struck Kedarnath town and washed off its upper parts [4]. Using a daily rainfall data set, Goswami et al. [5] have shown (i) significant rising trends in the frequency and the magnitude of extreme rain events and (ii) a significant decreasing trend in the frequency of moderate events over central India during the monsoon season [6].

Different reasons have been seen for this catastrophe. Some propose this the occasion happened because of flash floods and others are agreeable to a cloud burst. The primary reason behind such a voluminous stream is a penetrate in the snowmelt and rain fed Chorabari Lake (3960 m AMSL, roughly 400 m long, 200 m wide, 15–20 m profundity) otherwise called Gandhi Sarovar Lake, which was dammed by the moraines stored by the Chorabari glacier [4].

Despite the fact that all potential results of the environmental disaster are yet to be determined it has been considered as a critical factor for expanding the weakness of communities, particularly of poor people, ladies, and negligible networks living in disaster-inclined zone. There are a few factor adding to the flooding issue covering from geography, geomorphology, seepage, designing constructions and environment as indicated by the current work done by nearby scholars [7]. These segments of the social orders are generally more vulnerable to the account of more prominent reliance on atmosphere delicate areas like farming and forestry for their occupations or restricted versatile limit. In the current years, the constancy of natural and anthropogenic calamities has expanded in the whole Himalayan district making it one of the most powerless and delicate areas of the world.

Changes in rainfall pattern and bursting of high-height glacial lakes conjure flash floods and cloud burst in the upper Himalayas prompting flooding circumstances in the down-stream, the effect of large stream bodies outpouring might cause changes in downstream hydrology and geomorphology like change in absolute streams, change in occasional planning of streams, momentary vacillations in streams, change in outrageous high and low of stream, channel disintegration and statement, riverbed creation, and so forth [8]. Also, the expanded frequency of forest fires, soil disintegration, and changes in land use the example that is influencing the natural life and people similarly. A gigantic flash flood after hefty rainfall in the high height districts of Kedarnath valley in Uttarakhand in the year 2013 caused the greatest provincial natural disaster of the century that slaughtered a large number of individuals which incorporates local people and sightseers from everywhere India just as far off nations. The disaster was bothered because of the few disregarded developments in the delicate mountains and monstrous infringement of the stream beds. It has additionally brought about long haul socio-efficient and biological emergency in the area. The current paper subsequently means to underline and talk about the social and environmental weakness in the Kedar valley in the outcome of the 2013 disaster [9].

2 Rishi Ganga Valley Flash Flood

The Flash Flood in Rishi & Dhauli Ganga river, Uttarakhand Himalayan region occurred on 7th Feb 2021, Because of the collison, the Himalaya is seismically exceptionally dynamic, rough and delicate mountain framework. Further, slants in the higher Himalaya are by and large exceptionally steep and subsequently profoundly inclined to landslides, prompting weighty loss of both life and property [10]. There were casualty of 65 people and more than 141 people had missing report and Devastated 2 Hydropower projects. Flood was triggered by a combination of avalanche and debris flow. Glacier was melt due to global warming (climate change). Climate change is the subject of how weather examples change over many years or more, Climate change happens because of regular and human impacts and Since the Industrial Revolution (i.e., 1750), people have added to environmental change through the outflows of GHGs and vapor sprayers, and through changes in land use, bringing about an ascent in worldwide temperatures [11]. After Glacier was melt, it mixed with the Rishi & Dhauli ganga river. Since, the lakes are formed at higher elevations, the downstream impact is going to be severe due to extremely rapid debris flow. The breaching and associated flash flood are called GLOFs (Glacial Lake Outburst Floods). The study is not only for incorporating the disaster risk assessment in development planning of Himalayan region but also routine monitoring of potential area of geological failures in glaciated valleys along with supra-glacial lake [12].

3 Hydraulic and Hydrological Simulation

A significant test related to flash floods is the quantitative character of the forecast; the undertaking isn't simply to estimate the event of an occasion, which is troublesome enough without anyone else, however, to foresee the extent of the occasion. It is the measure of precipitation that changes a generally conventional rainfall into an unprecedented, dangerous circumstance, The increase the recurrence of outrageous precipitation, for example the quantity of occasions per unit time with force over a given limit, has commonly gotten substantially less consideration [13]. This test is exacerbated by the cooperation of meteorology with hydrology. Advances in flood estimating past the current situation with the-workmanship are to be accomplished, among others, based on broadening figure lead-time. This should be possible by climate gauging at different worldly and spatial goals. The downpour check is the most fundamental meteorological instrument for estimating the measure of precipitation at a specific area. Downpour measure records are helpful for meteorology, hydrology, agribusiness, and scientific and viable exercises. For the last mentioned, precipitation records can be utilized to assess the effect of the measure of precipitation on the plant water needs and its nonappearance in dry spell conditions [14].

The unremitting rainfall occurred at Uttarakhand region during 15–17 June 2013, caused annihilating floods and avalanches in the nation's most exceedingly terrible natural disaster since the 2004 wave. Specialists stated that it is another alert with respect to the effect of fast environmental change on the environment.

The extraordinary decimation in rainfall in Uttarakhand caused due to an exceptional meteorological occasion as because of informal formative exercises attempted in ongoing many years adding to the loss of lives and property. The satellite symbolisms show that gigantic avalanches happened in the upstream upper east district of the Kedarnath valley because of focused energy rainfall.

In the current work, a hydrological and hydraulic reenactment study was done on the Mandakini River to comprehend the occasions which occurred in the Kedarnath valley during 10–18 June 2013. The disaster was because of a coordinated impact of hefty rainfall power, the abrupt outburst of a lake (Chorabari), and steep geological conditions. The total situation was reenacted in the Geographic Information System (GIS) environment utilizing far off detecting information contributions through HEC–HMS and HEC–RAS hydrological demonstrating software.

As per the India Meteorological Department, combined rainfall during 14–18 June 2013 at Tehri, Uttarkasi, Tharali, and Jakoli was 381, 359, 326 and 390 mm separately. This high rainfall was because of a solid connection between an approaching box in the westerlies and the solid southeasterly storm wind stream in relationship with a rainstorm low-pressure framework over the North Indian district, bringing about the advancement of lower tropospheric wind combination over Uttarakhand and neighboring areas. Due to the solid collaboration between an approaching mid tropospheric box in the westerlies and the solid lower-tropospheric south easterly storm wind stream in relationship with a rainstorm low-pressure framework over the North

Indian locale, a lower tropospheric wind assembly zone created over Uttarakhand and its adjoining areas [15].

Because of the non-availability of adequate rainfall field information, satellite-based rainfall information was utilized in the current examination. Tropical Rainfall Measuring Mission (TRMM) 3 h rainfall information of $0.25 \times 0.25°$ spatial goal pictures were utilized in the examination for the period 10–18 June 2013, covering the Alaknanda and Bhagirathi catchments, the variability of release from both the catchments decides the stream system in the waterway Ganga. The worldwide idea of waterway Ganga and its colossal bowl makes its review muddled and its hydrological information hard to get [16]. Day by day aggregated rainfall was determined. The day by day rainfall dispersion fluctuates spatially from 50 to 200 mm during 15–17 June 2013. Gathered rainfall figured in the Bhagirathi and Alaknanda catchments during 10–18 June 2013 was discovered to be 550 and 530 mm separately. It was seen that substantial rainfall happened on 10 and 11 June 2013 also; these predecessor weighty rainfall occasions raised the dirt dampness to immersion level and the ensuing rainfall occasions brought about full run-off in the catchments. Transient dispersion of rainfall in these two bowls from 10 to 18 June 2013. Accordingly, the residue were contributed from two significant sources: the moraines and alluvial fans situated in the Trans and Higher Himalaya; and the avalanches in the Higher and Lesser Himalaya, Although the flood was the aftereffect of a focused energy precipitation occasion, the greatness was expanded because of the multiplication of settlements along the waterway and a progression of to some extent developed blasts on the river bed [17]. It tends to be discovered that rainfall which happened during 15–17 June 2013 was the primary setting off power behind the disaster. Remote satellite pictures of 28 May and 21 June 2013 showed that there was roughly 30% expansion in snow cover in the Alaknanda and Mandakini catchment territory.

It is an uncommon marvel to have a snowfall of this degree during June. As per the energy balance hypothesis, snowmelt during the snowfall time frame will be less. However, Spatial variety of rainfall in Bhagirathi and Alaknanda catchments as blocked by TRMM on 15, 16, and 17 June 2013 (in mm). But, because of the kinetic energy of extreme focus rainfall, dissolve might have quickened from the new day off. The stream in the Mandakini River preceding the occasion, as indicated by the Central Water Commission information, was of the request for 50–150 m/sec, which was an aftereffect of snowmelt as there was no critical rainfall before the occasion. Consequently, considering the pre-occasion snowmelt run-off and increment in the new snow cover during the occasion, an accepted normal snowmelt run-off of 150 m/sec was considered in the flood hydrograph calculations.

4 Geological Structure

The geological set-up of a zone assumes an indispensable job, which controls the whole lithological surface of the specific region. The region among Sonprayag and Kedarnath lies in Vaikrita Group [4] of Central Crystalline, which includes the

Munsiari, Joshimath, Pandukeshwar, and Pindari Formation. In Kumaon, these crystalline were incorporated as a fundamental piece of the "Focal Crystalline Zone" by Kazunori. Kedarnath sanctuary and the municipality are situated over the stones of Kyanite Sillimanite zone of Higher Central Himalaya. The geological statement of the zone proof that the higher focal glasslike mass overlying the lower Meta silt of Garhwal window arrangement alongside Thrust contact for example Fundamental Central Thrust (MCT).

5 Geomorphology Structure

Geomorphology of any hilly locale incorporates mostly four qualities that are significant for depicting the cycle following up on them. These are:

1. Height, regularly in outright terms;
2. Steep, even sharp slopes;
3. Rough landscape; and
4. The presence of day off ice.

The examination zone is exceptionally analyzed into various little edges and prods, wide U—molded to limit V—formed valleys, profound chasms to wide outwash fields with an assortment of inclines. The whole valley from Rambara up to the nose of the Mandakini waterway (Chorabari Glacier) is shrouded with frosty stores as the sidelong and average moraine. The region is covered by erosional just as depositional highlights including cirques, horns, edges, alluvial fans, morainic stores, bone cones, porches, and so on The absolute bottom in the Kedarnath region is Sonprayag for example the intersection purpose of Mandakini and Songanga Rivers, though the most noteworthy raised culmination is known as Kedarnath top (6940 m) Wadia establishment of Himalayan Geology, 2013, is the institute which has been set up to do investigate towards the advancement of new ideas and models for the geodynamic development of the Himalaya and foster investigation procedures and the executives of the regular holds like minerals, modern material, icy masses, and water [18].

6 Analysis of Weakness to Environmental Risks in Kedarnath Valley

Extraordinary atmosphere related fiascoes emerge because of the connection of atmosphere related unsafe occasions with the human and natural weakness, and their presentation and adapting capacity. Increasing paces of such negative changes in the atmosphere framework increment the danger of disasters, effects of which are by and large irreversible and inconvenient. It has now settled by an investigation of Wadia Institute of Himalayan Geology, that the impact of atmosphere inconstancy was one of the main considerations behind the disaster as the absolute terminal retreat of

Chorabari glacier somewhere in the range of 1962 and 2003 was around 262 m at a normal pace of 6.4 m per annum, and during 2003–2010 at a normal of 9.3 m per annum. It shows that about 9.6% territory of the glacier has been lost and which has been affirmed by the IPCC Synthesis Report too, which is the cryosphere coordinates climate varieties throughout a wide scope of time scales, making it a characteristic sensor of environment changeability and giving a noticeable articulation of climate change, Previously, the cryosphere has gone through enormous minor departure from many time scales related with ice ages and with more limited term varieties like the Younger Dryas or the Little Ice Age [19].

River Mandakini begins from the Chor Bamak glacier close to Kedarnath hallowed place and is generally taken care of by a few mountain streams. Mandakini joins Alaknanda at Rudraprayag and afterward Alaknanda joins Bhagirathi at Devprayag to frame the Ganga. The June 2013 catastrophe happened during the pinnacle traveler and journey season and not long previously the beginning of storm when by and large such hefty downpours are not anticipated. It was maybe the most exceedingly terrible disaster of the ongoing hundreds of years in which in excess of 6000 journey and local occupants were cleared away by the flash flood. It high measures of precipitation, joined with exceptionally productive and fast overflow on generally little catchments, are common of flash floods. A flash flood promptly follows the actuating storm occasion whereas the term ‘flash’ itself demonstrates an abrupt fast hydrological reaction of a generally little catchment, where water levels might ascend to their most extreme in no time or a couple of hours after the beginning of the downpour occasion [20]. The disaster harmed around 2174 streets, 85 engine spans, 140 harness spans, 995 public structures including school structures, emergency clinics, ladies and kid improvement focuses, block workplaces, and other private structures, 2540 houses either completely or then again incompletely harmed in the five influenced regions. It grabbed network to around 4200 towns and upset the whole correspondence framework. Radio semiconductor set is viewed as one of the significant public location frameworks around the world, which can be utilized effectively during any crisis, for example, environmental disaster, war, pestilence at least expense and support. In the examination zone, just 10.85% of families had a radio semiconductor set before the disaster, which was decreased to 3.89% after the misfortune. Nonetheless, about 90.28% of families in the influenced locale had TVs before the disaster, which was likewise decreased to 57.89% after the misfortune. It is assessed that the area endured an absolute loss of property worth 12 million including the loss of correspondence lines and other framework. Despite the fact that the accessibility of web association among the considered families was very poor before the disaster as just 12.33% of groups of Chandrapuri bunch had the web association yet it went to invalid after the fiasco. Presently 64% of individuals have cell phones while before the disaster practically 98% of individuals had it. The significance of cell phones as a wellspring of availability is notable to the individuals however because of helpless organization just as sporadic power individuals are being debilitated to have cell phones. Any appraisal of the effect of disaster-related weakness includes community cooperation at all levels. The recognizable proof of dangers and the elements liable for weakness are pivotal in the investigation and outlining techniques for adapting and transformation to the circumstance.

The customary method for dealing with stress to the weaknesses dependent on the conventional information framework should be converged with present day logical systems and techniques to accomplish better outcomes in the distant mountain regions. Considering this, community support, joint effort, ID of objectives, hazards, appraisal of community limit, combination of customary and current innovations, mindfulness, and expertise improvement of the community could be useful in lessening weakness in the valley.

7 Sequence of Happened Event in the Area of Mandakini River

There is no denying the way that there were hefty downpours in the region that prompted the expanded release of the streams and waterways. The equivalent is affirmed by rainfall and water level information.

Decimation in the Mandakini valley occurred in two flood occasions on 16 and 17 June 2013 and the last was related to the break of Chorabari Tal that had collected enough water to compel the moraine hindrance to give way. The previous occasion that washed off Rambara in the late night of 16 June 2013 could just occur by the bar of the course of Mandakini in the nearness of Kedarnath in order to (i) flood Kedarnath, (ii) power water into the relinquished eastern the channel of Mandakini and (iii) guarantee that enough water is seized to obliterate Bambara and Gaurikund with the unexpected evacuation of the hindrance. Hydro—geomorphic arrangement of the zone shows that Dudh Ganga is the just significant stream joining Mandakini among Kedarnath and Rambara that has the capability of cutting down enough trash in order to guarantee impoundment of Mandakini stream. Besides, the conversion of Mandakini and Dudh Ganga is situated at a spot bar over which could flood Kedarnath. Barricade at a downstream spot would not influence Kedarnath in view of the high angle of the stream in the zone.

It was this bar of Mandakini on 16 June 2013 that prompted the impoundment of the channel toward the west of Kedarnath. The dike on the left bank of the channel before long gave way and the relinquished channel of Mandakini toward the east of Kedarnath got dynamic. It was this occasion that brought about washing off of a few individuals in the late night of 16 June 2013 from Kedarnath, that in this way became water bolted. The rising degree of the landslide dammed lake constrained the boundary to give way furthermore, the resulting floods crushed Rambara and Gaurikund as additionally, the walker connect over Mandakini close to Kedarnath. All networks with the territory was along these lines snapped.

Constant downpours caused the degree of water in Chorabari Tal to rise. With the downturn of the glacier, the lake had a powerless moraine boundary that proved unable to withstand persistently rising hydrostatic weight. The stage was hence set for a significant disaster in Kedarnath and the boundary gave path around 0700 h on 17 June 2013. The volume of water was colossal and it conveyed with it enormous cold stones and outwash material that gagged the western channel of Mandakini and the progression of water and trash got redirected towards Kedarnath municipality that was in this manner attacked. There was definitely no admonition and a great many people were overwhelmed furthermore, had the opportunity to react. Other than Kedarnath, this occasion caused demolition in Bambara, Gaurikund, Sonprayag, and different spots.

8 Optical Remote Sensing of the Floods

9 Microwave Remote Sensing

Satellite information comparing to High Flood Levels (HFL) utilized (Figs. 1 and 2).

Mostly during 2003–2013 (around 100 satellite datasets) Important occasions remembered for expansion to 2003–13 information.

i. 1998-Assam
ii. 2000-West Bengal
iii. 2002-Bihar floods
iv. 1988-Delhi floods

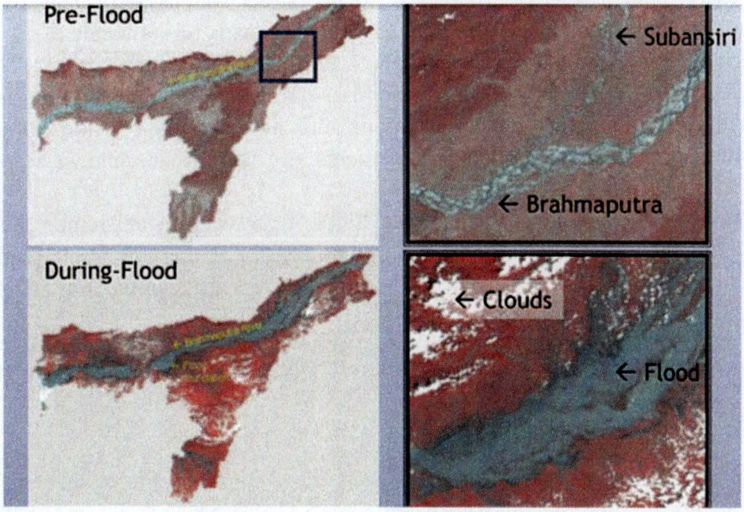

Fig. 1 Optical remote sensing map of pre-flood and during flood

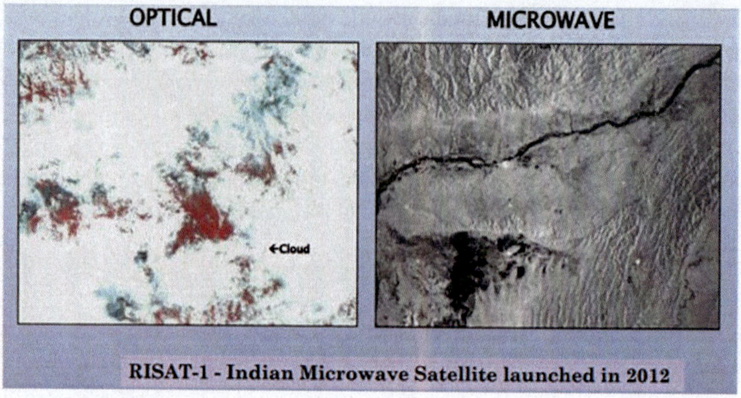

Fig. 2 Optical and microwave remote sensing map of Flood

Complete satellite-based greatest flood immersed region—Aggregated Flood Area—is about 10.934 MHA.

10 Role of Geographical Information System in Disaster Management

Disaster Mitigation and Management Center (DMMC) of Uttarakhand Government has given a brief description of drainage residence, street, water system, health framework, remote correspondence offices, and wellbeing maps for all locale of Uttarakhand arranged in GIS. NRSC has arranged Landslide danger maps for all the Char Dham courses utilizing GIS. For J&K, the Ministry of Environment and Forests as a team with the Indian Space Research Organization has gathered a National Wetland Atlas where point by point wetlands, settlements, streets, railroads, and waste guides have been made in GIS.

The above data ought to be made accessible to Disaster Management bodies, Public Administrative bodies and local individuals for arranging, asset assembly, and quick and composed reaction when natural disasters strike at the individual and community level [21].

11 Impact of the Disaster

The following disaster was statewide. Of all shapes and sizes waterways and mountain streams all through the state burst their banks unleashing devastation in close by towns. The weighty downpours destabilized mountain inclines causing landslides at a great many areas. The fundamental effect was felt close to the more delicate high ranges. It was given the primer appraisal information incorporated by various offices. In spite of the fact that a portion of the numbers was not clear, they uncovered the size of the disaster. The accompanying passages quickly portray the idea of the effects.

Floods: Flash floods are a typical event in the Himalayas however the obliteration at numerous areas this year was exceptionally unordinary. Huge streams like the Ganga, Kali, Saryu, and Ramganga (E) penetrated their risk marks. The Ganga immersed Rishikesh and Haridwar. Its feeder, River Bhagirathi, flooded pieces of Uttarkashi while the Alaknanda suffocated portions of Srinagar under 30 feet of water, mud, and residue. The Mandakini level rose 30–50 feet in its lower reach, close to Rudraprayag. Floods influenced all aspects of the state; country and metropolitan zones the same. The Yamuna immersed Vikasnagar.

Landslides: In a starter appraisal, the Indian Space Research Organization (ISRO) distinguished 2395 landslides in different pieces of the Mandakini, Alaknanda, and Bhagirathi watersheds (www.bhuvan-noeda.nrsc.gov.in). Just about 200 of them were among Kedarnath and Gaurikund. Street and media transmission joins were seriously influenced.

Loss of infra: Infrastructure in Uttarakhand was gravely hit. Streets, spans, electrical cables, irrigation trenches, drinking water flexibly frameworks, media transmission pinnacles and lodgings and houses were obliterated or harmed. Authorities esteemed the lost structures at several billion rupees (www.tehelka.com). The ensuing business misfortunes were comparable. The vast majority of the serious harm was in the northern locale of Uttarkashi, Rudraprayag, Chamoli, Bageshwar, and Pithoragarh. Government information indicated that 145 extensions had been cleared away and that streets were harmed at more than 2300 areas. A rampaging Mandakini stream cleared away most scaffolds across it. The more than 100-years of age connect interfacing India to Nepal at Jauljibi was washed away by the Kali waterway. In certain spots, individuals suffocated attempting to cross swollen mountain streams on shoddy extensions.

Life and Livelihoods: The human misfortune coming about because of the disaster is grimmer. Without homes, lands, and domesticated animals, the fundamental vocation resources of thousands of rustic families, reestablishing jobs will be a significant test. The sudden finish of the yatra season and its far-fetched resumption on this scale soon will ruin a great many families whose men administration travelers and sightseers on the yatra courses. They work taxis, transports lodges, dhabas, and slow down; some guide doormen conveying the youthful, old, and weak on their backs thousands of these individuals and the creatures were simply swept the Kedarnath.

12 Methodology

Information source.

- To satisfy the motivation behind this exploration paper computerized rise model (DEM),

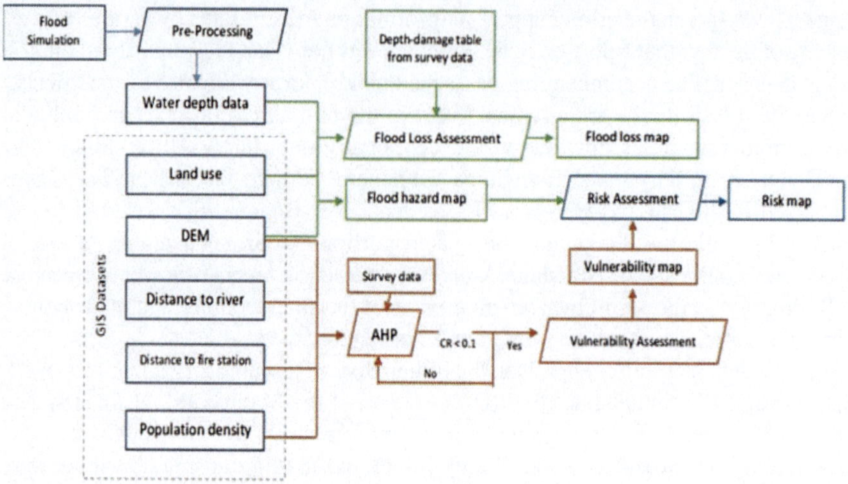

Fig. 3 Flow chart of risk map of flood

- Survey of India (SOI) toposheets,
- pre and post-disaster satellite picture of the investigation territory,
- CARTOSAT-1 and CARTOSAT-2 information could be used.

13 Technique

The land use and land cover (LULC) guide of the investigation zone were characterized utilizing the administered grouping procedure. To contemplate the harmed zone of Kedarnath, distant detecting and GIS methods were a utilized, and the harmed territory delegated high, moderate and low dependent on the force of the harms. The Built-up land, chilly/day off, slant, late moraine land, spring/tank, transport/foundation, valley, and water body are the LULC highlights characterized in this examination (Fig. 3).

The lessons learned while reacting to this disaster can help in fortifying the a disaster management framework.

14 Data Management

Interchanges having been disturbed not long after the disaster, there was no data coming to Emergency Operations Center (EOC) from the disaster influenced territory. Other than postponing the reaction this simultaneously added to the disarray as various adaptations were being circulated because of the absence of real data. EOC simultaneously was accepting countless solicitations for refreshed data from authorities, media, and the closest relative of the people who had approached Uttarakhand. To adapt up to the expanding heap of quarries various new phone associations were hastily enacted at the EOC. The emergency operations center (EOC) is a fundamental part of current emergency managements, previously understood as where authorities speak with the general population, support coordination, oversee tasks, create strategy, accumulate data, and host guests [22].

Many phone numbers were subsequently coursed through different modes and the equivalent added up to disarray. Earlier courses of action for media preparation and giving data to the closest relative of the people visiting Uttarakhand were additionally not set up. This brought about congestion at the EOC that frequently upset everyday practice working.

The consistent inflow of refreshed, true, and dependable data is fundamental to the powerful management of any disaster. There subsequently must be a component for guaranteeing the inflow of refreshed data from the disaster-influenced zone. The data got at the EOC from various sources is frequently voluminous and therefore there must be satisfactory data taking care of abilities with the goal that the data is immediately dissected and utilized for dynamic for reaction and asset assembly. The correspondence framework ought to be strong and solid with at any rate triple repetition to guarantee the accessibility of elective correspondence under all conditions. EOC ought to simultaneously have a solitary phone number with various lines and with the capacity of being upscaled during serious disaster occurrences. A committed four-digit complementary number of the EOC (1070) can be utilized for this reason. The equivalent can be promoted through different modes. The utilization of one number would be advantageous for all concerned. After any disaster media people are under gigantic strain to report the most recent updates and absence of data from the genuine sources frequently results in gossipy tidbits that add to disarray and injury of the influenced populace. This at the same time cripples the ones occupied with post-disaster tasks. Unique care, therefore, should be taken for a preparation of media at standard stretches by the properly approved individual approaching verified cutting-edge data and the people engaged with tasks ought not be overburdened with the duty of associating with media. Media ought to simultaneously be deterred from wandering into the EOC.

In the repercussions of any disaster closest relative of the influenced people are anxious to think about the government assistance of their friends and family. Separate courses of action have therefore to be made for reacting to such quarries. Unique consideration should be taken in cases where there is a chance of semantic contrasts between the ones answerable for reacting to public quarries and the expected guests.

In spite of cases of preemptive guidance with respect to the rate having been imparted to the government that was conveyed was nothing in a way that is better than a general forecast foreseeing hefty rainfall all through the state. Endeavors for the departure of the people prone to be influenced could therefore not be started ahead of time. Dependable admonition, with adequate lead-time, that is exact in reality and its powerful correspondence, in an understandable way, to the populace liable to be influenced by the occasion is the way to sparing human lives and moderating misfortunes.

15 Warning

With the current situation with logical information and the technological headway it is conceivable to produce and scatter admonitions of meteorological occasions well ahead of time. The adequately thick network of meteorological observatories with genuine time information transmission offices is anyway an essential for this, especially in the Himalayan landscape where climate boundaries are profoundly factored over short separations, which is keeping a limit among science an strategy in process of synthesizing science for strategy [23].

The network of meteorological observatories with ongoing information transmission abilities coordinated with rainfall—based landslide and flood forecast module can be used for creating dependable alerts well ahead of time. A framework skilled of promptly imparting these admonitions to the grassroots level, in a way that recommends activities to be started by individuals everywhere, must be an indispensable part of the notice framework. As the area is visited by vacationers and explorers in enormous numbers, portable informing administration with the arrangement of programmed conveyance of the notice to all dynamic cell phones in the zone prone to be influenced by the said cautioning could be considered for this. The admonition can simultaneously be shown at places where individuals assemble in enormous numbers and the equivalent can likewise be broadcasted through FM and other radio networks as additionally broadcasted through TV channels. Arrangements of the Disaster Management Act, 2005 can be used to guarantee that these messages are given high need by all media houses.

16 Relief and Rescue

In spite of the fact that profoundly specific and well prepared, the reaction powers brought in for search and salvage in the fallout of any disaster are regularly not familiar with the local landscape and climate conditions. They simultaneously don't have reasonable information on the elective courses, locally accessible resources and challenges, and difficulties that are probably going to be confronted while undertaking salvage. This is regularly basic, especially in the rugged landscape, and along these lines, the responders got from outside are regularly not excessively powerful.

The local individuals, therefore, appreciate the particular vital preferred position and in nearly all disaster occasions local individuals and different survivors are the people on the call. It is therefore suggested that local masses be prepared in search and save and be sufficiently prepared so they are more ready to confront the disaster and help their community. The particular reaction powers being raised by the states ought to simultaneously have an adequate number of people knowing about local ground real factors. This would guarantee their adequacy in case of any disaster.

17 Governance

Disaster management is a multi-departmental issue, coordination, and solidarity of order are basic to the accomplishment of the post-disaster reaction. SOPs and conventions relating to the equivalent must be therefore to be set down, coursed, and practiced well ahead of time to preclude the chance of breaches after a disaster has struck. As per the set down, SOPs senior individuals telling authority should assume responsibility for the issues not long after the disaster darning or sway.

Surveying the quantity of people engaged with the disaster was a significant test furthermore, there were differing claims from different quarters that additional to the disarray.

It is therefore necessitated that the journey, in regions where individuals need to travel, be managed and the main a predetermined number of people be permitted past a specific point after appropriately enlisting their subtleties.

Wellbeing registration should be made compulsory for all people wishing to wander past a given elevation. They ought to simultaneously additionally be informed on the territory conditions, climate, and related perils.

Bar of engine streets because of landslide and flash flood is normal in the slopes. It is therefore necessitated that elective engine streets be arranged and created to guarantee network during disaster rates. Considering the landscape conditions helipads are needed to be created in the slopes at vital areas. This would make disaster reaction quick and successful.

In the fallout of the disaster, correspondence with countless saved people turned into an issue because of etymological imperatives. It is suggested that the alleviation camps are enough to be set up with people who are acquainted with the language of the possible casualties.

Aside from the unfamiliar public, the disaster included individuals from more than 23 states of India and authorities from these states were deputed to deal with the explicit necessities of people from their state. There were anyway no past courses of action for the instructions of these authorities that necessary data on geology, territory, climate conditions, approach, and endeavors being made for search furthermore, salvage. This additional to crafted by the EOC. Explicit courses of action taking into account the necessities of the authorities rolling in from various states/countries relying upon the likely arrangement of the standard guests have this to be consolidated in the concerning SOP.

18 Conclusion

Flash floods are a natural phenomenon because of weighty rainfall in rugged territory like Kedarnath, which is constrained by the recognized geographic, orographic, and geomorphic boundaries of the Himalayan mountain ranges. In these boundaries, the stature of the specific territory, cirques, saddles, hanging valleys like geography, cumulus downpour clouds with stickiness, the arrangement of cold and warm convection current, and so forth assumes a significant job. These conditions furthermore, boundaries are constrained essentially. In this manner, the occasions of cloud burst can't be confined however the death toll and properties can be limited by taking certain careful steps. In such territories, development ought to be completed after exhaustive and point by point geologic, geomorphic, structural, land use, and risk zonation examines for the improvement of the locale by previously arranging.

References

1. B. Varghese and N. I. J. Paul, "Disaster Management : A Case Study of Uttarakhand A Case Study of Uttarakhand," no. September 2013, 2013, doi: https://doi.org/10.13140/2.1.4617.1521.
2. J.-S. Sun, T.-M. Han, K.-K. Lee, and U.-K. Kim, "A Study on the Measurement and Analysis of Whirling Vibration Behavior of Marine Propulsion Shafting System using Gap-sensors," *J. Korean Soc. Mar. Eng.*, vol. 39, no. 2, pp. 130–135, 2015, doi: https://doi.org/10.5916/jkosme.2015.39.2.130.
3. G. Ramalanjaona, "Impact of 2004 Tsunami in the Islands of Indian Ocean: Lessons Learned," *Emerg. Med. Int.*, vol. 2011, pp. 1–3, 2011, doi: https://doi.org/10.1155/2011/920813.
4. C. S. Dubey, D. P. Shukla, A. S. Ningreichon, and A. L. Usham, "Orographic control of the kedarnath disaster," *Curr. Sci.*, vol. 105, no. 11, pp. 1474–1476, 2013.
5. B. N. Goswamiv, S. Venugopald, S. Madhusoodananand, K. X. Prince, Increasing tren of extreme rain events over india in a warming environment. *Sci.*, Vol. 314–5804, pp. 1442–1445, 2006, DOI: https://doi.org/10.1126/science.1132027
6. M. Mohanty, K. Ray, and K. Chakravarthy, "Analysis of increasing heavy rainfall activity over western India, particularly Gujarat state, in the past decade," *High-Impact Weather Events over SAARC Reg.*, pp. 259–276, 2015, doi: https://doi.org/10.1007/978-3-319-10217-7_17.
7. A. H. Yahya, W. N. W. Noordin, and M. S. Chinnasamy, "Disaster Relief and Preparedness; Modeling Crisis Communication Disaster Intervention From 2014 Flood in Malaysia," *Researchgate.Net*, no. May 2019, 2017, [Online]. Available: https://www.researchgate.net/profile/Sara_Chinnasamy/publication/333516317_DISASTER_RELIEF_AND_PREPAREDNESS_MODELING_CRISIS_COMMUNICATION_DISASTER_INTERVENTION_FROM_2014_FLOOD_IN_MALAYSIA/links/5cf131374585153c3da94376/DISASTER-RELIEF-AND-PREPAREDNESS-MOD.
8. D. T. Thu Ha, S. H. Kim, and D. H. Bae, "Impacts of upstream structures on downstream discharge in the transboundary imjin river basin, Korean Peninsula," *Appl. Sci.*, vol. 10, no. 9, 2020, doi: https://doi.org/10.3390/app10093333.
9. R. K. Maikhuri *et al.*, "Socio-ecological vulnerability: Assessment and coping strategy to environmental disaster in Kedarnath valley, Uttarakhand, Indian Himalayan Region," *Int. J. Disaster Risk Reduct.*, vol. 25, pp. 111–124, 2017, doi: https://doi.org/10.1016/j.ijdrr.2017.09.002.

10. D. K. Dwivedi, "Dhauliganga, Uttarakhand, India Coalescent Natural Disaster: A Geoinformatics Perspective."
11. N. Morisetti, "Climate change and resource security.," *BMJ*, vol. 344, 2012, doi: https://doi.org/10.1136/bmj.e1352.
12. N. Rana *et al.*, "A preliminary assessment of the 7th February 2021 flashflood in lower Dhauli Ganga valley, Central Himalaya, India," *J. Earth Syst. Sci.*, vol. 130, no. 2, 2021, doi: https://doi.org/10.1007/s12040-021-01608-z.
13. G. Myhre *et al.*, "Frequency of extreme precipitation increases extensively with event rareness under global warming," *Sci. Rep.*, vol. 9, no. 1, 2019, doi: https://doi.org/10.1038/s41598-019-52277-4.
14. A. F. T.-R. and N. A. Mahyar Aboutalebi *,†, "Spatial and Temporal Analysis of Precipitation andEffective Rainfall Using Gauge Observations,Satellite, and Gridded Climate Data forAgricultural Water Management in theUpper Colorado River Basin."
15. S. D. Kotal, S. Sen Roy, and S. K. Roy Bhowmik, "Catastrophic heavy rainfall episode over Uttarakhand during 16–18 June 2013 - observational aspects," *Curr. Sci.*, vol. 107, no. 2, pp. 234–245, 2014, doi: https://doi.org/10.18520/cs/v107/i2/234-245.
16. V. Kamal, R. Sipolya, R. Kumar, and S. Mukherjee, "Morphometric and hydrological study of Alaknanda and Bhagirathi basins," *Proc. 16th Int. Assoc. Math. Geosci. - Geostatistical Geospatial Approaches Charact. Nat. Resour. Environ. Challenges, Process. Strateg. IAMG 2014*, pp. 468–471, 2014, doi: https://doi.org/10.1007/978-3-319-18663-4_125.
17. Y. P. Sundriyal *et al.*, "Terrain response to the extreme rainfall event of June 2013: Evidence from the Alaknanda and Mandakini River Valleys, Garhwal Himalaya India," *Episodes*, vol. 38, no. 3, pp. 179–188, 2015, doi: https://doi.org/10.18814/epiiugs/2015/v38i3/004.
18. K. Kumar and P. Srivastava, "Wadia Institute of Himalayan Geology," *Proc. Indian Natl. Sci. Acad.*, vol. 78, no. 3, pp. 565–576, 2012, doi: https://doi.org/10.1007/s12594-018-0919-5.
19. C. Change, "Synthesis Report: Contribution of Working Groups I, II and III to the Fourth Assessment Report of the Intergovernmental Panel on Climate Change Core Writing Team," *IPCC, Geneva, Switz.*, p. 104, 2007.
20. D. Loczy, S. Czigany, and E. Pirkhoffer, "Flash Flood Hazards," *Stud. Water Manag. Issues*, 2012, doi: https://doi.org/10.5772/28775.
21. Wikipedia the free encyclopedia, "2013_North_India_floods," *Wikipedia the free encyclopedia*, 2013.
22. P. Cavaliere *et al.*, "A research agenda to explore the emergency operations center," *J. Emerg. Manag.*, vol. 18, no. 6, pp. 525–534, 2020, doi: https://doi.org/10.5055/JEM.2020.0499.
23. G. Sundqvist, I. Bohlin, E. A. T. Hermansen, and S. Yearley, "Formalization and separation: A systematic basis for interpreting approaches to summarizing science for climate policy," *Soc. Stud. Sci.*, vol. 45, no. 3, pp. 416–440, 2015, doi: https://doi.org/10.1177/0306312715583737.

Disposal of Non-biodegradable Waste Using Eco-Friendly Methods

S. V. Mohanaprasadh, Pranshu Singh, and Kanchan Deoli Bahukhandi

Abstract From bags to computers non-biodegradable wastes found their way into everything. We knew little about how this is going to affect the environment when it was found. Many non-biodegradable wastes are complex polymers and are very hard to recycle. Usually, these non-biodegradable wastes end up in landfills and oceans. The world is moving towards a healthier and cleaner the best way to achieve it is by eradicating this waste, but it's a long way to go. At the present we slow the accumulation of these wastes and can concentrate on safe and eco-friendly ways to dispose this waste. In this article various methods are described to how improve waste management, use waste in as one of the consequent materials for other products or using waste for producing energy. All data is considered and recommended regarding improving waste management system in India.

1 Introduction

There are 195 countries in the world, 7.79 billion people in the world as of 2020. The plastic is inevitably one of the cheapest, lightweight, strong, durable, corrosion-resistant material, with high thermal and electrical insulation properties. The diversity of plastic polymers and the versatility of their properties will make plastic being heavily consumed, lead to production of a broad variety of products which have led to development in medical and technological industries, contributed to energy savings

S. V. Mohanaprasadh
Masters of Technology in Health Safety and Environment, University of Petroleum and Energy Studies, Bidholi, Dehradun, India

P. Singh
B.Tech Fire and Safety Engineering, University of Petroleum and Energy Studies, Bidholi, Dehradun, India

K. D. Bahukhandi (✉)
Sustainability Cluster (HSE and Civil Engineering), University of Petroleum and Energy Studies (UPES), Bidholi, Dehradun, India
e-mail: kanchan@ddn.upes.ac.in

K. D. Bahukhandi et al. (eds.), *Environmental Pollution and Natural Resource Management*, Springer Proceedings in Earth and Environmental Sciences,
https://doi.org/10.1007/978-3-031-05335-1_20

and other various technological advances, this has led to massive consumption of plastics and production of plastic waste which has led to massive negative impacts on the world [1].

Recent population growth and urbanization have led to a rise in non-biodegradable garbage, which has had a negative impact on our ecology. Since the 1950s, when the first plastic was produced, the world has been contaminated by more than 8 billion tons of plastic. This massive amount of plastic has ended up in landfills and seas. According to scientists, around 7–12 million metric tons of plastic enter our seas each year [2]. This destroys marine life and releases poisons into the water body. Only a tiny portion is recycled and reused. According to researchers, just 9% of the world's 8 billion tons of plastic garbage makes it to recycling centers [3].

Biodegradable waste has no negative impact on the environment. Natural forces such as fire, water, air, microorganisms, or soil easily decompose them [4]. Furthermore, these are all good for the environment in some way. This category includes plant and animal waste, dead animals and plants, vegetables, paper, food, and other items. Waste that does not decompose naturally is called "non-biodegradable waste". Landfill sites are being utilized as open dumping grounds, with much too much trash being dumped without resource recovery, resulting in leachate and methane gas. The global quantity of municipal solid trash was created in 2016, with predictions for 2030 and 2050. Globally, it is estimated that 3.4 billion metric tons of municipal solid trash would be created by 2050 [5]. Waste generation is increasing day by day at a very high rate and if we don't do something about this, the world will become a huge dump yard of non-biodegradable waste, and this is not very far away from our current scenario.

Segregation of the wastes is the initial stage in handling these non-biodegradable wastes. When trash is segregated into basic categories such as wet, dry, and metallic, it has a significant improvement potential, and as a result, recycling and reusing [6]. Moreover, failure to maintain waste segregation obstructs proper recycling, efficient bio-methanation, and safe waste-to-energy plant operation, resulting in the release of hazardous chemicals into the atmosphere.

1.1 Plastic Waste Production in the World

According to UN environment, it is estimated that more than 8.3 million tons of plastic has been generated since the 1950s, where only 60% of the plastic waste is dumped in the environment or it is dumped in the landfills. Most of the world uses the rivers to get rid of the plastic, the plastic is dumped into the rivers and led to the ocean by the flow of river. 10 rivers in the world alone carry about 90% of the plastic in the world. [UN environment] The ten rivers are (Fig. 1 and Table 1).

According to 2010 data, China manufactured the most plastic, totaling 59.08 tons of plastic. Despite being the world's greatest producer of plastic garbage, the average daily plastic waste per person in the country is one of the lowest in the world, at.12 kilos. China intends to outlaw single-use, non-biodegradable plastic bags in all major

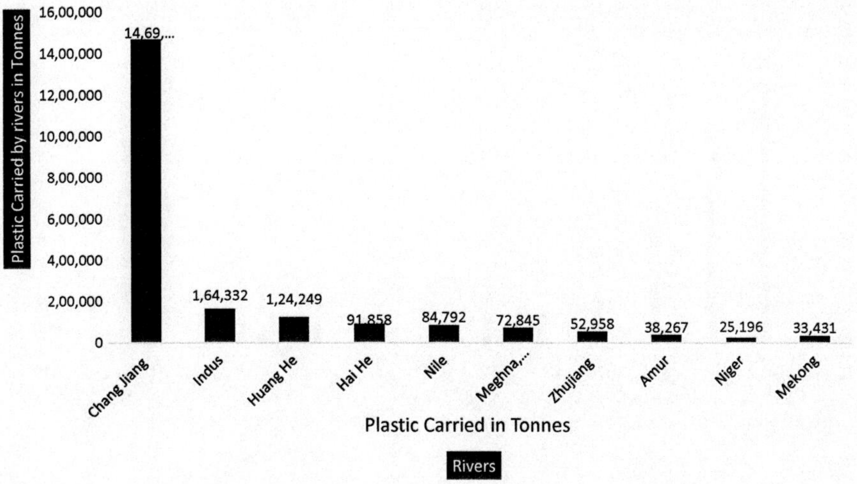

Fig. 1 Showing plastic carried by rivers in Tons

Table 1 *Source* Data from [6]

Rivers	Plastic Carried by Rivers in Tonnes
Chang Jiang	1,469,481
Indus	164,332
Huang He	124,249
Hai He	91,858
Nile	84,792
Meghna, Brahmaputra, Ganges	72,845
Zhujiang	52,958
Amur	38,267
Mekong	33,431
Niger	25,196
Grand Total	2,157,409

cities by the end of 2020, and in all cities and villages by 2022. Furthermore, eateries will be prohibited from using single-use straws by the end of 2020 [7].

The United States is the world's second-largest producer of plastic trash. In 2010, the United States generated around 37.83 million tons of plastic trash and over 275,000 tons of plastic litter. The United States burns almost six times as much plastic garbage as it recycles. The United States, along with other nations such as Canada and the United Kingdom, is notorious for exporting plastic waste to Asian countries in order to dump their problem onto another country [8].

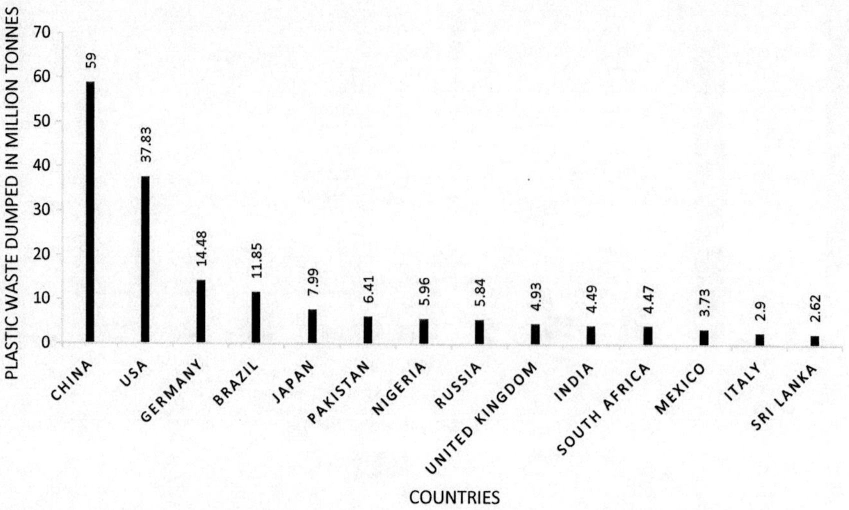

Fig. 2 Plastic waste dumped by countries. *Source*https://www.science.org/doi/abs/10.1126/science.1260352

Germany generates 14.48 million tons of plastic trash every year, with 31,239 tons of plastic litter potentially entering rivers. At.46 kilos per person per day, Germany has one of the highest rates of plastic trash in the world. In 2018, Germany's Environment Ministry unveiled a five-point strategy to minimize the country's plastic waste [9].

Brazil, the world's fifth-largest country, is the fourth-largest producer of plastic trash. Brazil produces around 11.85 million tons of plastic trash each year. Brazil is expected to recycle just 1.28% of its total plastic trash, resulting in 7.7 million tons of garbage ending up in landfills. Every year, Japan creates around 7.99 tons of plastic trash and over 143,000 tons of plastic litter.

Japan has a coastline that stretches for almost 18,000 km. Some argue that because of Japan's preoccupation with sanitation, many items are wrapped, re-wrapped, and packaged in many layers of plastic. The Japanese government has set a target of decreasing plastic consumption by 25% by 2030 [10] (Figs. 2 and 3).

1.2 Plastic Waste in India

The volume of waste in India is projected to increase from 64 to 72 million tons at present to 125 million tons by 2031. Untreated waste from Indian cities lies for months and years at dumpsites where land was originally allocated for developing landfills for safe disposal [11].

India is a fast-developing nation, with growing urbanization and it has 17.7% of the total world population. India generates around 9.46 MT plastic waste annually and was the third-largest consumer of plastic waste in 2009 with a total consumption

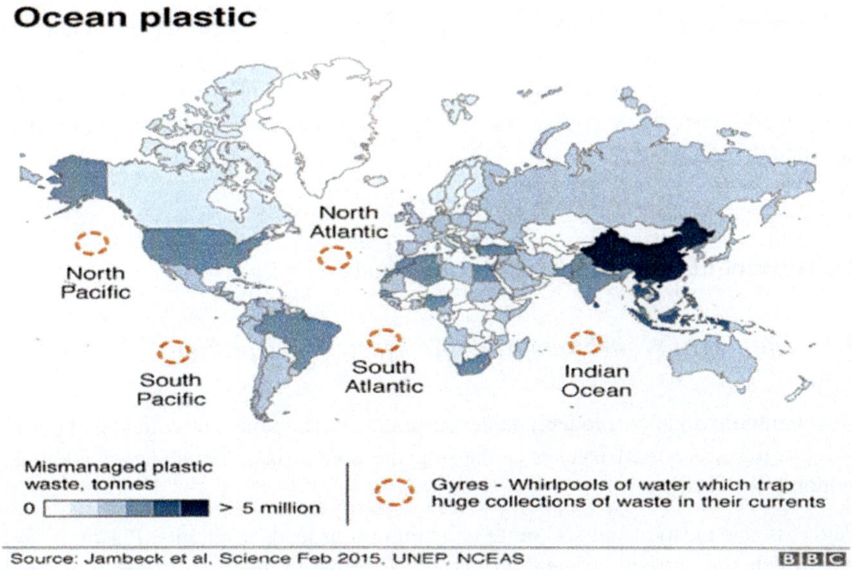

Fig. 3 Ocean plastic. *Source* Jambeck et al. science February 2015, UNEP, NCEAS-BBC

of 12.5 MT. It is very hard for a country with such a large area and population to remove waste immediately. It is commendable that several Indian governments have outlawed single-use plastics, but we still have a long way to go to decompose the other wastes that have been piling up in our dump-yards for years. The total amount of solid waste created in 2018–2019 is 152,076 TPD, of which 149,748 TPD is collected, equivalent to 98.5% of the total waste. 50,161 TPD was deposited of in a landfill, whereas 55,759 TPD was handled as MSW. According to the research, the country now has 3159 dumpsites, which are a major cause of air pollution and ground water contamination. According to a study conducted by ASSOCHAM and PwC, "an estimated amount of 7.46 million metric tons (MT) of hazardous waste is generated by the country's 43,936 industries, of which 3.41 million MT (46%) is landfill waste, 0.69 million MT (9%) is waste that can be burned in an incinerator, and 3.35 million MT (45%) is recyclable hazardous waste" [12].

The Ministry of Environment, Forest and Climate change and Government of India have distinguished waste in the following type:

1. Municipal solid waste: Commercial and Residential waste, including treated bio-medical waste.
2. Bio-Medical waste: Generated by treatment of human beings or animals, research, or in the production or testing of the biological.
3. Plastic Waste: waste generated by disposal of plastic.
4. E-Waste: electrical or electronic equipment.
5. Hazardous Waste: Waste which causes corrosion, reactivity or toxicity.
6. Industrial Waste: Waste Material produced during Industrial activity.

Unfortunately, no city in India can claim to have 100% trash segregation at the household level, and on average, only 70% of waste gets collected, with the other 30% being mixed up and lost in the urban environment, out of which only 12.45% of the trash collected is technologically treated, while the remainder is dumped in the open dumping [13].

2 Impact of Non-biodegradable Waste

2.1 Impact of Non-biodegradable Waste Environment

Soil contamination non-biodegradable wastes such as chemical insecticides (D.D.T) and fertilizers are used in excess, causing the soil to become acidic or alkaline, reducing soil fertility.

Cattles ingest plastic from soil, after consumption it leads to fatality. Plastic in soil also affects the nutrient efficiently of soils, a decline in quality of crops is noticed because of this. Plastic contaminates water bodies; water gets clogged as well as water gets contaminated.

Ecological imbalance, air pollution, water pollution, and land pollution are some of the negative effects of non-biodegradable waste. They pollute soil and water resources because they don't break down naturally. Stacks of these wastes serve as breeding grounds for disease carriers such as mosquito's and Flies, which transmit a wide range of infectious illnesses.

2.2 Impact of Non-biodegradable Wastes in Drainage Systems

Due to no segregation or improper dumping of plastics, most end up in drainage system. Flooding on the roads is caused by poorly designed drainage systems combined with reckless waste dumping. It leads to various diseases and acts as breeding ground for mosquitoes [14].

2.3 Impact of Non-biodegradable Waste on Human Being

An average person's day starts with a packet of milk, which he cuts it open pours the milk into the cup and disposes the plastic cover in trash can. We have started to coexist with plastic and plastic has become an unavoidable. The US, for instance, generates more than 250 million tons of waste every year. In India, it is estimated that the same as 62,000,000 tons. This, according to experts, will increase to 150,000,000 tons by 2030 [op]. India alone produces 40 million tonnes urban municipal waste annually and most of it are sent to unsanitary open landfills, due to poor infrastructure for recycling plastic. Rapid urbanization, which increased by a ratio of 8.9 in Mumbai, 13.6 in Bangalore, and 20.4 in Delhi between 1990 and 2000, is the main cause of failure. From 1950 to 2015, the population of the United States grew at an exponential rate in India [15].

2.4 *Impact of Plastics on Ocean*

Marine plastics are the result of poor trash disposal infrastructure and management, as well as a lack of public awareness about their environmental impact. The financial impact of marine plastics on coastal communities is also significant, particularly for fisheries and to maintain tourism [16]. Currently, half of the world's population lives within 60 km of the sea, and three-quarters of the world's major cities are located on coasts. Very little information on causes aesthetic issues. It endangers marine operations such as fishing as well as tourism [17].

3 Basic Plastic Waste Management Techniques

3.1 *Segregation*

At the home level or in the communal bin, there is no coordinated and scientifically designed segregation of MSW. Waste sorting is primarily done by the unorganized sector and is rarely used. There is no organized and scientifically planned segregation of MSW. Segregation and sorting take place under very unsafe and hazardous conditions. Lack of segregation deprive proper scientific receivership of MSW from being solicited by waste producers.

3.2 *Collection*

Household waste is generally collected in community containers made of metal, concrete, or a mix of the two. Street sweepings are also collected and deposited in communal bins. Rag pickers also collect certain amount of waste.

3.3 *Reuse/Recycle*

Recycling includes actions such as gathering items from garbage that may be profitably recovered and used to create new goods. Non-segregated waste is deposited into community landfills, its optimal recycling is not feasible due to the lack of containers. Rag-pickers, on the other hand, typically sifted and took what they wanted to sell. The main aspects influencing the recycling of plastics are identified as: polymer cross contamination, additives, non-polymer impurities, and degradation [18].

3.4 *Transportation*

Modes of transportation for MSWM practiced in India are bullock carts, hand rickshaws, compactors, trucks, tractor, trailers, and dumpers. Trucks with a capacity of 5–9 tonnes are employed in smaller towns. MSW is carried via stationary compactors, mobile compactors, and tarpaulin-covered trucks, with roughly 65, 15, and 20% of trash transported through these compactors, respectively. In certain urban areas, such as Mumbai, there are just a few transfer stations [19].

4 Recycling and Reuse

4.1 *Recycling*

Non-biodegradable plastics that enter municipal garbage streams may complicate existing plastic recycling facilities. Recycling is not always cost efficient, it's good to find ways to reuse it other products which does not affect the environment. It's better to reuse the waste in different forms (Fig. 4).

4.2 *Chemical Recycling*

Chemical recycling is a process where the polymer is broken down into smaller molecules that can be easily separated from impurities. Chemical recycling is a process in which polymers are broken down into smaller molecules that may be readily separated from contaminants such as waxes and paraffin. Cracking and hydrogenation are two examples of such processes [20].

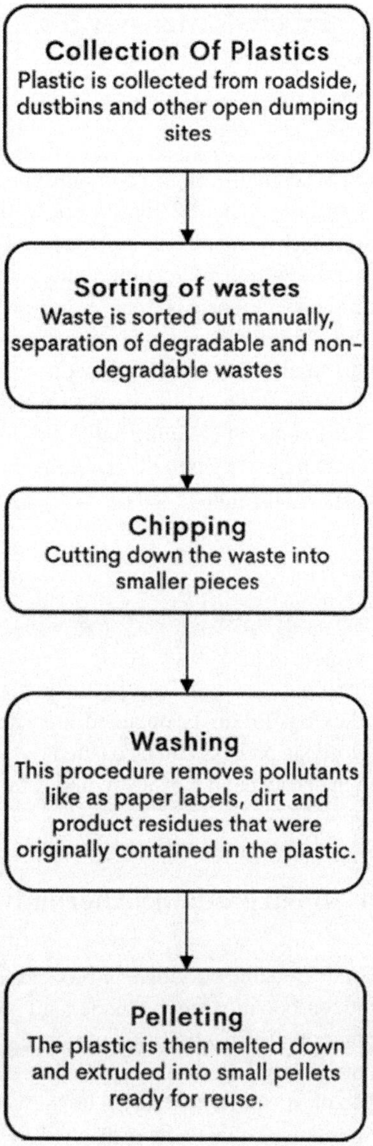

Fig. 4 Process involved in recycling

4.3 Reuse

It is very hard to eradicate non-biodegradable waste totally from this planet. These waste very hard to separate from other wastes only way is to separate it while disposing and reuse it, until we find a permanent solution for eradicating these wastes. These are few ways to reusing it smartly.

5 Pyrolysis Methods

5.1 Incineration or Combustion Method

Plastic combustion is a complex chemical reaction. Plastics can be micro-molecular or macro-molecular substances, depending on their structure. Warming, deterioration, flash over and combustion are all present at the same time during plastic combustion. Low-molecular compounds can be evaporated directly in the air and, depending on their composition, can either produce a flammable mixture or oxidize in solid form. To start the combustion process, macro-molecular plastics must breakdown into small molecule components.

Incineration of plastic waste in an open field is a major source of air pollution. Most of the times, the Municipal Solid Waste containing about 12% of plastics is burnt, releasing toxic gases like Dioxins, Furans, Mercury and Poly-chlorinated Biphenyls into the atmosphere [21].

5.2 Thermal Cracking

Breaking down of long polymer plastics into smaller useful components is called cracking. Products obtained are used for making fuels and other chemicals. The pyrolysis process can be carried out with or without a catalyst. It is thermal cracking in the absence of a catalyst, and catalytic pyrolysis in the presence of a catalyst [22].

6 Biodegradable Alternatives over Conventional Plastics

Biodegradable polymers or biodegradable plastics are polymers that can break down into carbon dioxide, methane, and other compounds. Water, or biomass in inorganic compounds Enzymatic activity is the prevailing mechanism of standardized measurable germs. Tests that reflect in a specific time frame Condition of disposal available. Bio plastic polymers stay relatively much more costly than their conventional plastic alternatives, this is the restriction for its widespread use. Biodegradable alternatives have less performance, less durable and life of the product is less compared to its counterparts. These are the few cons in using these alternatives, but if you see the bigger picture, we are helping our planet [23] (Table 2).

Table 2 Non-biodegradable and their eco-friendly alternatives

Non-biodegradable products	Eco-friendly alternatives
Bottles	Bamboo cans
Metal scraps	Stainless steel
Batteries	Rechargeable batteries
Grocery bags	Cloth bags
Styrofoam	Beeswax wrap
Aluminum cans	Reusable cans

7 Reuse of Non-biodegradable Waste into

7.1 Using Non-biodegradable Waste for Making Road

Plastic being used in road construction is not a new concept, as PVC (Polyvinyl chloride) or HDPE (High-Density polyethylene) are already in use for road construction. PVC and HDPE are used to form plastic mats. The plastic roads include transition mats to ease the passage of tires up to and down from the crossing. Both PVC and HDPE help in the protection of wetland haul roads from rutting by distributing the load across the surface [24].

7.1.1 Use of Tire and Plastic Waste

Plastic and tire are two of the non-biodegradable waste, used tires waste and plastic waste is increasing day by day at a tremendous amount. Therefore, one of the methods suggested is to use this waste for construction of roads [25].

Plastic is mixed with bitumen; plastic increases the melting point of bitumen and makes the road retain its flexibility during winters leading to its long life [26]. As per Singh and Sahu there are two methods for using plastics in road construction Dry Process and Wet Process, in dry process the plastic is separated, washed, shredded and mixed with bitumen at 160–170 °C. In the wet process, after washing the plastic is ground into powder form. 6–8% of the new plastic powder is added to the bitumen beforehand the aggregate is added. The Bitumen and the powdered plastic are mixed, they should be thoroughly and evenly mixed at the temperature of 155–165 °C [27]. The LDPE (low density polyethylene) plastic can also be used in construction of by using it in the dry process. Rakode used LDPE in road construction in using the dry process to generate the aggregate, he also suggests that rubber from Tire waste can be used as well, by utilizing the wet process [28]. Rakode used the Marshal method, marshal method was developed by Bruce Marshal in 1939, the method involves the selection of asphalt binder content with a suitable density which satisfies minimum stability and range of flow values.

7.2 *Non-biodegradable Wastes to Fuel*

There are two methods in this to obtain fuel from plastics.

- Incineration
- Thermal cracking

7.2.1 Incineration Method

In this method energy is recovered from incineration of plastic waste, Hydrocarbons or obtained and they can be used as substitute for fossil fuels. By this Carbon-dioxide burden is reduced on the environment. Polyethylene has a calorific value close to fuel oil, and the energy produced by combustion of polyethylene is equal to the amount created by the manufacturing process, which makes it an acceptable alternative [29].

Disadvantages on this method

- Produces greenhouse gases
- Produces toxic pollutants
- Yield is low

We can overcome these disadvantages using scrubbers and other filters which stop these toxic gases from entering the atmosphere.

7.2.2 Thermal Cracking

Thermal cracking is breaking down of large polymers by heating at very high temperature. Heating is carried out in in absence of oxygen to make sure oxidation does not occur. (Temp range 300–900 °C). The generated products comprise carbohydrates and a percentage of volatility. Some volatile fractions may be condensed to produce paraffins, is paraffins, olefines, naphthene and aromas, while remaining portion is a no condensable high calorific value gas. The produced products and their precise composition depend on the kind of plastic waste and the circumstances of the process.

7.2.3 Catalytic Cracking

In catalytic cracking, the same process is carried out in the presence of a catalyst. Catalyst presence decreases the temperature and duration of the process. Another benefit is that the breakdown of the polymeric chain in thermal cracking forms a wide range of products, while the distribution of products is much narrower, at peaks with lighter hydrocarbons, in catalytic degradation [26].

7.3 Using Non-biodegradable Waste to Make Bricks

Clay bricks are the most used as the building materials around the world. In the process of manufacturing clay bricks a lot of non-renewable resources are required as well as the process also contributes to global warming, by releasing greenhouse gasses and chemical pollutants. The clay bricks are fired up for achieving the required strength, the heating is done on coal, 24 million tons/year coal is consumed globally for the process. The consumption of coal and the entire process of manufacturing bricks has catastrophic consequences, it leads to smog, global warming, climate change, and other health issues. Other than coal being a non-renewable source, the materials which the bricks are made of also not available in abundance, as well as the 60 to 70% of cost of construction is associated with the material. On the other hand, plastic waste is available in abundance, with 300 million tons of plastic waste being produced globally. The plastic is either dumped on the landfill or in the ocean, either way they contribute to pollution, harm the environment and acquire unnecessary space. A common solution to waste management and the Brick problem can be to use plastic waste as the consequent material for making bricks [30].

As in the methodology used and described R. S. Kognole, Kiran Shipkule, Manish Patil, Lokesh Patil, Udaysinh Survase. Plastic waste must be dried post collection, the moisture content from the waste must be removed. The dried plastic is crushed into fine particle sizes. The fine particle sized plastic waste is subjected to massive heat in a furnace, where it can be liquidized, stone dust is added and mixed to the melted plastic, the mixture is then poured into molds and left to dry to finally form bricks [31]. A variety of plastic waste can be used in the process, such as polythene, high density polythene (nylon 66), plastic bottles, plastic composite. In the process conducted by Frank Ikechukwu Aneke and Celumusa Shabangu scarp plastic waste was used collected directly from the Mariannhill landfill, the SPW consists of long chains of hydrocarbon compounds containing Carbon, Hydrogen and Oxygen [32].

With addition to reduction in usage of non-renewable resources, a productive method of waste management, reducing the manufacturing cost and most importantly reducing the negative impacts on the environment in the process of manufacturing bricks, bricks made from plastic and sand have additional advantages over clay bricks as well [33]. The water absorption of plastic sand bricks is zero percent. In the test conducted by Frank, the results show that the bricks recorded compressive strength of 38.14 MPa and tensile strength of 9.51 MPa, showing that the compressive strength of plastic strength was 2.5 and 3 times higher than the clay fired bricks [34]. Clay brick samples fired in the muffle furnace at a reduced temperature of 6000 °C were tested for compressive strength after 3 days of firing which showed that the value ranges from 11 to 21 N/mm2. Experimental results inferred that glass powder and quarry dust in the ratio 1.5:2.5 gave higher compressive strength of 21.31 N/mm2 with a standard deviation of 1.28 [35].

7.4 Tire Waste as Fuel

The fossil fuel sources in the world are limited. According to the BP statistical review of World energy Coal reserves will be exhausted by the year 2169, Natural Gas by 2068 and Crude Oil by year 2066. Other than the supply of these sources being limited there are also negative impacts of consuming these fuels. Considering these issues there is a need to find alternative sources of energy. Annually there are 1.5 billion tires produced in the world which leads 17 million of used tires or tire waste being produced.

European Tire and Rubber Manufacturers Association, Brussels, Belgium. Tires are mainly made from rubber, carbon black, steel and textile components which also gives an option to recycled and reuse them. As well as their chemical and physical properties make them an extremely valuable resource Tire is waste either dumped in landfills or used to provide energy for cement kilns.

Pyrolysis is the process of chemically decomposing solid waste by heat in an atmosphere whose oxygen levels has been reduced. It is also defined as a thermal process of degrading residues into gases after combustion as taken place [34].

Pyrolysis of tires produces a variety of products, in particular oils and gasses, which are largely depended upon the type of reactor used, the temperature and heating rate. A chemically complex oil is produced by the pyrolysis of tires which contains aliphatic, aromatic, hetero-atom and polar fractions. The main aromatic compounds found are benzene, toluene, xylene, styrene, limonene, and indene and their alkylated homologous and polyclinic aromatic hydrocarbons from 2 to 5 rings. Dai and team conducted the pyrolysis and reported a tire oil composition from the pyrolysis of tires in a circulating liquidized bed of 26.77 wt. % of alkane, 42.09 wt.% aromatics, 26.64 wt.% of non-hydrocarbons and 4.05 wt.% as asphalt. Conesa and team reported a 39.5 wt.% aliphatic fraction, 19.1 wt.% aromatic fraction, 21.3 wt.% hetero-atom fraction and 20.1 wt.% polar fraction for oils produced from a laboratory scale liquidized bed reactor at 700 °C. Aylón et al., used a screw kiln reactor to pyrolyze tires at 600 °C reported an alkane fraction of 6.7 wt.%, aromatic fraction of 65.6 wt.% and polar fraction of 27.8 wt.%. Laresgoiti and team reported an increase in 15 aromatic content of tire pyrolysis oils from a fixed bed reactor from 53.4 to 74.8 wt.% as the temperature was raised from 300 to 700 °C [36]. Gas from the pyrolysis of scrap tires in a liquidized bed reactor, Energy and Fuels. Waste tire pyrolysis: Comparison between fixed bed reactor and moving bed reactor. The oil produced shows like a gas oil, Diesel fuel or light fuel oil and have successfully been tested and combusted in test furnaces or blended with diesel fuel or burned with Diesel engines [37].

8 Waste Recovery Methods Around the World Which Can Be Implemented in India

8.1 Germany

Germany is one of the leading countries in waste recovery, having highest waste recovery rate in the world. The government of Germany has made laws which ensures provisions on the obligation of the producers and distributors to take back used packaging. Retailers participate in the system of collection and recycling of the packaging materials. There are various provisions made to recover paper, packaging materials and recovering biodegradable waste. Most of the packaging materials is used for storing and containing food and water. For example, Germany has machines built on the streets to recover packaging of the food from the consumers. In 2010, Germany has recorded packaging waste recovery rate at almost 85%, which implies that the population of Germany also participates effectively in the programs established by the government. Many countries are trying to have an effective waste management system, the reason why Germany is successful in this is because they have established an effective system by using methods such as separating waste bin, not just as biodegradable or non-biodegradable but also different method for collecting waste glass, wastepaper, cardboard and used water bottles individually and establishing laws which oblige the producers and distributors to participate in the process of waste recovery [38].

8.2 Columbia

Colombia's municipalities generate about 28,800 tonnes of solid waste each day, with the major cities of Bogotá, Cali, Medellin, and Barranquilla each producing 10,000 tonnes. To tackle their significant waste issue, they came up with a solution called Eco-bot.

Eco-bot

This is a national recycling project that promotes the culture of recycling. No, not by merely informing its inhabitants about the benefits of recycling, but by encouraging and rewarding every item that is recycled. ECOBOT stands for "reverse vending machine," which may be seen in shopping malls, institutions, and public areas and encourages PET bottle recycling.

How does this work?

Every time you deposit plastic bottle or cap you found, the machine gives you restaurant coupon or movie ticket. This encourages many youngsters collect plastic wastes and deposit in this vending machine, so that this these plastics don't end up in landfills and they are sent to recycling plants [39] (Fig. 5).

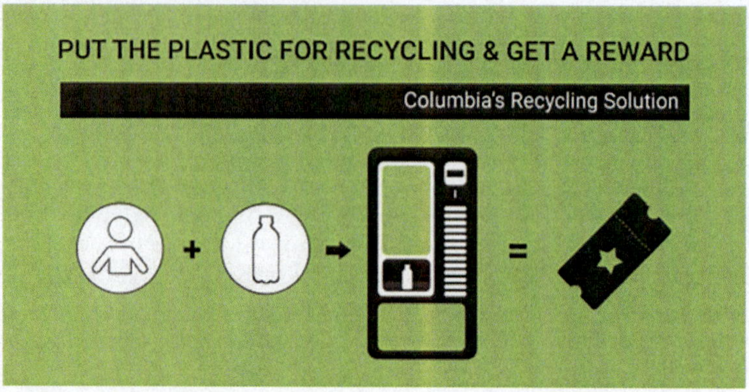

Fig. 5 https://www.ndtv.com/photos/news/5-countries-that-redefined-waste-management-techniques-23853

8.3 Sweden

Sweden is teaching the whole how deal with the trash present in fact they run out of trash and asking other countries for their trash so that they can run their recycling plants. In Sweden, less than 1% of residential trash is disposed of in landfills. In Sweden, 32 waste treatment plants provide heat to 810,000 homes and power to roughly 250,000 private residences. Every year, almost a million tonnes of trash are recycled. Sweden's first district heating system has been in operation since the 1940s. The city of Gothenburg is home to one of the world's largest waste-to-energy incinerators, with a capacity of more than tonnes per year.

Waste incineration in Sweden produced as much energy in 2007 as 1.1 million m^3 of oil. The waste sector will reduce its emissions of greenhouse gases by 76% during the years 1990–2020, according to the Climate Committee's forecast [40].

9 Biodegradable Plastics

In recent years, biodegradable plastics have been extensively studied and have been commercialized in the manufacture of a variety of products, including garbage bags, compost bags, poly bags, and agricultural mulch films. After being disposed of in the environment, biodegradable plastics can be degraded by microorganisms to create CO_2 and H_2O. Recycling of plastics and other non-degradable waste are usually time consuming, expensive and requires lot of processes. Researchers have come up with a new biodegradable plastic. The negative influence of synthetic polymers on the natural environment causes several issues with waste deposition and consumption. The major benefit of biodegradable polymers is that they may be composted with organic waste and returned to the soil to nourish it. So, they are parsed by nature, their usage not only reduces the threats to animals posed by dumping conventional plastic, but it also reduces the expense of labor for removing plastic waste from the environment. By decreasing the amount of waste that can be recycled into useful monomers and oligomers by microbial and enzyme treatments, decomposition will help landfills last longer and be more sustainable [41]. Using biodegradable polymers instead of synthetic materials in a number of sectors can assist to conserve the environment considerably. Biodegradable plastics can be the future of plastics, but it has a long way to go to be produced in large scale.

Biodegradable plastics give best solution against the synthetic plastics, because of its easily degradable nature. There are many types of biodegradable plastics with different degrees of biodegradability, among them polyhydroxybutyrate (PHBs) are the only 100% biodegradable ones.

9.1 Polyhydroxybutyrate

Polyhydroxybutyrate are macro molecules produced by bacteria and inclusion bodies that accumulate as reserve material when bacteria develop under various stress situations [42]. They are polymers with characteristics that are similar to synthetic thermoplastics such as polypropylene. This makes them suitable for a wide range of uses for potential commercial mass manufacturing of biodegradable polymers that can replace petroleum-based plastics [43].

10 Conclusion

Non-biodegradable waste problems are increasing daily and it's a never-ending fight against it. In the present what we can do is stall the non-biodegradable waste from entering landfills treat it and reuse. This can be done only by educating people and changing their attitude towards non-biodegradable wastes. Many countries have come with different solutions by converting these waste into something useful. Producing energy by using tire fuels is also a good recycling method; fuel obtained from the pyrolysis of tires was producing energy in comparisons of diesel fuel. Pyrolysis of tires is also very good method for recycling waste, but it will also lead to toxic gas emissions. In this paper we have discussed various methods such as laying roads with plastic infused tar can be very effective because construction of roads in any country is inevitable and roads made with plastics provide better strength and durability. Making of bricks out of non-biodegradable waste is a every eco-friendly method we can adopt for country with big population such as India. It will help in reducing the cost of construction as well as solve the problem of the abundance of plastic waste. Construction as well as solve the problem of the abundance of plastic waste. Provided the plastic waste bricks must not deform under pressure and withstand hot and cold temperatures as clay bricks do Waste management techniques adopted by Germany can be a good option for India, but for that there is need of proper infrastructure, laws and developing a mode of communication to show the importance of recycle and reuse. Bio-degradable plastic and Polyhydroxybutyrate plastic can be the future of plastics, but they are very far from mass production. So, there is a long way to go but it's now or never with increase in non-biodegradable waste rates.

References

1. Andrady Anthony L. and Neal Mike A. 2009 Applications and societal benefits of plastics Phil. Trans. R. Soc. B3641977–1984 https://doi.org/10.1098/rstb.2008.0304 Thompson, Richard C et al. "Plastics, the environment and human health: current consensus and future trends." Philosophical transactions of the Royal Society of London. Series B, Biological sciences vol. 364, 1526 (2009): 2153–66. https://doi.org/10.1098/rstb.2009.0053
2. Plastic Waste Problem and Education for Plastic Waste Management. Cheuk-Fai Chow, Wing-Mui Winnie So, Tsz-Yan Cheung and Siu-Kit Dennis Yeung, Feb 2017
3. https://www.nationalgeographic.org/article/whopping-91-percent-plastic-isnt-recycled/
4. https://sdg.iisd.org/news/world-bank-report-warns-global-solid-waste-could-increase-70-percent-by-2050/
5. www.statista.com/topics/4983/waste-generation-worldwide
6. Christian Schmidt, Tobias Krauth, and Stephan Wagner "Export of Plastic Debris by Rivers into the Sea", published in Environmental Science & Technology (2017)
7. Plastic pollution, Hannah Ritchie and Max roser, September, 2018
8. worldpopulationreview.com/country-rankings/plastic-pollution-by-country
9. Waste Management in Germany—Development to a Sustainable Circular Economy? M. Nellesa, b,*, J. Grünesa, G. Morschecka, International Conference on Solid Waste Management, 5IconSWM 2015
10. Plastic Waste Problem and Education for Plastic Waste Management Cheuk-Fai Chow, Wing-Mui Winnie So, Tsz-Yan Cheung, and Siu-Kit Dennis Yeung, May 2009
11. A Comprehensive Study on Waste Segregation Techniques
12. Murlidhar Gangwani, Madhuresh Pandey, Nikhil Punjabi, Prateek Khatwani, Sunita Sahu
13. CPCB. (2013). Status report on municipal solid waste management. Retrieved from http://www.cpcb.nic.in/divisionsofheadoffice/pcp/MSW_Report.pdf, http://pratham.org/images/paper_on_ragpickers.pdf
14. https://www.thestreet.com/world/countries-most-plastic-waste
15. Isher Judge Ahluwalia and Utkarsh Pate, Solid Waste Management in India. An Assessment of Resource Recovery and Environmental Impact
16. https://cpcb.nic.in/annual-report.php
17. Status and challenges of municipal solid waste management in India: A review. Rajkumar Joshi1 and Sirajuddin Ahmed
18. J. H. Song, R. J. Murphy, R. Narayan and G. B. H. Davies, Phil. Trans. R. Soc. B (2009) 364, 2127–2139, Biodegradable and compo-stable alternatives to conventional plastics
19. Chua, Dan Hoover U. Lim, Luis Antonio Sj. Ceniza, Nicole Renee N. Chio, Kim Deborrah G. Go, Kendra Kirsten L.Along Don Jose Avila Street, Disposal of Non-Biodegradable Waste by Small Business Owners
20. Florian Thevenon, Chris Carroll and João Sousa, The Characterization of Marine Plastics and their Environmental Impacts, Situation Analysis Report, ICUN
21. Gregory, 2009, Environmental implications of plastic debris in marine settings—entanglement, ingestion, smothering, hangers-on, hitch-hiking and alien invasions
22. J. Hopewell, Robert Dvorak and Edward Kosior. "Plastics recycling: challenges and opportunities". Phil. Trans. R. Soc. B (2009) 364, P. 2115–2126

23. VANNESSA GOODSHIP, Science Progress (2007), 90(4), 245–268, Plastic recycling
24. Rinku Verma*, K. S. Vinoda, M. Papireddy, A. N. S Gowda* Procedia Environmental Sciences 35 (2016) 701–708
25. Dinish Chacko, Anirudh. P, Anuj. K, Abhijith Mohan and Akshay M.K, International Journal of Scientific & Engineering Research, Volume 7, Issue 3, March-2016,Plastic Waste to Fuel: A Sustainable method for Waste Management
26. Ahmed Trimbakwala, Plastic Road use of waste plastic in Road construction, 2017
27. R S Deshmukh, L S Kendre, L C Chaudhari, S S Hirve, S P Shinde, Use of non-Biodegradable material in Bituminous Pavements
28. S Rakode Use of waste plastic and waste rubber Tyre in highway pavements, 2012, Patricia Nana Ama Asare, Francis Atta Kuranchie & Eric Antwi Ofosu
29. R.K Singh and A.K Sahu, Application of plastic materials in road construction, 2016
30. B M. A. J Chavan, use of plastic waste in flexible pavements, 2013
31. https://www.businessinsider.in/retail/adidas-has-sold-1-million-pairs-of-sneakers-made-from-ocean-trash-and-reveals-a-new-normal-in-footwear/articleshow/63349815.cms
32. Roy C, Chaala A, Darmstadt H (1999) Vacuum pyrolysis of used tyres end-uses for oil and carbon black products. J. of Anal. and App. Pyrol. 51, 201-221, Murillo R, Aylón E, Navarro MV, Callén MS, Aranda A, Mastral AM (2006) The application of thermal processes to valorise waste tyre. Fuel Proc. Technol. 87, 143-147.
33. R. S. Kognole I Kiran Shipkule I Kiran Shipkule I Manish Patil I Lokesh Patil I Udaysinh Survase "Utilization of Plastic waste for Making Plastic Bricks" Published in International Journal of Trend in Scientific Research and Development (ijtsrd), ISSN: 2456-6470, Volume-3 I Issue-4, June 2019, pp.878–88
34. Frank Ikechukwu Aneke, Celumusa Shabangu, Green-efficient masonry bricks produced from scrap plastic waste and foundry sand, 2021
35. Sellakutty, Dinesh. (2016). UTILISATION OF WASTE PLASTIC IN MANUFACTURING OF BRICKS AND PAVER BLOCKS. International Journal of Applied Engineering Research
36. Energy efficient production of clay bricks using industrial waste, Mary Lissy P.N., Carolin Peter, Kavya Mohan, Shone Greens, Sneha George, Heliyon, Elsevier, October 2018 WWS
37. Roy C, Chaala A, Darmstadt H (1999) Vacuum pyrolysis of used tyres end-uses for oil and carbon black products. J. of Anal. and App. Pyrol. 51, 201-221. 2
38. Dai X., Yin X., Wu C., Zhang W., Chen Y., 2001. Pyrolysis of waste tires in a circulating fluidized-bed reactor. Energy 26, 385–399, Conesa J.A., Font R., Marcilla A., 1996
39. M. LEWANDOWSKI1*, Katarzyna JANUSZEWICZ1, Wojciech KOSAKOWSKI;International Conference on Solid Waste Management, 2019
40. M. Nellesa b, J. Grünesa, G. Morscheck."Waste Management in Germany—Development to a Sustainable Circular Economy", Procedia Environmental Sciences 35 (2016)
41. Ágnes NAGY, Rajmund KUTI2 ,Vol. 15, No. 3 (2016) 231–237, The Environmental Impact of Plastic Waste Incineration
42. Sameer Maithel, Energy Utilization in Brick Kilns, PhD. Thesis, Indian Institute of Technology Bombay, Department of Energy Science & Engineering, 2003
43. Waste Management, Volume 33, Issue 8, 2013, Pages 1714–1728, ISSN 0956–053X, https://doi.org/10.1016/j.wasman.2013.05.003, ETRMA, 2011. End of life tires: A valuable resource with growing potential; 2011 Edition

UAS and Sustainable Development

Sudhir Kumar Chaturvedi and Surya Prakash Tiwari

Abstract Unmanned Aerial System (UAS) is an aerial vehicle that operates without human intervention using a ground-based controller. UASs are expected to observe and protect very large areas as well as control and monitor remote areas that are beyond the reach of humans. Drones or unmanned aerial vehicles are changing the way we communicate with the world and earth environment. These are, essentially, flying robots that can be remotely controlled or fly self-governing by programming-controlled flight designs in their installed frameworks, working related to locally available sensors and its general components include body frames, propellers, flight controllers, ESC (Electronic speed controller), batteries, motors and sensors required for use. Applications of UASs are widely increasing around the world for civilian, commercial as well as military purposes. We have examined some of the applications of UASs and the possible futuristic applications of UASs in India. This paper explores the use of UAAs in a variety of fields, including forest fire, land detection, coastal surveillance, agriculture, and a variety of commercial applications, since these flying machines are fast and do not endanger human life. Until now, UASs in India are primarily used for observation, reconnaissance and air navigation in military and defense zones. It is therefore important to establish more UAS-based research on these topics.

Keywords UAS · Drones · Forest fire · Coastal surveillance · Agriculture · Land detection

S. K. Chaturvedi (✉)
School of Engineering, UPES, Dehradun 248007, India
e-mail: sudhir.chaturvedi@ddn.upes.ac.in

S. P. Tiwari
King Fahd University of Petroleum and Minerals, Dhahran 31261, Saudi Arabia

K. D. Bahukhandi et al. (eds.), *Environmental Pollution and Natural Resource Management*, Springer Proceedings in Earth and Environmental Sciences,
https://doi.org/10.1007/978-3-031-05335-1_21

1 Introduction

Technologies are those inventive concepts that reflect progressive creativity in a sector to achieve a competitive edge. UAS is one of the most creative systems used in the world. UAS are basically unmanned aerial vehicles that are remotely controlled or equipped to operate independently. An autonomous flying machine frame is defined as a powered aerial vehicle that does not convey a human operator, uses aerodynamic forces to lift the vehicle, and can operate independently or remotely. Recently, UASs have made rapid progress due to their varied applications in the area of gadgets, optics, digital engineering, vitality storage, etc. Technical advances such as Global Navigation Satellite Systems, Inertial Measurement Units, Light Detection and Range, Synthetic Aperture Radar, Imaging Sensors and Mechanical Autonomy have contributed to the growth of UAS. The UASs are grouped into six categories: aim and decoy, observation, logistics, battle, research and development, and for economic and corporate purposes. Popular and company UASs are used by professionals in applications such as photography and film, precision agriculture, infrastructure examination, etc.

Figure 1 shows the UAS classification tree. Details on the drone surveying is presented in subsequent section of this study.

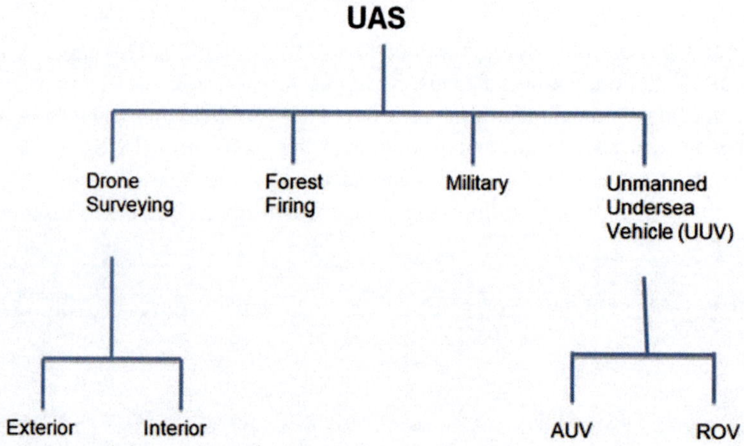

Fig. 1 UAS classification tree [1]

2 Study of Drone Surveying Method

2.1 Background of UAS

Drone surveying is something that many countries are adapting but the use of UAS by civilians is relatively a new concept. Hence one should need to be licensed for making use of UAS and should know the rules and regulations before using it. The best and most precious benefit of drone surveying is at once obvious even to individuals who are not professionals within the discipline. Using a drone to take aerial pictures of full-size portions of land makes the entire technique a lot simpler rather than using traditional techniques.

Other benefits include:

1. It is much safer
2. They provide excellent detail
3. It is faster and cheaper
4. It consolidates your toolkit

2.2 Drone Surveying Method

There are two types of drone surveying:

(1) *Interior drone surveying:* Inside looking over comprises of giving countless estimations to deliver a geometric portrayal of a structure anytime. There are many applications: industrial buildings, public buildings (sports halls, exhibition hall), historic buildings (museums, castles, and churches), archaeological sites, etc. Demonstrating structures empowers the creation of plans, models and warm adjusts which help in checking and reporting the development of a structure. Good knowledge of the structure allows for the implementation of evacuation or contingency plans.

(2) *Exterior drone surveying:* Exterior drone overviews include estimating all components recognizable from outside. A variety of survey techniques are available, including terrestrial photogrammetric, aerial photogrammetric, laser scanning, LiDAR, tachometry, and GNSS rover. Flexible tracking and mapping, rambles, VTOL aircraft, knapsack, and so on are examples of survey vehicles (see kinds of looking over).

2.3 Drones in Agriculture

The long-term aim is to create drones for farmers that can carry 20 L of pesticides for larger farms. Farmers can buy Agri-drones, but maintaining them is expensive, both in terms of fuel innovation and maintenance. Drones today operate on batteries, and the cost of maintaining the batteries is high. If the rancher chooses, automaton will check the harvest on a regular basis. Weekly, regular, or even hourly photographs will display the changes in the plants over time, showing possible "problem spots". After finding these trouble spots, the farmer will work to strengthen crop management and manufacturing. With the era constantly improving, imaging of the crops will want to enhance for better analysis. With the facts that drones gather, store and visualize information from the crops the farmers are able to investigate their vegetation and make educated decisions on how to proceed given the correct crop facts. Software programs for studying and correcting crop manufacturing have the capability to develop on this marketplace. Farmers will fly a drone over their crops, accurately perceive a difficulty in a specific area, and take the important moves to accurate the trouble. This offers the farmer time to awareness on the huge photograph of manufacturing instead of spending time surveying their crops. Additional uses include maintaining track of cattle, surveying fences, and tracking for plant pathogens. Examples of agricultural drones: Fixed-wing Ag drones, eBee Ag UAV, Precision Hawk's Lancaster fixed wing UAV.

A few estimates required in Indian agribusiness are:

1. Prevention of natural harm
2. Viable checking of crop wellbeing
3. Getting bits of knowledge into the soil wellbeing
4. Arranging water system and the correct utilization of compost

2.4 Drones in Land Surveying

UAS remote sensing has an immense potential for landslide mapping. Multi-temporal images can be used to examine the sequence of the slides. In fact, it also requires advanced image processing technologies. In the case of category drone images, there is a broad scale mapping such that it will not be paintings of daily form algorithms. Often segmentation algorithms are used for the type of algorithm. They are also used to have public-earthquake, landslide maps: it is useful that we didn't have to wait for satellite imagery, so we could get a high decision map shortly after the earthquake. One can also render multi-temporal maps by flying drones again daily (e.g., after a main rainfall or aftershock to correlate those inputs with landslide remobilization or the prevalence of extra landslides). Thanks to the rapid organization of Fox6 or Foxy Slim and the centimeter accuracy of the Drone Box RTK, the essence of the calculation is the same as the accuracy of the summary charts.

2.5 Drones in Mining Industry

With the potential to map much as in land surveying, the UAS has also found its use in the mining industry. Mines are the areas from which different minerals are mined and these sites are risky and time-consuming for the user to carry out inspections of stocks or dumps or mines with a risk of soil or rockslides. Drone surveying is very effective and can offer a more detailed outcome. Some UAS, such as Fox6, FoxySlim VTOL and FoxyPro VTOL, can hold more than 100 ha and FoxyWalk is capable of measuring stocks while FoxBathy can survey underwater stocks with both tools and DroneBox RTK vehicles compatible.

2.6 Drones in Smart Cities

UAVs are engaged with a wide scope of utilizations and capacities in smart cities. A portion of these applications are observing traffic, key foundations and checking improvement at a normal premise. UAVs can give a few administrations and openings that can profit keen urban areas. Some of the applications of drones in smart cities include traffic and crowd management, natural disaster control and checking, smart transportation. What could be simpler than to fly an elevated review automaton, for example, Fox4 or Fox6 (in consistence with current guidelines), or to walk or roll a Foxy Walk earthly study apparatus which in under a half-day permits the organizer access to precise and refreshed plans in regard to consider territories- consistently with centimeter exactness on account of Drone Box RTX (Table 1).

3 Major Applications of Drone

3.1 Application of UAS in Forest Firing

Forest Fires has been part of the Earth's natural cycle for thousands of years. Forests have a variety of essential functions in general. They can refine water, balance dirt, cycle supplements, moderate the environment, and store fuel. They also provide ecosystems for biodiversity and nurturing ecosystems rich in biological diversity. Economically, forests sustain the forest goods sector, which provides hundreds of thousands of jobs and adds billions of dollars to the country's national wealth. Unfortunately, every year millions of hectares of land are burned by forest fires, and hundreds of millions of dollars are spent on this burning.

Early identification and containment of forest fires is essential to minimize the damage that fires can cause as a result of their rapid spread of convection and long combustion period. Generally blackwood fire checks and locations use either mechanical gadgets or environmental scanners, but both methods may be both dangerous and exorbitant in that remote sensing technology has been one of the most commonly employed resources for successful forest survey and management. The Use of Small Drones for Environmental Mapping and Monitoring. As with small

Table 1 Agriculture and land surveying examples

Application	Agricluture	Land detection, mining industry
Example	Fixed-wing ag drones, eBee Ag UAV	Hex copter
About	This uses propelled, flight arranging, flight management and image preparing software, incorporates their restrictive eMotion Ag flight arranging and ground station software, making the way toward making a flight plan extremely straightforward	This enables the real-time production of drawings and georeferenced 3D models. They provide precise mapping and ensure operator safety
Information that can be captured	Vegetation indices (NDVI, CCCI, NDRE, MCARI, CWSI, et al.) Plant counts Soil H20 levels Soil temp Topography / 3D mapping (limited/not survey quality)	Emergency mapping of a territory in response to disasters The user gets a clear picture of the disaster area and can therefore manage relief and repairs Estimate the extent of damage. For long range sites such as earthquake sites, Fusion VTOL is a very effective tool
Weight	1.1 kg/2.4 lbs	2.5 kg
Sensors	Sequoia multi-spectral sensor provides more accurate data than NIR options	Thermal, multispectral, optical, hyper spectral or LIDAR
Flight time	55 min	30 min
Area covered	Up to 500 acres per flight (400 ft elevation, 12 cm/pixel resolution)	2 km(height 2 km)

drones, technologies such as GPS are originally used for military uses but are gradually being used in commercial applications. While these drones are not commonly used for this reason, the market for them in the future is very strong. Drones are mainly used for mapping and surveillance in this area, but some of those drones are also used for forest stand monitoring. India, which has seen a 46% rise in forest fires over the last 16 years (2003–17), has seen a 125 percent increase (from 15,937 to 35,888) in such fires in just two years (2015 to 2017). In 2017, the largest amounts of forest fires were reported in Madhya Pradesh (4,781) followed by Odisha (4,416) and Chhattisgarh (4,373). In fact, 23 out of 33 states and associations. In Punjab, such instances of fire rose sevenfold, led by Haryana and Rajasthan, who saw four, and three times a leap in numbers (Fig. 2).

Some of the drones used for this purpose are:

(1) *Matrice 200 V2:* The final stage for airborne profitability consolidates the difficult structure and simple configurability to act as a solution to a variety of mechanical applications. Upgrades to the M200 Series V2 boost astute control

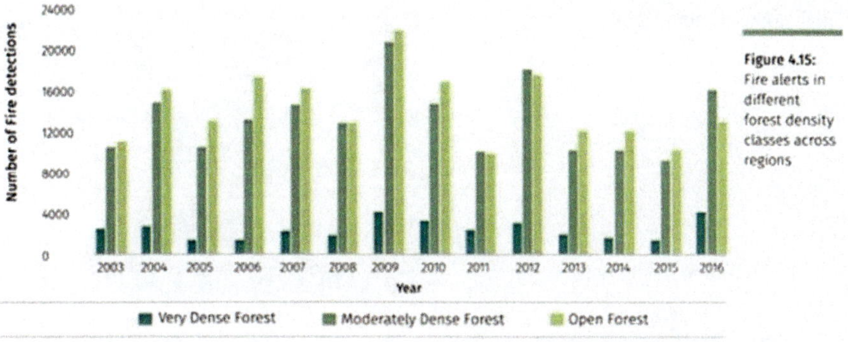

Fig. 2 State of forest report (ISFR) 2017

Fig. 3 Matrice 200 V2

frameworks, flight execution, which provide flight well-being and information security highlights.

(2) *UAV Marcus zephyr:* Marcus UAV systems offer on-demand real-time video and high-resolution aerial images. Unmanned aircraft is a safer choice than flying a conventional aircraft for etheric reconnaissance and research. They send images of elevated modeling, prosecution, catastrophe relief, and a vast variety of various occupations (Figs. 3 and 4, Table 2)

3.2 Application of UAS in Military

The way the wars are fought has changed fully when we reach the new era. Today's war technology is very modern, and it is mainly concerned with defending allies while causing more damage to those who are not. Certain unmanned aerial systems are used to carry out vital missions, such as protecting the protection of human life. So, what exactly is an Unmanned Aerial Vehicle (UAS)? In principle, it could be anything from a $5 child-controlled toy to the $100 million Global HAWK by the United States Army. The UAS is used for vital missions such as collecting key enemy

Fig. 4 Marcus zephyr [2]

Table 2 Specifications of UAS used in forest firing

Drone	Sensors	Payload capacity	Flight time	Range	Dimensions
1. Matrice 200 V2	Obstacle sensing External GPS module	1.34 kg	38 min	8 km	Unfolded propellers and landing gears included, 883 × 886 × 398
2.UAV MARCUS ZEPHYR			1 h		

camp information, storage centres, surveillance of some HVTs and, in certain cases, destroying various sensitive HVT and military camps. These UASs are often used for very critical missions.

Many of the world's best-known UAVs are:

Those are the numerous UAVs that are primarily used by the air force, the navy. But soldiers in many very critical missions often use UAV'S to find out about the location and number of enemies in different small UAVs. They are so fragile that they remain undetected (Figs. 5 and 6).

In India for the past two decades the air force, navy, army are using UAVs. Mainly the DRDO (Defense research and development organization) is making these UAVs

Fig. 5 MQ9 reaper [3]

Fig. 6 Black hornet [4]

Fig. 7 DRDO Rustom2 [5]

for the forces. But India has also bought various UAV's from its ally countries such as Israel, Russia (Fig. 7).

The key components of each UAV are based on the same concept as the power grid, transmitter and receiver, etc. But when it comes to uses, the tools mounted on them differ since, for example, in military UAVs, they are mostly mounted with infrared cameras, very accurate HD cameras, or weapons mounted on them.

UAVs are mostly used for reconnaissance missions in India. It is basically used to gather the Intel from the hiding places of various terrorists, without endangering human life. But this is not enough. We can use these UAVs more than we are currently using by making some changes.

The weapon can be mounted on UAVs by increasing its payload capacity and making other modifications. And it can be used for missions where the jets are used. It is very expensive to use fighter jets instead of UAVs. And we are also putting human life at risk. These drones can be used for the border patrolling by mounting some sensors, such as motion detection, infrared, thermal and others. These UAVs can also be used to provide supplies to the ground troops under harsh conditions.

In India, there are a range of drones with the following specifications (Table 3).

Table 3 Specifications of UAS used in military

Name	Wingspan	Payload	Service ceiling	Range	Endurance
DRDO rustom	7.9 m	95 kg to 350 kg depend on variants	7925 m	250 km	About 24 h
DRDO nishant	6.5 m	45 kg	3962 m	100 km	4 h 30 min
DRDO netra	0.9 m	High resolution CCD camera	300 m	2.5 km	30 min
DRDO imperial eagle	1.6 m	250 g	4572 m	15 km	60 min

3.3 Application of UAS in Unmanned Undersea Vehicle (UUV)

These days UAS are getting more attention by private and public sector companies that are working in the defense domain. Be that as it may, the skies are not the main playing field for unmanned vehicles. Humming underneath the tide is a school of unmanned undersea vehicles (UUVs) gaining popularity as those of their flying partners. Unmanned Underwater Vehicles (UUV), known as submerged machines, is any vehicles that can work submerged with no human impedance.

These vehicles might be classified into two categories:

(1) *Remotely Operated Underwater Vehicles (ROVs)*

These are restricted by a human administrator who is based on the other side of the planet. ROVs are connected to ships through the umbilical or progressive interface connections which send a sign to the operator.

(2) *Autonomous Underwater Vehicles (AUVs)*

Autonomous Underwater Vehicles (AUVs) function independently of human interference. AUVs are customized to accomplish a given task or mission, use built-in sensors to determine their areas of operation and returning to a pre-modified location for data handling.

Suitable for business, scholastic, rational and leisure purposes, two types of UUVs may perform a variety of tasks. They can, for example, test oil rigs, help researchers analyze and collect living beings at a much deeper level than manned vehicles can accomplish, and they can also help track down hidden fortunes. UUVs may typically be fitted with a variety of sensors such as side-scan sonar, inertial navigation systems (INS) and GPS support and an acoustic Doppler current profiler (ADCP) in areas other than protection. These devices enable UUVs to conduct overview missions, such as the detection and mapping of underwater wrecks, rocks and other obstructions that pose a challenge to the path of commercial and recreational vessels.

Unmanned undersea vehicles are the latest in a long line of vehicles that have started to take the place of divers in the sea. They have been designed for marine research, science, research, rescue, recovery and overview.

Table 4 Specifications of some of the UUVs [7]

Vehicle	Endurance	Depth (m)	Speed	Weight/Payload	Dimension	Initial operating capability (FY)
SAHRV	6 h @ 4 kt	100	5 kt	36 kg	0.2 × 1.6 m	03
LDUUV	5 h @ 6 kt	200	12 kt	2,450 kg	0.7 × 8 m	13
BPAUV	TBD	300	6 kt	400 kg	0.5 × 3 m	06

Despite having a booming economy and some of the world's brightest minds, India has not completely devoted its capital to the production of UUVs. IURS (Indian Underwater Research Society) has recently been formed, an independent body of students interested in AUVs and ROVs. They gave India its first low-cost AUV named BhAUV (Bharat Autonomous Underwater Vehicle) [6]. But India is still lagging behind compared to other countries. AUV-150, MAYA AUV are some of the AUVs that have been developed in India. DRDO, NIO, the Indian Navy and many such organizations have shown interest in, and are currently working on, the development of this new technology.

Some of the main applications of Unmanned Underwater Vehicles [7] are as follows (Table 4):

Semiautonomous Hydrographic Reconnaissance Vehicle System (SAHRV)

PMS 325 J Expeditionary Warfare has created a semi-autonomous hydrographic surveillance vehicle (SAHRV) platform, an extension of a framework called the Remote Environmental Monitoring Unit System (REMUS), produced for routine independent surveys at seaside locations for the inspection network. In the examination network, the SAHRV system is commonly used. Sensors for measuring conductivity, temperature, water depth, and optical backscatter are included. It has side-filter sonar, much like an acoustic Doppler current profiler that looks up and down. A short-standard acoustic system is used to present the road. Control is carried out via a PC phone that executes the specific path of the waypoint directions.

Long-range Mine Reconnaissance System (LMRS)

The Long-range Mine Reconnaissance System (LMRS) is a UUV designed to be propelled and retrieved from a submerged submarine while it is traveling at a very slow speed. The LMRS main objective is to expand the submarine's ability to lead mine observation in a stealthy manner. The LMRS dispatch system necessitates a dedicated torpedo tube, and the recovery system necessitates an additional torpedo tube, suggesting a major lack of adaptability for various submarine missions. The LMRS is a follow-up to the Near-Term Mine Reconnaissance System (NMRS), which was developed in response to the submarine network's need for a semiautonomous vehicle to conduct surveillance in front of submarines [8].

Multi-Reconfigurable Unmanned Undersea Vehicle (MRUUV)

The multi-reconfigurable unmanned undersea vehicle (MRUUV) is the next step in the evolution of UUVs. It is expected to continue with the LMRS. ISR sensors, as well as ASW capabilities, will be included in the MRUUV. By switching out modules, the MRUUV would be able to reconfigure for different missions. The modules will allow the MRUUV to perform various tasks, such as ocean surveillance, underwater hunting and study, interchange and route assistance, and the underwater trail and track [9].

Large Diameter Multi-Reconfigurable Unmanned Undersea Vehicle (LD MRUUV)

The large, multi-reconfigurable UUV (LD MRUUV) diameter is designed as a large transport, suitable for reconfiguration to convey distinctive sensor and task packages. LD MRUUV could express and send a few smaller, more precise UUVs to a difficult terrain and fill the docking station as a resurrection of vitality and knowledge download. It will extend the U-boat venture into the regions in question. This system could fulfill a vast range of occupations envisaged in the threat debate and future maritime practices, as discussed above, in particular ASW; coastal anti-surface warfare; Special Operations Forces; clandestine insight, monitoring, surveillance, and mine-survey emphasis [10, 11].

4 Conclusion and Future Prospects

Since the use of UAS would be substantially more beneficial based on many factors. Carrying extinguishers to a forest fire in a helicopter or rotorcraft is very expensive and difficult, whereas an unmanned aerial vehicle has a slight weight and height advantage: not needing to carry human pilots also avoids any human loss. This adds a significant amount of weight to the retrieval, stretches the travel time or raises the payload limit. Being more heat-resistant will also have some leeway, as the specialization can be flew closer to the fire. Moreover, the use of UAS in the defense army saves both lives and manpower. It can now be seen in many ways that can continue to save time, money and human life. In India, there is also a need to expand the use of these UAVs in other fields. They can also be used for refueling and other uses.

This study has summarised the possible use of UAS in land surveying, forest fire and underwater applications and contrasts its availability and use in India with other countries. For various implementations, the changed components must be fixed in UAVs to execute the operation. Creating UUVs that are self-sufficient, long-lasting and self-sustaining are critical needs for the Navy's vital missions. It is about being determined and making a strategy where we need to go for unmanned vehicles. It is just a matter of applying and enhancing the ingenuity that we already have to carry out in a variety of missions and have the capacity that we need. While formatively behind their aerial kinfolk, UUV development is ready to accelerate.

Acknowledgements We would like to thank the University of Petroleum and Energy Studies, Dehradun, India and Center for Environment & Water at King Fahd University of Petroleum & Minerals, Dhahran, Saudi Arabia.

References

1. https://dronebelow.com/2018/12/06/drone-classification-applications-and-challenges/
2. https://www.aviationtoday.com/2014/11/20/5-potential-commercial-uses-of-unmanned-aircraft-systems/
3. https://www.washingtontimes.com/news/2015/jun/16/russian-microwave-gun-can-disable-warheads-6-miles/
4. https://www.defenceweb.co.za/industry/industry-industry/desert-wolf-marketing-black-hornet-2-handheld-uav-in-sa/
5. https://www.hindustantimes.com/india-news/drdo-s-rustom-2-drone-flight-tested-india-goes-for-armed-heron-uavs/story-CZ5jd9tRo6Ph2jcq2HOpmM.html
6. Indian Underwater Robotics Society; Sehgal, Anuj & Kadarusman, Jason & Chawla, Gaurav & Garvit, Kumar & Neville, Michael & Jolly, Prabhnoor & Agarwal, Aabhas & Arjun, (2005). BhAUV: India's First AUV.
7. Unmanned Surface and Undersea Vehicles: Capabilities and Potential.'National Research Council.2005. Autonomous Vehicles in Support of Naval Operations. Washington DC: The National Academics Press.doi: https://doi.org/10.17226/11379
8. Hardy, Tim, and Gavin Barlow. "Unmanned Underwater Vehicle (UUV) deployment and retrieval considerations for submarines." *International Naval Engineering Conference and Exhibition 2008*. 2008.
9. Mancini A. E Frontoni, P Zingaretti, 2016. A multi/hyper-spectral imaging system for land cover/land use using unmanned aerial systems. IEEE International Conference on Unmanned Aircraft Systems (ICUAS), Arlington, VA, USA, pp. 1148–1155.
10. Nebiker S, N Lack, M Abacherli, S Landerach, 2016. Light-weight multispectral UAV sensors and their capabilities for predicting grain yield and detecting plant diseases. International Archives of the Photogrammetry, Remote Sensing and Spatial Information Sciences, Volume XLI-B1, XXIII ISPRS Congress.
11. Proctor C, Y He, 2015. Workflow for building a hyperspectral UAV: Challenges and opportunities. UAV-g 2015 International Archives of the Photogrammetry, Remote Sensing and Spatial Information Sciences, Vol. XL-1/W4: 415–419

Impact of Discharge of Wastewater and Used Oil from Motor Service Stations on Groundwater Quality of Dehradun City of Outer Himalaya Region

Kanchan Deoli Bahukhandi, H. Tonsana, Mayuri Deka, Shalini Vohra, and Mohini Singh

Abstract The effluent discharge from automobile service stations is a matter of deep concern as it is directly released to the environment without any treatment particularly in developing countries. The different types of used oil and fluids are prominent contaminants from automobile sector which are being consistently let-out risking the environment and health of living beings. The disposition of oily wastes on land surface effects the soil quality and can infiltrate the sub-soil impacting the groundwater quality. The deterioration of groundwater quality has challenged the survival of the ecosystem and in the forefront need to take urgent steps to revive and sustain the groundwater. The current research work is based on assessment of effluent quality discharged from different Motor Service stations by drawing comparison of the physiochemical wastewater quality parameters against the BIS standard of drinking water quality. The analysis showed that 7 out of 15 parameters measured were off permissible standard limits and exhibited maximum values. High levels of Total Dissolved Solids (TDS) 666.65 mg/l, Dissolved Oxygen (DO) 6.8 mg/l, Alkalinity 457.5 mg/l, Calcium (Ca) 142.9 mg/l, Magnesium (Mg) 69.86 mg/l, Total Hardness 550 mg/l, Oil & Grease 76 mg/l were indicated in the collected samples which does not imply a positive scenario of the water ecosystem. High concentrations of Oil & grease at the different automobile service centers were found in the range of 20–76 mg/l. Thus, this study has been done to provide meaningful insights on the effects on aquatic biota and the importance of effluent treatment particularly in automobile sectors. Further, the study can be used as a source of reference to do research studies in this topic.

Keywords Water quality · Groundwater · Effluent discharge · Used oil · Waste oil · Hazardous waste

K. D. Bahukhandi (✉) · H. Tonsana · M. Deka · S. Vohra · M. Singh
Sustainability Cluster (HSE and Civil Engineering), University of Petroleum and Energy Studies (UPES), Bidholi, Dehradun, India
e-mail: kanchan@ddn.upes.ac.in

K. D. Bahukhandi et al. (eds.), *Environmental Pollution and Natural Resource Management*, Springer Proceedings in Earth and Environmental Sciences,
https://doi.org/10.1007/978-3-031-05335-1_22

1 Introduction

Growth of industrialization and urbanization has increased the necessity of fresh water for various needs and purposes which has resulted in environmental problems. In developing countries mainly, groundwater is the primary source for drinking, irrigation sector and industrial sector [3–5]. In today's scenario, there is an urgent need to find solution on groundwater sustainability because of improper treatment and discharge of effluent and sewage facility which has led to deterioration of quality of groundwater [2, 5–7, 14, 16–19, 21, 26]. As per India's annual inventory record of 2021, sewage generation from urban areas is estimated to be around 72,368 MLD (Millions of Litre per Day) which is based on water supply @185 lpcd (litres per capita per day) wherein rate of sewage generation is 80%. There is a total set-up of 1631 Sewage Treatment Plants (STPs) in whole of India of which 1093 STPs are operational with a capacity utilization of only 20,235 MLD and the difference of 52,133 MLD quantity of untreated sewage is let-out into the natural drainage system [13] which either flows towards nearby water bodies or it is accumulated in the immediate environ forming cesspools [14, 15].

In recent times, the automobile industry is one of the demand growing industries where India was ranked fifth largest in the year 2020 and was ranked seventh in 2019 in terms of manufacturing commercial vehicles. With growth in population and modernization, the sale of domestic vehicles between financial Year 2016–2020 increased at 1.29% CAGR (Compound Annual Growth (Foundation, n.d.). This has led to increase in set-up of automobile service centers which undertake different jobs such as workshop, servicing, repairing, car-washing, etc. In general, for car-washing 70L of fresh water is required and 200L of water in automatic car wash station. The amount of usage of water may differ in stations but conclusively, a huge volume of water is used for washing cars where the wastewater is directly discharged into the drainage system rather than recycling and treating and then reusing it. Therefore, all types of contaminants are directly released into the environment [3, 31]. Similarly, numerous activities in car-servicing and workshop stations can be source of pollutant for contaminating the environment. These activities release various effluents and hazardous wastes such as oil & grease, lube oil, spent oil, heavy metals, detergents, paints, weld rods, plastic, rubber, etc. [4, 32].

One of the most common inorganic groundwater pollutants is used oil from automobile industries. The used oil otherwise known as waste oil or spent oil is categorized as hazardous waste and it is listed in Schedule I under Rule 3(1) 17(i) of the Hazardous and Other Wastes (Management and Transboundary Movement) Rules, 2016 [12]. Used oil from automobile industries includes different types of oils and fluids—oil filters, lubricants, motor oil, oil leaks from cars, etc. are improperly discarded on the ground as well as are washed off to the nearby drainage system without proper treatment. Consequently, it contaminates the groundwater and the soil [4]. Burns et al. [11] had researched that used oil takes more than six years to decompose in the environment causing long-term problems for the biota. And, even it takes weeks to months to decompose the oil by microorganisms under controlled laboratorial conditions. [1, 33] studied that the Polyaromatic hydrocarbons, one of the components of oil has been identified as carcinogenic in nature by the International Agency for Research on Cancer and, they are more harmful to living beings when present in bigger concentration in its composition. Nowak et al. [28] highlighted that water is considered contaminant in the presence of 1 ppm oil. The mineral oils are not easily biodegradable in nature and hence, it biodegradability is very low thereby, posing a threat to the aquatic ecosystem.

Vazquez-Duhalt [33] analyzed and described the impact of used motor oil on soil and aquatic organism. It also discussed the mutagenic and carcinogenic health effects of used motor oil and provided information on the biodegradation of lubricating motor oil. These studies have highlighted the harmful effects of improper disposal of used oil and its consequence on the environment [29]. Therefore, there is a dire need to take urgent mitigation steps for prevention and abatement of environment pollution to ensure sustainable development.

The current research paper evaluates the effluent discharged from different Motor Servicing Centers of Doon valley to determine the extent of contamination of ground water. The generated data can be used as a future reference to do research studies in this area. And, finally, acknowledging the importance of installation wastewater treatment and mitigate contamination of groundwater and its surrounding environment.

2 Methodology

The study was carried out in the district of Dehradun, the capital city of Uttarakhand which lies at 2200 feet altitude above sea level in India's northern region. The city covers a total geographical area of 67.0 km^2 and lies between 29°50'–30° 30' N Latitude and 77° 35'–78° 20' E Longitude having population of about 578,420 [30]. The Dehradun, known as Doon valley, have two main rivers Song and Asan which receives water from different perennial and non-perennial streams which originated in lesser north Himalaya and south Sub Himalaya. These rivers finally discharge its water from the city into the Ganga river in the East and Yamuna river lying in the west of Doon Valley [6, 8, 9, 25].

Field survey was carried out for collection of primary data of the effluent samples from selected 10 different Automotive stations of Doon valley. The effluent samples were collected from the outlet source of the stations. On-field pH, Temperature, TDS parameters were measured, and rest other parameters were analyzed in the Health Safety & Environmental Engineering Lab at UPES as per the APHA Standard 2012 (APHA Standard 2012) of wastewater and drinking water quality. pH of effluent was measured using pH meter. Alkalinity of water was determined by titration where acid of known strength was added to a volume of treated water sample. Chloride was analysed by using titration method with standard silver nitrate using potassium chromate as an indicator. Total hardness, Calcium concentration and calcium hardness were analyzed by titration method using standard EDTA (Ethylene-diamine tetra acetic

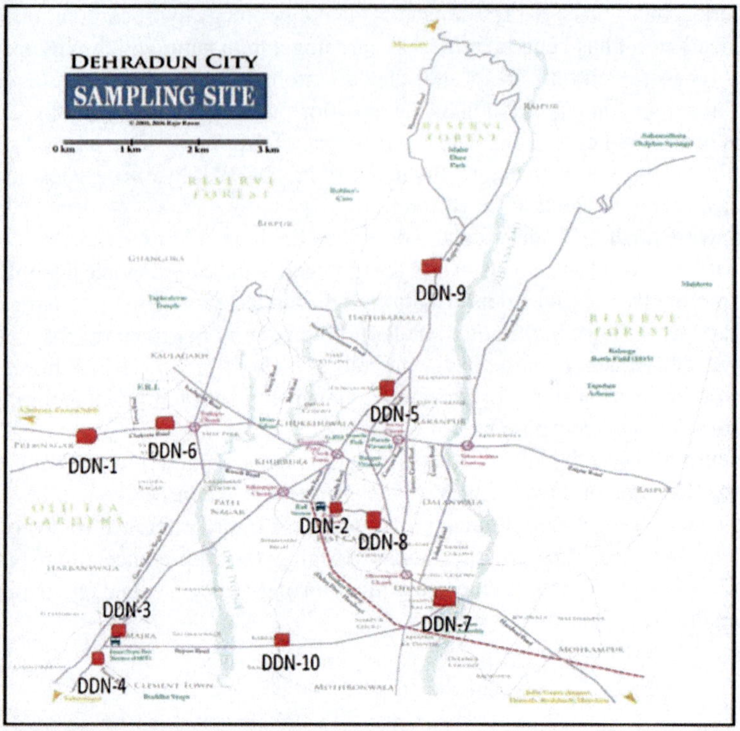

Fig. 1 Sampling location map of Dehradun City

acid) solution. Magnesium hardness was determined by analyzing the total hardness and calcium hardness of the wastewater. The presence of Sodium and Potassium was based on emission spectroscopy. Used oil & grease were extracted from water samples by intimate contact with petroleum ether (Fig. 1).

3 Results and Discussion

The physiochemical water quality parameter has been presented in Table 2 and comparison was drawn against the BIS standards of drinking water quality [10]. The temperature ranged from 33.4 to 33.8 °C in all the sampling locations. The conductivity varied from 643 μSimen/cm to 995 μSimen/cm in all sampling locations where DDN 3 station showed highest value (Table1; Fig. 2). The TDS concentration had crossed maximum permissible limit (500 mg/l) at three sampling locations namely- DDN 2, DDN3 and DDN 5 where DDN 3 showed highest TDS value of 666.65 mg/l (Table1; Fig. 3). The pH and TSS (total suspended solid) were found under permissible limit of BIS standard of drinking water quality. The parameters

Table 1 Sampling locations from different servicing station

Sl. no	Code	Name of the servicing center	Location
1	DDN–1	Bajaj servicing center	Premnagar
2	DDN–2	Hyundai service center	Prince Chowk
3	DDN–3	Oberoi motors workshop	Majra
4	DDN–4	Chevrolet workshop	Transport nagar
5	DDN–5	Mohit auto service center	Rajpur road
6	DDN–6	Rohan motors	Chakrata road
7	DDN–7	Hyundai workshop	Haridwar road
8	DDN–8	Roadways workshop	Old ISBT
9	DDN–9	Mayur auto	Rajpur road
10	DDN–10	Mahindra and Mahindra	Kargi Chowk

DO 6.3–6.8 mg/l and Alkalinity 353.8–457.5 mg/l of effluent exceeded than the normal standard range. High concentration values of Calcium 105–142.9 mg/l and Magnesium 46.8–69.86 mg/l were found in all collected water samples. Similarly, the total hardness of water was found far exceeding in the range of 520–550 mg/l from the permissible limit standard of 200 mg/l. Further, the concentration of oil & grease in DDN-8 showed 76 mg/l highest among all the collected samples and in fact, all the samples showed ranged 20–76 mg/l against the standard limit of 10 mg/l (Table 1; Figs.3, 4, 5, 6, 7, 8, 9,10, 11and12).

4 Conclusion

The automobile sector is huge and it is one of the fasted growing industry in today's modern era. As there is increase in demand and supply, it has resulted in generation of waste and accompanied by environmental pollution. The growing market of automobile sector is alarming and is becoming major polluters leading to detrimental effect on the ecosystem. With increase in number of automobile service stations, wastes are dumped or disposed off directly to the environment without being treated and/or properly stored and sent to authorized recycle vendors. They also have the same fate as domestic wastes in the city. A stringent regular monitoring of discharged water quality should be done in the motor servicing centers as majority of the wastes—different types of fluids and oils, lead acid batteries, fuel filters and other materials falls within the category of hazardous wastes [27]. Proper design facility of collection, treatment of wastewater and ensure effluent is brought to permissible standard limit before disposal at point source. No stagnation of wastewater should be allowed to avoid percolation of contaminants in groundwater. The wastes water generated by

Table 2 Water quality analysis

Sl. no	Parameters	Range	BIS: 2012
1	Temperature, °C	33.4–33.8	NA
2	Conductivity, µ Simen/cm	643–995	NA
3	Total dissolved solids (TDS), mg/l	587–666.65	500
4	Total suspended solids (TSS), mg/l	11 0.0–27.1	100
5	pH	7.2–7.9	6.5–8.5
6	Dissolved oxygen (DO), mg/l	6.3–6.8	5
7	Alkalinity, mg/l	353.8–457.5	200
8	Calcium (Ca), mg/l	105–142.9	75
9	Magnesium (Mg), mg/l	46.8–69.86	30
10	Total hardness, mg/l	520–550	200
11	Ca hardness	262–357	NA
12	Chlorine (Cl), mg/l	34.75–54.59	84
13	Sodium (Na), mg/l	9.7–24	NA
14	Potassium (K), mg/l	1.3–11	NA
15	Oil & Grease	20–76	10

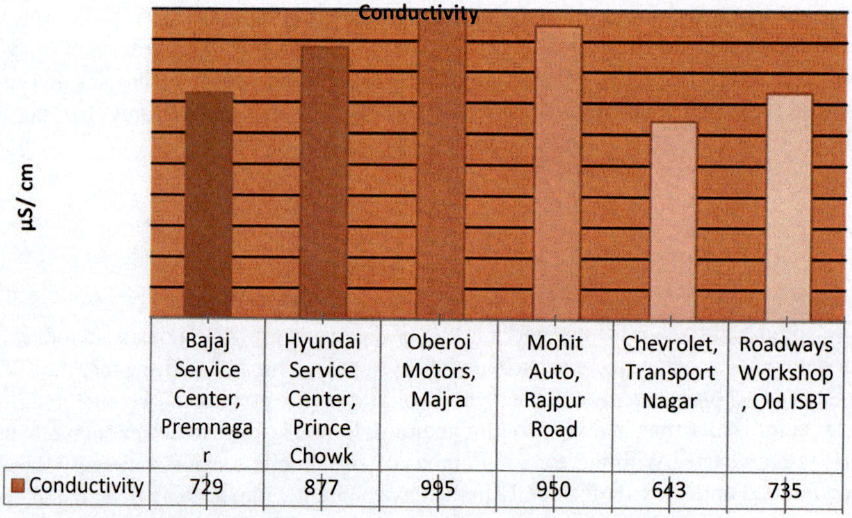

Fig. 2 Conductivity concentration in waste oil generated from different service center

Motor Servicing Center get mixed with domestic wastes and pollute the groundwater. The Oil & Grease content in the samples were not conforming to the desirable limit of BIS which is 10 mg/L [10]. **Roadways Workshop**, Old ISBT has the highest Oil & Grease content with **76 mg/L** (Fig. 10). All the samples from the Motor Servicing

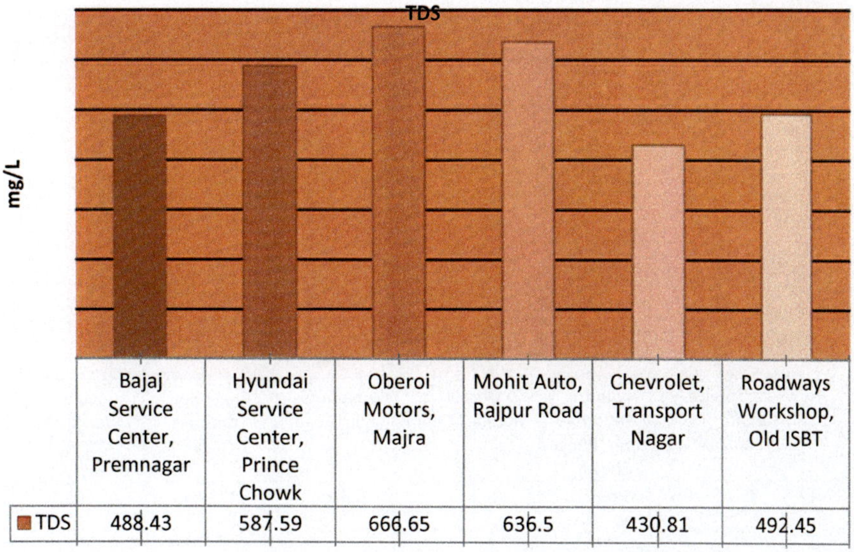

Fig. 3 TDS concentration in waste oil generated from different service center

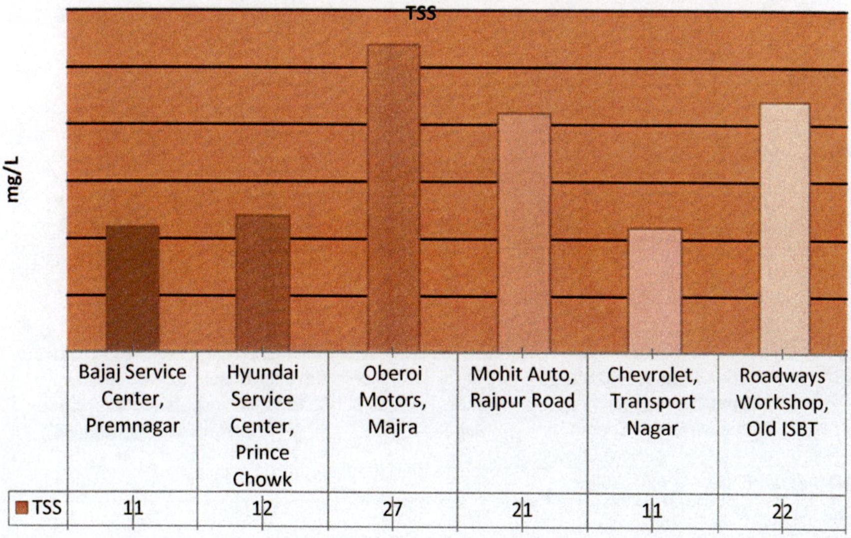

Fig. 4 TSS concentration in waste oil generated from different service center

Centers had Cation % in the order of **Ca > Mg > Na > K**. Ca % contributes almost 50% of the total Cation analyzed. Disposal of discharged water from the Servicing Centers should be either prohibited or limited in the city to avoid any leaching process into the groundwater or to provide sewerage system if it is within the city limit. It is

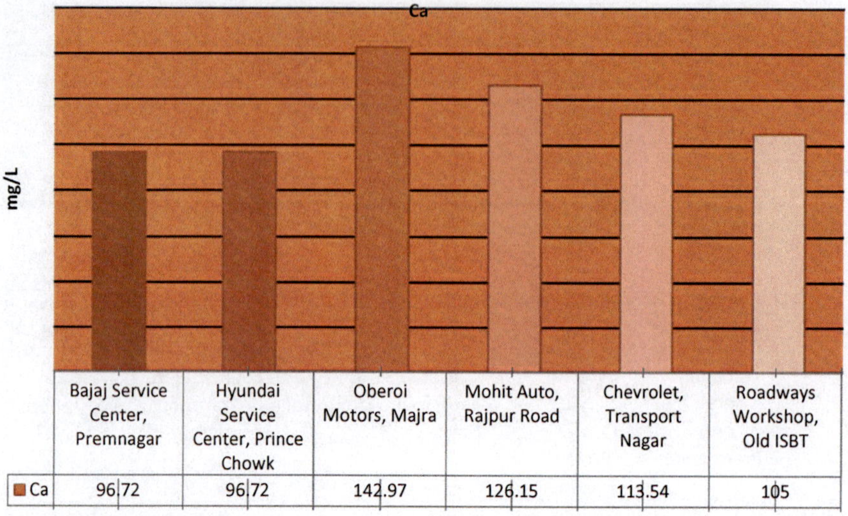

Fig. 5 Ca concentration in waste oil generated from different service center

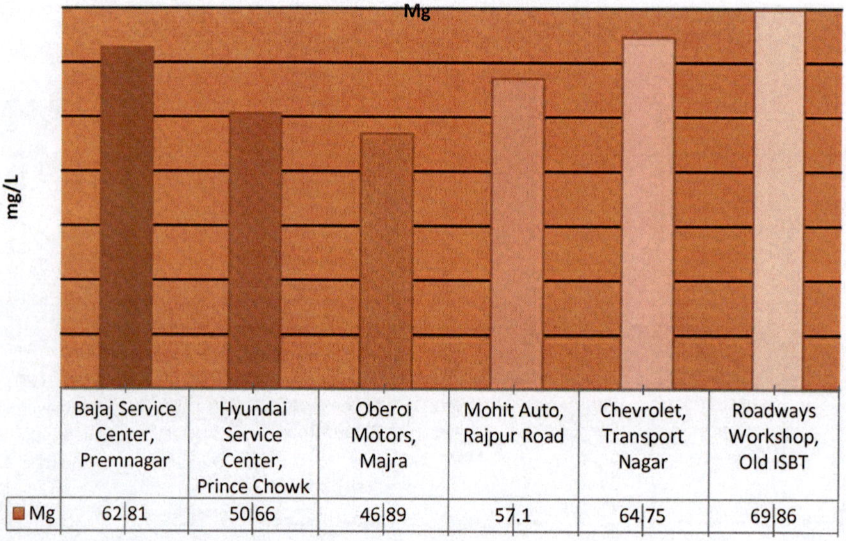

Fig. 6 Mg concentration in waste oil generated from different service center

suggested that some low cost and easy to implement techniques may be provided to the automobile service centers to avoid groundwater contamination. Further, more in-depth studies need to be done in this topic and provide efficacy solutions to these emerging problems.

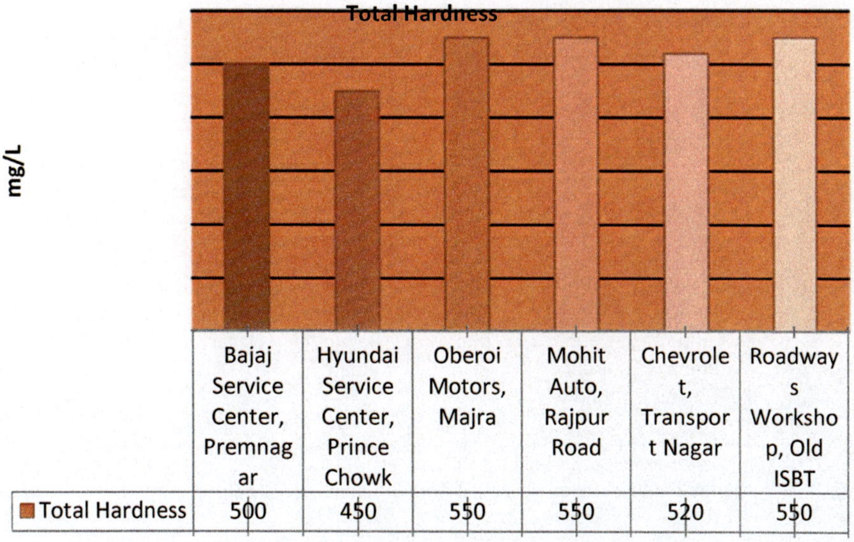

Fig. 7 Total Hardness concentration in waste oil generated from different service center

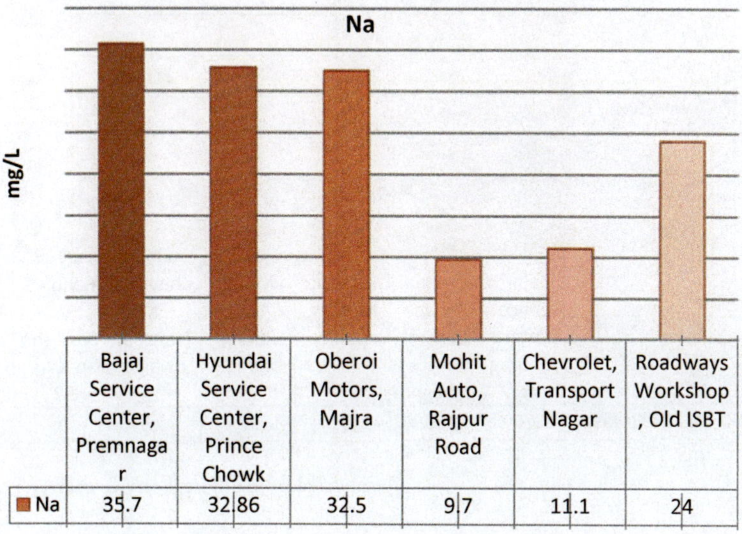

Fig. 8 Na concentration in waste oil generated from different service center

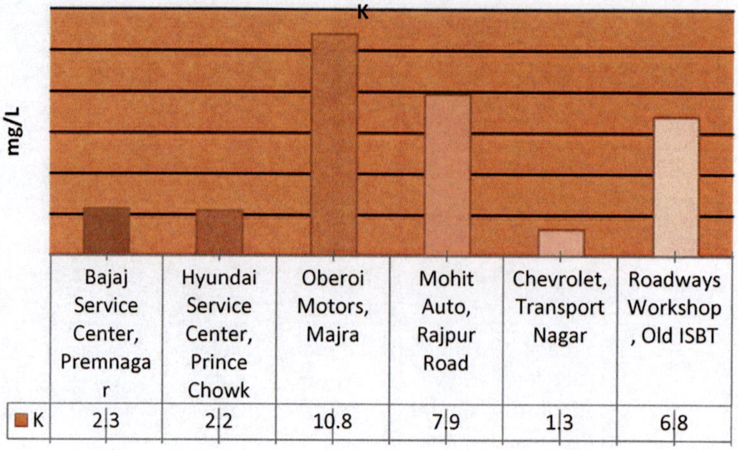

Fig. 9 K concentration in waste oil generated from different service center

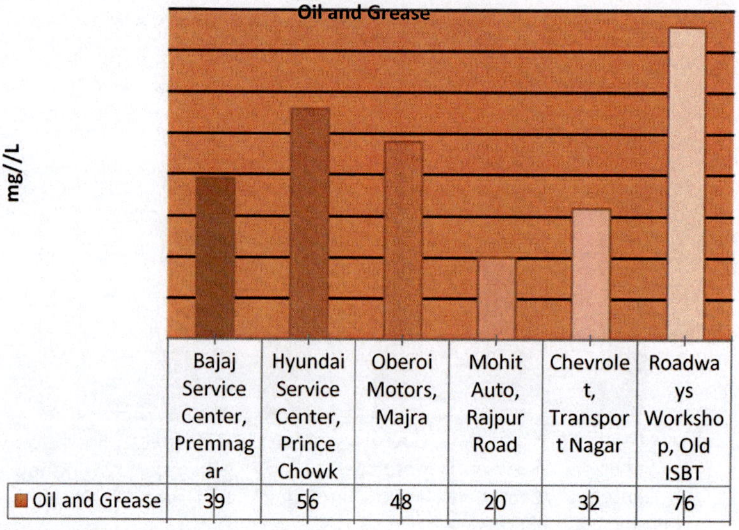

Fig. 10 Oil & grease concentration in waste oil generated from different service center

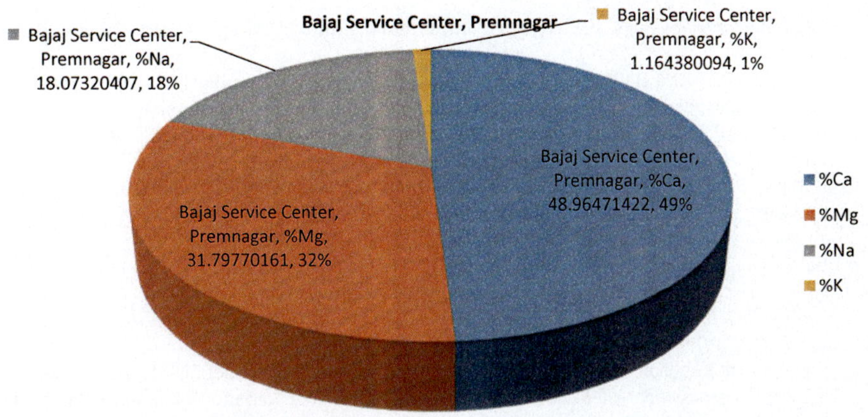

Fig. 11 Cation % distribution of each motor servicing center

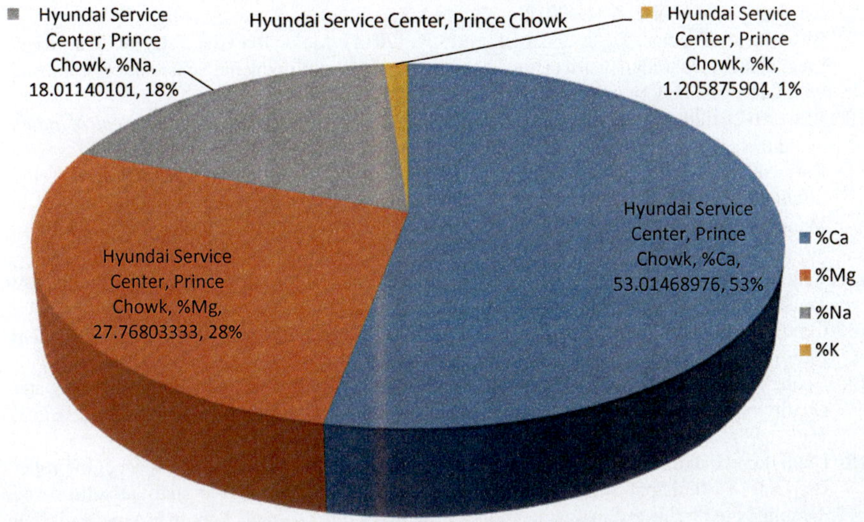

Fig. 12 Cation % distribution of each motor servicing center

References

1. Abdel-Shafy, H. I., & Mansour, M. S. M. (2016). A review on polycyclic aromatic hydrocarbons: Source, environmental impact, effect on human health and remediation. *Egyptian Journal of Petroleum*, *25*(1), 107–123. https://doi.org/10.1016/j.ejpe.2015.03.011
2. Arora, S., Bhaukhandi, K. D., & Mishra, P. K. (2020). Coronavirus lockdown helped the environment to bounce back. *Science of the Total Environment*, *742*, 140573. https://doi.org/10.1016/j.scitotenv.2020.140573
3. Asha, M. N., Chandan, K. S., Harish, H. P., NikhileswarReddy, S., Sharath, K. S., & Liza, G. M. (2016). Recycling of Waste Water Collected from Automobile Service Station. *Procedia*

Environmental Sciences, 35, 289–297. https://doi.org/https://doi.org/10.1016/j.proenv.2016.07.009

4. Ashraf, S., Rizvi, N. B., Rasool, A., Mahmud, T., Huang, G. G., & Zulfajri, M. (2020). Evaluation of heavy metal ions in the groundwater samples from selected automobile workshop areas in northern Pakistan. *Groundwater for Sustainable Development, 11*, 100428. https://doi.org/10.1016/j.gsd.2020.100428
5. Bartarya S. K. and Bahukhandi D. K. (2012). Impact Assessment of Urbanization and Industrialization on Surface and Ground Water Quality. *Global Journal of Engineering, Design & Technology, 1(1)*(11–22). https://doi.org/10.24018/ejfood.2020.2.4.79
6. Bahukhandi, K. D., Saini, R., Ahuja, N. J., Thakur, D., & Siddiqui, N. A. (2021). *Assessment of the impact of the municipal solid waste on Hydrochemistry of the Surface and Groundwater quality of Dehradun district of Uttarakhand : Seasonal variation in water quality. 42*(1), 175–188.
7. Bahukhandi Kanchan Deoli Siddiqui NA, Singh Rajesh, Arora Shefali, Mondal Prasenjit, S. (2018). Assessment of Major Ion Chemistry and Spatial Variation of Water Quality of Ganga River of Uttarakhand, India. *Journal of Chemical and Pharmaceutical Research, 10*(7), 145–150.
8. Bahukhandi Kanchan Deoli Bahukhandi, K. D., Bartarya, S. (2012). Hydrochemistry of Surface and groundwater of Doon valley. *Proceeding Volume "Fifth International Groundwater Conference (IGWCC), 3*(74), 2012.
9. Bhaukhandi Kanchan Deoli and Bartarya S.K. (2014). Major Ion Chemistry of Asan River Catchment of Dehradun district (Impact Assessment of Anthropogenic Sources). *Octa Journal of Environmental Research, 2(2).*
10. BIS. (2012). Indian Standard Drinking Water Specification (Second Revision). *Bureau of Indian Standards, IS 10500*(May), 1–11. http://cgwb.gov.in/Documents/WQ-standards.pdf
11. Burns, K. A., Garrity, S. D., Jorissen, D., MacPherson, J., Stoelting, M., Tierney, J., & Yelle-Simmons, L. (1994). The Galeta Oil Spill. II. Unexpected Persistence of Oil Trapped in Mangrove Sediments. In *Estuarine, Coastal and Shelf Science* (Vol. 38, Issue 4, pp. 349–364). https://doi.org/10.1006/ecss.1994.1025
12. Central Pollution Control Board. (2016). Hazardous Waste (Management, Handling & Transboundary Movement) Rules, 2016. *The Gazette of India*, 68.
13. Central Pollution Control Board (CPCB). (2021). *National Inventory of Sewage Treatment Plants March 2021. March*, 183.
14. Deoli, B. K., Bartarya, S. K., & Siddiqui, N. A. (2017). Assessment of surface and ground water Quality of Haridwar district of Uttarakhand. *International Journal of Chem. Tech. Research, 10*(10), 95–118.
15. Deoli Bahukhandi, kanchan, Kanchan Bahukhandi, D., & Aaron, A. (2016). Impact of Improper Disposal of Municipal Solid Waste on Groundwater Quality in and around the solid waste dumping site of Vishakhapatnam, Andhra Pradesh, India. *Original Research Article Journal of Global Ecology and Environment, 5*(3), 133–143. https://www.researchgate.net/publication/342338363
16. Bahukhandi, K. D., Siddiqui, N.A. (2008). Impact assessment of Urbanization on water quality. Environmental Pollution Control Journal, 12 (1).
17. Bahukhandi, K. D., Mondal, P., Singh, S. (2015). Water Quality Assessment of River Ganga Health Hazard Identification and Control Singh. International Journal of Scientific and Research Publications, 5 (5), 1-8.
18. Bahukhandi, K.D., Aaron, A. (2017). Impact of Improper Disposal of Municipal Solid Waste on ground water quality in and around the solid waste dumping sites of Visakhapatnam, Andrapradesh. Indian Journal of Global Ecology and Environment, 5(3), 133-143.
19. Bahukhandi, K.D, Siddiqui, N.A., Arora S. (2020). Assessment of Heavy Metal and Physio-chemical Parameter in Surface and Groundwater Quality of Dehradun District of Uttarakhand, Springer Transactions in Civil and Environmental Engineering, Advances in Water Pollution Monitoring and Control, Proceedings from HSFEA 2018. https://doi.org/10.1007/978-981-32-9956-6_16

20. Bahukhandi, K. Agrawal, S. Singhal, S. (2020). Impact of lockdown Covid-19 pandemic on Himalayan Environment" in International Journal of Environment Analytical Chemistry" , Tailor and Francis, Vol 100, Issue 15, https://doi.org/10.1080/03067319.2020.185775. (Impact factor 1.67, Q3) (SCI), 1–5 , 10290397, 03067319
21. Ritesh Saini, Nihal Anwar Siddiqui, Kanchan Deoli Bahukhandi, Neelu J Ahuja (2018) : "Study of effects of Municipal solid waste on water quality of Dehradun City" Journal of Environmental Science, Computer Science and Engineering & Technology , An International Peer Review E-3 Journal of Sciences and Technology, JECET; June 2018- August 2018; Sec. A; Vol.7. No.3, 409–416, E-ISSN: 2278–179X, PP 409–416, [DOI: https://doi.org/10.24214/jecet.A.7.3.40916
22. Bahukhandi D. K, Ahuja J.K (2012), Building Block Methodology Assisted Knowledge-Based System for Environmental-Flow Assessment of Suswa River of Dehradun Dist., India: A Reminiscent Framework. International Research Journal of Environment Sciences 2(12), pp 74–80
23. Federation, W. E. (2012). APHA, AWWA, WEF. "Standard Methods for examination of water and wastewater." *Anales de Hidrología Médica, 5*(2), 185–186–186. https://doi.org/10.5209/rev_ANHM.2012.v5.n2.40440
24. Foundation, I. B. E. (n.d.). *Automobile Industry in India.* Oct, 11 2021. https://www.ibef.org/industry/india-automobiles.aspx
25. Kanchan Deoli Bahukhandi, Nihal Anwar Siddiqui, S. K. Bartarya, A. S. (2020). Assessment of Heavy Metal and Physiochemical Parameter in Surface and Groundwater Quality of Dehradun District of Uttarakhand. *Springer Transactions in Civil and Environmental Engineering. Springer, Singapore.* https://doi.org/10.1007/978-981-32-9956-6_16
26. Kumar, M., Goswami, R., Kumar, A., & Srivastava, M. (2020). Chemosphere Scenario , perspectives and mechanism of arsenic and fl uoride Co- occurrence in the groundwater : A review. *Chemosphere, 249*(336), 126126. https://doi.org/https://doi.org/10.1016/j.chemosphere.2020.126126
27. Ministry of Environment, Forest & Climate Change, G. of I. (2019). *Guidelines for Environmentally Sound Facilities for Handling , Processing and Recycling of End-of- Life Vehicles (ELV).*
28. Nowak, P., Kucharska, K., & Kamiński, M. (2019). Ecological and health effects of lubricant oils emitted into the environment. *International Journal of Environmental Research and Public Health, 16*(16), 1–13. https://doi.org/https://doi.org/10.3390/ijerph16163002
29. Rajesh, S., Kanchan, B., Prasanjeet, M., & Satendra, S. (2015). *Water Quality Assessment of River Ganga Health Hazard Identification and Control.* 5(5), 1–8.
30. Registrar General & Census Commissioner, I. (2014). The Census of india 2011 report pdf. *Government of India, July.*
31. Rodriguez Boluarte, I. A., Andersen, M., Pramanik, B. K., Chang, C. Y., Bagshaw, S., Farago, L., Jegatheesan, V., & Shu, L. (2016). Reuse of car wash wastewater by chemical coagulation and membrane bioreactor treatment processes. *International Biodeterioration and Biodegradation, 113*, 44–48. https://doi.org/https://doi.org/10.1016/j.ibiod.2016.01.017
32. Sharma, P., Sharma, A., Sharma, A., & Srivastava, P. (2016). Automobile Waste and Its Management. Rjces, 4(2), 1–7
33. Vazquez-Duhalt, R. (1989). The purpose of this review is to gain an insight into the production and fate of used crankcase oil ; the effects of dispersion and spillage of used motor oil on the soil and aquatic environment ; and the environmental importance of waste oil combustion . *The Science of the Total Environment, 79*, 1–23.
34. Bahukhandi, K. D., Bartarya, S. K., Siddiqui, N.A. (2017). Assessment of surface and ground water Quality of Haridwar district of Uttarakhand. International Journal of Chem Tech Research,10(10), 95-118.

35. Barakat, A., Baghdadi, M.E., Rais, J., Aghezzaf, B. and Slassi, M.M. (2016). Assessment of spatial and seasonal water quality variation of Oum Er Rbia River (Morocco) using multivariate statistical techniques. Int Soil Water Conserv Res. 4(4), 284-292.
36. Bartarya, S.K. (1993). Hydrochemistry and Rock weathering in a sub tropical lesser Himalayan river Basin in Kumaun India. Journal of Hydrology, 146, 149- 174.

Implementation of an Integrated Management System in Automobile Industry

Reyya Pujitha, Kanchan Deoli Bahukhandi, and Rahul Silori

Abstract Presently the international standards are in priority for Environment, Sustainability and Business Development as well as for Federal Laws, State Laws and other laws are focusing on their regional development. The implementation of IMS based on the foresight and vision for the organization, these can increase the efficiency, transparency as well as rationalization of common approaches, ideas and tools. This paper discuss about the integration of a different management systems into one cohesive system. The primary focus of this is to implement the five international standards-ISO45001:2018, ISO14001:2015(E), ISO28000:2007(E), ISO22301:2012 and ISO50001:2018(E) in Automobile Industry. It discusses about the levels of integration, starts with the adoption stage of Standards to the final audit certification. It also discuss about the challenges that may face while integrating the standards and the benefits of integration to an organization. The objective of this paper associated with the journey and the practical consideration of IMS in the stage of implementation.

Keywords Integrated management system (IMS) · International organization of standardization (ISO) · Implementation · Standards

1 Introduction

Integrated Management system (IMS) is the collaboration of two or more standards into one cohesive system with a holistic set of documents, policies, procedures and processes. In order to acquire an ISO certification, certain organizations certify more than one certification [1]. For example an organization is certified for Quality and Health and Safety management. Subsequently, the quality department centers only on quality, while the EHS department on Health and Safety. In doing so, they do not emphasize on the other standards. By integrating differing standards into one system,

R. Pujitha · K. D. Bahukhandi (✉) · R. Silori
Department of HSE and Civil Engineering, University of Petroleum and Energy Studies, Dehradun, India
e-mail: kanchan@ddn.upes.ac.in

K. D. Bahukhandi et al. (eds.), *Environmental Pollution and Natural Resource Management*, Springer Proceedings in Earth and Environmental Sciences,
https://doi.org/10.1007/978-3-031-05335-1_23

we can increase the efficiency, transparency as well as rationalization of common approaches, ideas and tools [2].

The global realities of current business strategies emphasize individuals and organization to change and adapt IMS. Besides assisting certification, IMS helps in establishing the foresight and vision, thereby governing the present to build a brighter forth coming. A brighter future of an organization depends upon the maturity of present management systems and effectiveness through its core objective and process for the business continuity [3]. The objectives of IMS are towards the organizations which show case a single management system and has its focus on more than one management systems which are commercially available, based on International Standards [4]. This effort is to rationalize the organizations design and development to approach IMS. Despite having initial challenges, it was a reasonable effort. The objective of this paper is associated with the IMS journey and the practical consideration for effective development and implementation of the IMS approach. While adopting a system, one should be sensible towards business structure to the management system because normally business and financial challenges are apart and the other barriers are in the form of personality, employee's behavior and work culture.

This paper helps to implement the IMS by identifying standards, tasks, recommended approaches and processes (Fig. 1). The bottom line for this paper would be a guideline for a more efficient, effective and productive business that has rationalized the logic of its business practices, markets and risk with the goal of a healthier balance sheet. In this paper an attempt has been made with respect to the importance of IMS in Implementation of Environment, Energy, Business Continuity, Occupational Health& Safety, and Security supply chain Management Systems.

2 Levels of Integration

The implementation of IMS is divided into 10 levels:

1. Awareness training
2. Policy and Objective
3. Internal Gap Analysis
4. Documentation/Process Design
5. Documentation/Process Implementation
6. Internal Audit
7. Organization of Management System Review Meeting
8. Thorough Gap analysis of Implementation System
9. Corrective Action
10. Final Certification Audit.

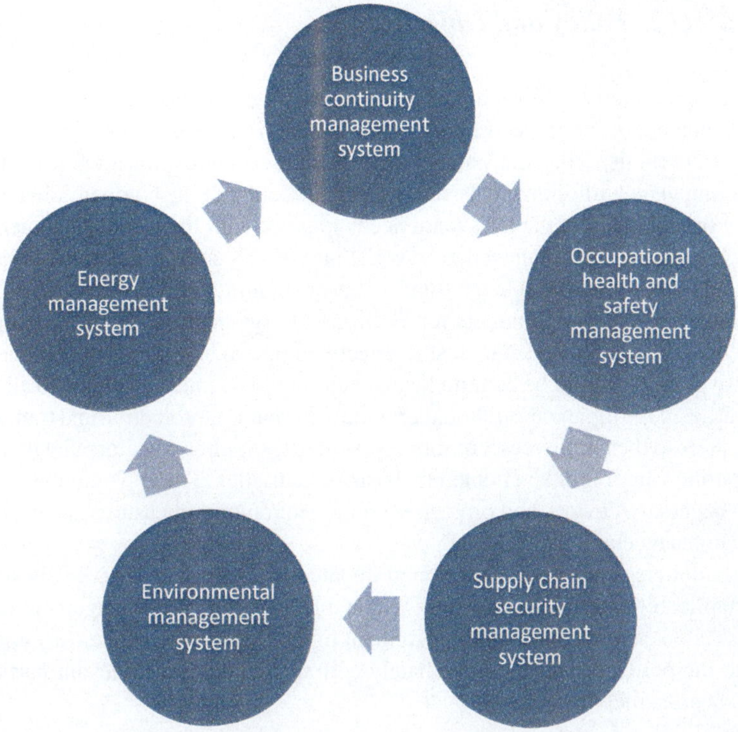

Fig. 1 Integrated management system

2.1 Level 1: Awareness Training

When an organization adopts a new system either technical or management, relevant employer should be aware of that system as well as IMS [5]. It should be aware to every employee respective to their job role and surroundings. In the initial stages of the implementation, one has to meet different challenges and the gaps remain while it gets implementing for awareness of IMS towards the employers [6]. It can be subdivided into various categories. The first one is **Management and supervisor level**- In this level they should be well trained for the respective courses to follow, to supervise and for internal audits, second is for rest of the employees to be aware of the IMS towards their work and the last category is for the new employees. After the recruitment of a new employ the awareness training has to be conducted and it should be documented for verification.

2.2 Level 2: Policy and Objectives

The implementation of IMS is done by collaborating five international standards for the development of Integrated Energy, Environment, Health and Safety, Security and Business Continuity. The policy and objectives for the organization are formulated by considering vision, mission, goals and strategic direction of the company. To develop a policy or an objective the IMS team needs to work with the top management and middle level functional management. As the part of this commitment, the organization should take all reasonable measures to ensure continual improvement by setting measurable objectives and targets for all levels of organization [7]. All appropriate resources should be provided to ensure effective implementation of the policies.

These policies should be communicated and should be made available to all interested parties. The organization should provide information, protection and training on IMS matters to the employees, contractors, vendors, suppliers, visitors and members of the public who might be effected by business activities. These would make available all necessary devices and protective equipment/counter measures and supervise their maintenance and use.

These policies shall be reviewed and updated if necessary, periodically for ongoing applicability, effectiveness and sustainable as part of the management review process and to ensure continual improvement of the IMS performance. Any changes made to the policies shall be immediately communicated to all relevant parties by emails, boards, meetings etc.

2.3 Level 3: Internal Gap Analysis

The identification of gaps depends up on location, type of industry, amount of production and other factors. A gap analysis is mainly a review process or a type of minor auditing process or "show me the evidence" type which comes in the form of a record or a document [2, 8].

2.4 Level 4: Documentation/Process Design

Overall process design must be documented as per the requirement of the adopted standards. The procedures should be written and implement to the organization like functional procedures, work instructions, system procedures, associated forms, organization charts etc. [9].

2.5 Level 5: Documentation/Process Implementation

Process Implementation is not only the process for documentation as it turns the strategies and plans into actions. In order to implement the successful plan in the organization, IMS should be treated into one of the work culture [10]. These implementing process should be done along with the level 4 across the organization so that all departments are covered.

2.6 Level 6: Internal Audit

Internal audits are independent activity designed to add value in an organizations operations. It helps to increase the effectiveness of an organization objectives by bringing in a disciplined and systematic approach for the effectiveness of risk management, control and governance processes [7].

2.7 Level 7: Organize a Management System Review Meeting

The purpose of the management review is to evaluate the effectiveness of the existence system. It is mandatory to review the various aspects like policy, strategic directions of the organization, setting of the objective, reviewing of the existence objectives, the extent they have reached, results of process performance, system performance if already exist, results of complaints, feedbacks, suggestions, previous incidences, results of an internal audit, legal and other compliance reports that related to the standards for integration [11, 12].

2.8 Level 8: Thorough Gap Analysis of Implementation System

Once again conduct a formal gap analysis and also thorough with the previous gap analysis to assess the compliance of the IMS and its effectiveness for the final certification audit [13].

2.9 Level 9: Corrective Actions

On the basis of gap analysis, internal audits, compliance reports nonconformities are discovered for corrective actions. These are usually a set of actions which require law and standard to eliminate an undesirable situations. For example, non conformity may also be a machine failure, market complaint, customer complaint, misinterpretation of written instructions to carry out a work. These non-conformities must be related to the adopted standards.

A corrective action is designed by a team including personnel who are involved in the actual observation and the quality assurance in the point of non-conformance to implement systematically and observed to their abilities to eliminate for the future occurrence [14]. In this process if there any similar chances to get a nonconformity those all should be also corrected. To ensure the corrective action for an investigation root cause of the incident is pivotal [15, 16].

2.9.1 Level 10: Final Certification Audit

Final certification audit will be conducted by the external person. This auditing can also say as self-examination to organization and these may also help for the business standard to show about the state of organization and statutory purpose [17, 18].

3 Challenges to Integrate

- The development, implementation and maintenance of a management system are time-consuming, and frequent challenges can be encountered.
- The main challenge to implement IMS is to integrate the people, time, Data and delivery then the people will increase their competence and motivation, time will turn into predictive, data will be modeled and finally delivery will be mauled into interactive dashboards.
- Having gone through this process by implementing one by one and reach to a certain comfort level then another storm will occur from other side.
- The most common fears encountered is the perception that defining processes and advocating consistency will stifle entrepreneurship, creativity, and flexibility.
- By historically we can also say that management system based on written procedures and multiple of paper works that ultimately increased into directorate, sometimes simply doesn't make any sense.
- The development of any new system carries with it certain costs, likewise integrating systems requires resources which consumes cost [19].
- If IMS is not well planned or implemented, the integrated system can quickly become a major cost area and a source of frustration and if not managed well, the whole exercise of integrating management systems and processes can contribute to "turf wars" due to overlap or gaps.
- In the process of integrating a system, the person tasked for this was not a competent nor experienced in this work the effort will be plagued by a false start, unnecessary paperwork and templates and other works.

4 Benefits of an Integration

- **Improving performance:** By implementing the system in a planned manner it definitely shows a positive impact on the system by improving its performance of their standards by showing some productivity [20].
- **Eliminating redundancies**: When the organization following different management system then the similarities will be there in all systems like policies, processes, documents, trainings etc. when you integrating a system these redundancies will reduce and can save your organization a great amount of time, and in return money.
- **Accountability**: When you align the different system into an integrated system by collaborating policies, objectives and all will show the improvement in accountability.
- **Establish consistency**: Integration will create best consistency among the system that will reduce the complex and be easier to understand.
- **Cost Reduction**: When you have different system each system will have to do audits and assessments individually, when you integrated up to certain activities you can reduce time, eliminate the interrupted time and reduce the cost.
- **Reducing maintenance**: When you have an integrated management system you can maintain the requirements concurrently, streamlining the process and allowing the organization to focus on improvements rather than maintaining multiple systems when that is unnecessary.
- **Facilitating decision making**: Eliminating layoffs and creating consistency within the organization of the functional needs and performance. This integrated approach also allows the organization to analyze functional and department barriers and therefore improve communication and decision making [3].

5 Conclusion

The concept of an integrated management system is collaborating the occupational health and safety, energy, environment, security and business continuity Management systems into one cohesive management system in an automobile industry. It integrates all the set of documents, procedures, policies, processes, auditing etc. The results after the implementation of IMS will not only increases efficiency, transparency, rationalization of idea, tools but also shows some benefits in cost (Table 1).

Table 1 Documents requires for five international standards

Integrated management system											
S. No.	Documents required for ISO	ISO 28000:2018		ISO 14001:2015		ISO 45001:2018		ISO 50001:2018		ISO 22301:2012	
		Req?	Clause	Req?	Clause	Req?	Clause	Req?	Clause	Req?	Clause?
1	Legal and regulatory requirements up to date	Req	4.3.2	x		Req	6.1.3	x		Req	4.2.2
2	Scope	x		Req	4.3	Req	4.3	Req	4.3	Req	4.3.1
3	Policy	Req	4.2	Req	5.2	Req	5.2	Req	5.2	Req	5.3
4	Objectives	Req	4.3.1	Req	6.2.1	Req	6.2.2	Req	6.2	Req	6.2
5	Competence	Req	4.4.2	Req	7.2	Req	7.2	x	7.2	Req	7.2
6	Risk Assessment procedure include establish, implement and maintained	x		Req	6.1.1	Req	6.1.1	x		Req	8.2.3
7	Explanation of any such exclusion shall not effect the organizations ability and responsibility	x		x		x		x		Req	4.3.2
8	Procedure to ensure continuity of activities and management of an disruptive incidents	x		x		x		x		Req	8.4.1

(continued)

Table 1 (continued)

Integrated management system											
S. No.	Documents required for ISO	ISO 28000:2018		ISO 14001:2015		ISO 45001:2018		ISO 50001:2018		ISO 22301:2012	
		Req?	Clause	Req?	Clause	Req?	Clause	Req?	Clause	Req?	Clause?
9	Procedures and management structures to respond to a disruptive incident using personnel with the necessary responsibility authority and competence to manage a incident	x		x		x		x		Req	8.4.2
10	The decision of an organization using life safety as a first priority and in consultation with relevant interested parties, whether to communicate externally about its significant risks and impacts	x		x		x		x		Req	8.4.2
11	Implementation of an audit program and audit results	x		x		x		Req	9.2.2	Req	9.2

(continued)

Table 1 (continued)

Integrated management system											
S. No.	Documents required for ISO	ISO 28000:2018		ISO 14001:2015		ISO 45001:2018		ISO 50001:2018		ISO 22301:2012	
		Req?	Clause	Req?	Clause	Req?	Clause	Req?	Clause	Req?	Clause?
12	Nature of the non-conformity and any subsequent actions taken and the results for corrective action	x		x		x		Req	10.1	Req	10.1
13	Top management shall demonstrate leadership and commitment with respect to continual improvement, scope, targets, requirements	x		x		x		Req	5.1	x	
14	Boundaries and targets	Req	4.3.1	x		x		Req	4.3	x	
15	Procedures for how to achieve its objectives, targets and action plans	x		x		Req	6.2.2	Req	6.2.3	x	
16	Methods for determining and any updating the energy performance indicators (EnPIs)	x		x		x		Req	6.4	x	

(continued)

Table 1 (continued)

Integrated management system											
S. No.	Documents required for ISO	ISO 28000:2018		ISO 14001:2015		ISO 45001:2018		ISO 50001:2018		ISO 22301:2012	
		Req?	Clause	Req?	Clause	Req?	Clause	Req?	Clause	Req?	Clause?
17	Reviewed EnPIs values and compared to their respective Energy baseline (EnBs) values	x		x		x		Req	6.4	x	
18	EnBs information, Relevant Variable data and modifications	x		x		x		Req	6.5	x	
19	Measurement, monitoring and other means of establishing accuracy and repeatability	x		Req	9.1.1	Req	9.1.1	Req	6.6	x	
20	Emergency preparedness and response for the emergency situations	x		Req	8.2	Req	8.2	x		x	
21	Process needed to the extent necessary as planned	x		Req	8.1	x		x		x	
22	Aspect, associated impacts and criteria used to determine its significant aspects	x		Req	6.1.2	Req	6.1.2.2	Req	6.3	x	

(continued)

Table 1 (continued)

Integrated management system											
S. No.	Documents required for ISO	ISO 28000:2018		ISO 14001:2015		ISO 45001:2018		ISO 50001:2018		ISO 22301:2012	
		Req?	Clause	Req?	Clause	Req?	Clause	Req?	Clause	Req?	Clause?
23	Compliance obligations related to environmental aspects and how these apply to organization	x		Req	6.1.3	x		x		x	
24	Internal and external communications relevant to the system	x		Req	7.4	Req	7.4.1	x		x	
25	Compliance evaluation results	x		Req	9.1.2	Req	9.1.2	x		x	
26	results of performance evaluation and on maintenance, calibration or verification of measuring	Req	4.5.2	x		x		x		x	

(continued)

Table 1 (continued)

Integrated management system											
S. No.	Documents required for ISO	ISO 28000:2018		ISO 14001:2015		ISO 45001:2018		ISO 50001:2018		ISO 22301:2012	
		Req?	Clause	Req?	Clause	Req?	Clause	Req?	Clause	Req?	Clause?
27	Security management requirements of the design, specification and installations, identification of adequate resources including staffing levels, training needs and skills, development of operational controls, organization overall threat and risk management framework	Req	4.3.1	x		x		x		x	
28	The means and time—scale by which security management objectives and targets are to be achieved	Req	4.3.5(b)	x		x		x		x	

(continued)

Table 1 (continued)

Integrated management system											
S. No.	Documents required for ISO	ISO 28000:2018		ISO 14001:2015		ISO 45001:2018		ISO 50001:2018		ISO 22301:2012	
		Req?	Clause	Req?	Clause	Req?	Clause	Req?	Clause	Req?	Clause?
29	Any changes in the documented procedures resulting from corrective and preventive action and shall include required training where necessary been recorded	Req	4.5.3	x		x		x		x	
30	Management review results	Req	4.6	x		x		x	9.3.4	Req	9.3
31	Operational planning, implement and control relate to significant energy use (SEU's) and requirements needed to implement the actions	x		x		x		x	7.1	x	
32	Design activities to the energy performance	x		x		x		x	8.2	x	

(continued)

Table 1 (continued)

Integrated management system											
S. No.	Documents required for ISO	ISO 28000:2018		ISO 14001:2015		ISO 45001:2018		ISO 50001:2018		ISO 22301:2012	
		Req?	Clause	Req?	Clause	Req?	Clause	Req?	Clause	Req?	Clause?
33	top management shall assign the responsibilities and authorities to the management team for ensuring that conforms to the requirement	Req	4.3.5 (a)	x		Req	5.3	Req	5.3	x	
34	Suggested improvements to the EnMS or the energy performance from employee	x		x		x		Req	7.4	x	
35	Evaluated compliance with legal and other requirements related to energy efficiency use and consumption and any actions taken	x		x		x		Req	9.1.2	x	

References

1. Holdsworth, R., Practical applications approach to design, development and implementation of an integrated management system. Journal of Hazardous Materials, 2003. **104**(1–3): p. 193–205.
2. Garengo, P. and S. Biazzo, From ISO quality standards to an integrated management system: An implementation process in SME. Total Quality Management & Business Excellence, 2013. **24**(3–4): p. 310–335.
3. Rajkovic, D. and M. Aleksic, Corporative motives on implementation of integrated management system (IMS). International Journal for quality research, 2009. **3**(3): p. 1–5.
4. Jackson, S.L., The ISO 14001 implementation guide: creating an integrated management system. Vol. 3. 1997: John Wiley & Sons.
5. Rumane, A.R., Quality management in construction projects. 2017: CRC Press.
6. López-Fresno, P., Implementation of an integrated management system in an airline: a case study. The TQM Journal, 2010. **22**(6): p. 629–647.
7. Arifin, K., et al., Implementation of integrated management system in Malaysia: The level of organization's understanding and awareness. European Journal of Scientific Research, 2009. **31**(2): p. 188–195.
8. Barafort, B., A.-L. Mesquida, and A. Mas, Integrating risk management in IT settings from ISO standards and management systems perspectives. Computer Standards & Interfaces, 2017. **54**: p. 176–185.
9. Zutshi, A. and A.S. Sohal, Integrated management system: the experiences of three Australian organisations. Journal of manufacturing technology management, 2005. **16**(2): p. 211–232.
10. Beckmerhagen, I., et al., Integration of management systems: focus on safety in the nuclear industry. International Journal of Quality & Reliability Management, 2003. **20**(2): p. 210–228.
11. Zeng, S.X., P. Tian, and J.J. Shi, Implementing integration of ISO 9001 and ISO 14001 for construction. Managerial Auditing Journal, 2005. **20**(4): p. 394–407.
12. Ionescu, G., et al., Implementation of integrated management systems and corporate social responsibility initiatives—a Romanian hospitality industry perspective. Sustainability, 2018. **10**(10): p. 3684.
13. Zeng, S.X., J.J. Shi, and G. Lou, A synergetic model for implementing an integrated management system: an empirical study in China. Journal of cleaner production, 2007. **15**(18): p. 1760–1767.
14. Nunhes, T.V., L.C.F.M. Barbosa, and O.J. de Oliveira, Identification and analysis of the elements and functions integrable in integrated management systems. Journal of cleaner production, 2017. **142**: p. 3225–3235.
15. Moumen, M. and H. El Aoufir, Quality, safety and environment management systems (QSE): Analysis of empirical studies on integrated management systems (IMS). Journal of Decision Systems, 2017. **26**(3): p. 207–228.
16. Salomone, R., Integrated management systems: experiences in Italian organizations. Journal of cleaner production, 2008. **16**(16): p. 1786–1806.
17. Domingues, J., et al. Integrated versus non-integrated perspectives of auditors concerning the new ISO 9001 revision. in 2016 IEEE International Conference on Industrial Engineering and Engineering Management (IEEM). 2016. IEEE.
18. Wilkinson, G. and B. Dale, System integration: the views and activities of certification bodies. The TQM Magazine, 1998. **10**(4): p. 288–292.
19. Khanna, H.K., S. Laroiya, and D. Sharma, Integrated management systems in Indian manufacturing organizations: Some key findings from an empirical study. The TQM journal, 2010. **22**(6): p. 670–686.
20. Santos, G., et al., Implementing and certifying ISO 14001 in Portugal: motives, difficulties and benefits after ISO 9001 certification. Total Quality Management & Business Excellence, 2016. **27**(11-12): p. 1211–1223.

Development of Accident Prediction Model for Low Frequency and High Severity (LFHS) Industrial Accidents

N. N. Pisharody, Kanchan D. Bahukhandi, Prashant S. Rawat, and R. K. Elangovan

Abstract *Back Ground*: Industrial safety is a multi-disciplinary subject involving interaction of human, equipment and work environment. It is difficult to design equipment and work environment for each individual as each individual is unique. Therefore, accident control measures should be implemented by safety through design, supplemented by administrative controls and personal protective equipment. In order to implement an effective accident control program, all accidents should be investigated and analyzed to identify significant contributors. *Methods*: Regression analysis is one of the tools for identification of significant contributors through development of accident prediction models. The paper illustrates multiple linear regression analysis of Low Frequency High Severity (LFHS) accidents occurred in a Public Sector Power Company in India during the period of 10 years from 2006 to 2015. The independent variables of the model are 'Types of LFHS Accidents' and dependent variable is man-days loss attributed to LFHS accidents. *Results*: The significant contributors of the LFHS accidents are 'Exposure to or Contact with Electric Current', 'Fall of Persons from Height' and 'Stepping on Striking Against or Struck by Object'. The paper also gives accident control measures to be taken by 'Safety through Design', 'Administrative Controls' and 'Personal Protective Equipment'. *Conclusion*: Effective implementation of these measures will support organization in accident prevention.

Keywords Accident prediction model · LFHS accidents · Regression analysis · Safety through design

N. N. Pisharody (✉)
Nuclear Power Corporation of India Limited (NPCIL), 12-N-01, V.S. Bhavan, Anushaktinagar, Mumbai 400094, India
e-mail: nnpisharody1965@gmail.com

K. D. Bahukhandi (✉) · P. S. Rawat
University of Petroleum and Energy Studies (UPES), Dehradun, India
e-mail: kanchan@ddn.upes.ac.in

R. K. Elangovan
Directorate General of Factory Advice and Labor Institute (DGFASLI), Mumbai, India

K. D. Bahukhandi et al. (eds.), *Environmental Pollution and Natural Resource Management*, Springer Proceedings in Earth and Environmental Sciences,
https://doi.org/10.1007/978-3-031-05335-1_24

1 Introduction

Accidents are unexpected events resulting in injury to persons or loss of property or production [1]. Accidents are classified as Near Miss, First-Aid Injury, Non-Reportable Injury, Reportable Injury and Fatality depending upon the consequences. Near misses, first-aid injuries and non-reportable injuries are accidents of High Frequency and Low Severity (HFLS); whereas reportable injuries and fatalities are accidents of Low Frequency and High Severity (LFHS) [2–4]. Frequency and severity are estimated in terms of number of accidents and man-days loss per million man-hours respectively [5].

There were 113 LFHS accidents comprising of 95 reportable Injuries and 18 fatalities; at 11 twin unit Operating Plants and 3 twin unit Construction Projects in a Public Sector Power Company in India during the period of 10 years from 2006 to 2015. The number of LFHS accidents and man days lost are given in Table 1.

Accidents may recur, if root causes are not identified and corrective measures are not taken. Therefore, all accidents shall be investigated to address the causes to prevent events of similar kind in future [4, 6–8]. Even though all accidents are analyzed and corrective measures are taken, accidents are recurring in the company. Therefore, it is felt necessity to further reinforce accident preventive measures. Accident prediction models help to identify dominant contributors of accidents to take decisions on accident preventive measures [9, 10].

This paper gives details on development of an accident prediction model based on the analysis of LFHS accidents occurred in the Public Sector Power Company during the period from 2006 to 2015 with reference to 'Type of LFHS Accidents' [5]. The paper also brings out safety measures to be further reinforced to minimize consequences of accidents as low as reasonably practicable.

Table 1 Number of LFHS accidents during from 2006 to 2015

Year	Reportable injury	Fatality	Total No. of accidents	Man days loss
2006	7	3	10	18,830
2007	18	3	21	18,357
2008	12	1	13	6312
2009	10	1	11	7135
2010	11	2	13	12,355
2011	6	1	7	6109
2012	14	1	15	8620
2013	8	3	11	18,450
2014	4	2	6	18,991
2015	5	1	6	10,724
Total	95	18	113	125,883

2 Materials and Methods

2.1 Taxonomy

The paper comprises of 9 sections. Section 1 is Introduction, Sect. 2 is Materials and Methods, Sect. 3 gives Results, Sect. 4 gives Discussion, Sect. 5 is Conclusion, Sect. 6 gives Future Research Directions, remaining Sections give Conflicts of Interest, Acknowledgement, References, Tables and Figures.

2.2 Literature Survey

Literature survey on 'Accident Prevention and Control' and 'Accident Prediction Models' shows that immense literature on 'Accident Prevention and Control' at chemical process industries are available, whereas literature on conventional industries is relatively less. Similarly, large quantity of literature is available on 'Accident Prediction Models' on road accidents and crash models, however literature on accident prediction models on conventional industrial accidents is limited.

Accident Prevention and Control

Accident prevention and control is a multi-disciplinary subject comprising of vide spectrum of topics on administration, program implementation, engineering and technology [11, 12]. The first step for administration and implementation of safety program is identification of safety professionals with requisite qualifications and experience; and defining their role. Safety professionals facilitate establishment of a safety organization by development of safety and occupational health programs to meet statutory requirements as per the national laws, with the objective to safeguard safety, health and welfare of workers and to protect them from exploitation [3].

An organization shall establish Occupational, Health Safety Management System (OHSMS) for the effective implementation of safety programs. At the outset, the top management shall issue an OHS policy, expressing its commitment on OHS responsibilities. Implementation of OHSMS in an organization is based on Plan-Do-Check-Act (PDCA) cycle. It involves review of existing OHS programs and establishment of programs for hazard identification and risk assessment, hazard control, accident prevention, accident reporting and investigation, surveillance of work environment and industrial hygiene; work place ergonomics, safety education, emergency preparedness, corrective and preventive action, development of safety culture, enforcement of safety legislation, maintenance of records, verification of system effectiveness through safety auditing and periodic management review for continual improvement [11, 13–18].

Hazard identification and risk assessment facilitate implementation of hazard control programs as per the hierarchy of accident control in Fig. 1 [19, 20]. In this hierarchy, elimination, substitution and engineering controls assume the upper

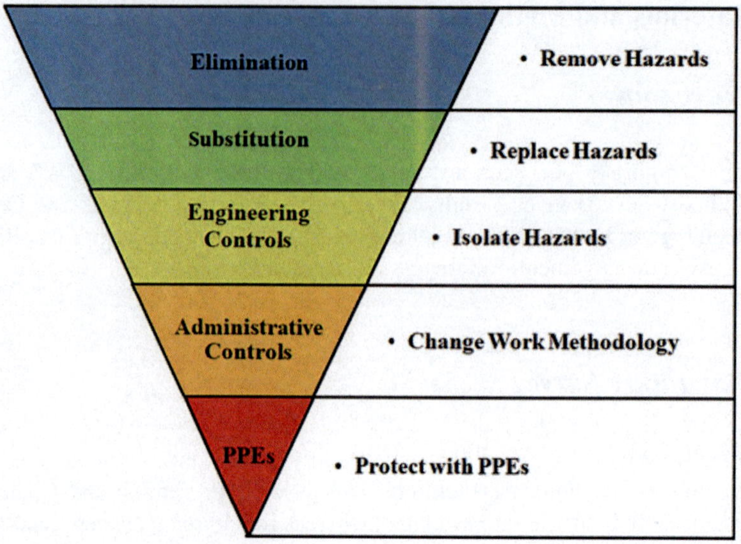

Fig. 1 Hierarchy of accident control

tiers. Next in the hierarchy are administrative controls such as enforcement, training, job hazard analysis, safe work procedures, work permit system, access control etc. Personal Protective Equipment (PPE) is the lowest in the hierarchy. PPE do not prevent accidents; however it mitigates the consequences of the accident [21].

A study carried out on the cause of industrial accidents in aviation, railway and nuclear industries validates that about 50–60% of the root causes of the accidents are attributed to the design and hence design improvement should be considered as one of the major strategies for accident prevention [21]. Accident control by design can be achieved by 'Safety through Design' in structures, systems, equipments and components by judicious application of elimination, substitution and engineering controls. 'Safety through Design' integrates hazard analysis and risk assessment right from the design stage. It has an encompassing effect on the facilities and layout. Though initial investment for 'Safety through Design' could be relatively high, cost is comparable with subsequent retrofitting efforts. At later stage, if safety related deficiencies are identified in the design, it may be impossible, impracticable and difficult or expensive to implement. In such cases, 'Administrative Controls' and 'Personal Protective Equipment' can supplement 'Safety through Design' [11, 12]. Despite all these efforts, there could be accidents and injuries due to shortfalls in program implementation. Therefore, it is essential to investigate each accident to identify causes and contributors, whether it is HFLS or LFHS accident [2].

An accident is preceded by numerous incidents and near misses. Even near misses are preceded by many precursors, which are anomalous events without any immediate safety impact; however may indicate unknown or insufficiently understood significant risks. Therefore, all precursors shall also be analyzed to identify potential safety risks, to take corrective measures and hence to prevent more severe accidents. Therefore, organizations shall have a frame work for identification, reporting, investigation, causal analysis, solution identification, dissemination and resolution of precursors and near-misses [1, 22–24].

Organizations may adopt a 7 point protocol for identification of precursors to direct their resources to improve safety. It includes selection of area, identification of precursors, prioritization of precursors, identification of ineffective control processes, identification of latent conditions, identification of affected safety barriers and derivation of conclusions As long as precursors and near misses are kept under control, either by preventive or by controlling actions; it may not escalate to serious accidents. In fact, investigation of precursors and near misses shall be integral part of organizational safety management system. There should be a systematic mechanism to interrupt and prevent precursors and immediate factors (PaIFs) through real-time tracking from historical accident records [4, 6–8]. However, organizational and management factors such as weak safety culture and ineffective root cause analysis could lead to failure of near miss reporting system. A near-miss reporting system should be based on blame free and forgiveness culture. Identification of human errors as causes for accidents should be based on man–machine interface and system induced errors [25].

A dangerous situation with potential for accident is formed due to technical failure, human failure and organizational failure. If adequate defenses are available, dangerous situation could return to normal, else it will develop into an incident. With adequate human recovery actions, the incident could reduce to a near miss; else it may escalate to an accident. Hence human recovery has an important role in accident prevention as it reduces a possible accident into a mere near miss [26].

Work place injuries are caused by both individual and organizational factors. Individual factors are employees' educational level and organizational role. Organizational factors are organizational culture, organizational support, job satisfaction, and perception on workplace safety. Workers' perception on workplace safety has very high impact on accident frequency. Workplace injuries are the result of human and work environmental factors. An effective organizational management by paying attention to social structures at the workplace can reduce accident frequency. Organizational and cultural precursors such as leadership, oversight and scrutiny; and organizational learning play critical role in accident prevention [27, 28].

Accident models based on theories of accident philosophy, accident sequence, and accident causation help to understand and analyze accidents and assist accident prevention. Accident models are used in both reactive and proactive safety management [28, 29]. Most of the accidents models are based on chemical process industries. The System Hazard Identification, Prediction and Prevention (SHIPP) model developed by Rathnayaka et al. uses a combination of event and fault tree concepts based on precursor information to establish cause-consequence relationship [30, 31]. Similarly, Dynamic Sequential Accident Models (DSAMs) based on fault trees and event trees by utilizing precursor data such as near-misses, mishaps, incidents and accidents are also available to estimate posterior risk profile and support decision making [32].

There were major accidents with fire and explosion in chemical process industries world over at hazardous material storage facilities, where large number of highly flammable and toxic substances is stored in huge quantities. When highly flammable Liquid Petroleum Gas (LPG) is stored under pressure, it involves risk of explosion due to Boiling Liquid Expanding Vapor Explosion (BLEVE). Therefore, such tanks shall be installed at optimum safe distance such that in case of an event in one tank, other thanks are not harmed. Therefore, it is mandatory that chemical process industries are designed and operated by complying with requirements of applicable standards to prevent major accidents involving fire, explosion and toxic release [33]. In a tank farm where highly flammable liquids are stored in two or more tanks and if one of the storage tanks meets with an accident and leak leading to fire and explosion; adjacent tanks are also likely to fail due to heat. Inter-utility distances between hazardous facilities within the periphery of chemical process industry and risk of accidents can be minimized by using 'Risk and Distance Minimization in Process industry Units Siting' (RIDIMPUS) methodology [34]. The safe and optimum distance between two or more hazardous storage tanks in a tank farm can also be determined by accident models such as Point Source model, Shokri and Beyler's model, Mudan's model or Baldwin and Thomas model [35].

A template based quantitative risk assessment of hazardous material storage facilities to identify and reduce risks could also prevent accidents in chemical process industries [36]. The Chemical Accident Simulation Tool (CAST) software can generate various scenarios of different types of accidental fires and explosions in chemical process industry with likely consequences in terms of area impacted and types of impacts. The CAST is developed with an integrated mapping tool to display damage zones around accident center and it makes the application useful in decision making [37].

Studies on occupational accident models, human factors and safety culture by Attwood et al. [38] indicate that holistic quantitative models, capable of predicting accident frequencies in conventional industries are scarce. Therefore, more research should be made in this direction. This paper is such an effort to develop an accident prediction model on 'Type of LFHS Accidents'.

Accident Prediction Models

Accident prediction models are developed to identify significant contributors for the accidents to take appropriate corrective actions. Generally adopted methods for accident prediction are scenario analysis, regression method, time-series method, Markov chain method, grey model, neural networks and Bayesian networks. Sometimes a combination of different methods is also used to bring a realistic prediction model [13]. Amongst these methodologies, multiple linear regression and Analysis of Variance (ANOVA) is used for the prediction of a dependent variable with two or more independent variables. Multiple linear regression analysis is a statistical method for estimating relationship among variables for defining relationships between the variables in the form of an equation. It is used to estimate a dependent variable using certain independent variables. It is widely used for prediction to understand which among the independent variables are significantly related to the dependent variable.

With some limited conditions, regression analysis is used for inferring causal relationships between independent and dependent variables also. Statistical significance of the model is determined by the values of coefficients of determination i.e. R^2 and Adjusted R^2 values, F-Test, T-Test and *p*-values for a specific values of significance [39–41].

Multiple linear regression models are used in academics to analyze student's final grade as dependent variable and measurement and evaluation, educational psychology, curriculum development, guidance, teaching methods student's scores in tests, quiz and final examination etc. as independent variables. Prediction capabilities of these models are adequate and the results are statistically significant [42, 43].

Multiple linear regression is extensively used to predict road accidents, although its application on prediction of occupational accidents is less. In road accident prediction models, dependent variable is number of road accidents and independent variables are quantitative and qualitative parameters such as road cross-section dimensions, traffic volume, speed, road shoulder width, lighting conditions, traffic signs and traffic signals etc. Validity of these models is checked by coefficient of determination and by comparing predicted values with actual. Coefficient of determination shows satisfactory goodness fit of the models [14, 44].

Multiple linear regression is also not free from controversies. One common issue with multiple linear regression is multicollinearity as sometimes it may make interpretation of results difficult. However, Kraha et al. [45] opines that multicollinearity will not affect prediction capability of the model. They also propose to use multiple indices such as correlation coefficients, beta weights, structure coefficients, all possible subsets regression, commonality coefficients, dominance weights, and relative importance weights etc. to understand contributions of predictors in a regression model.

2.3 Rationale of Model Selection

An accident prediction model is an algorithm of pitting a dependent variable against several independent variables, each of which is assigned a constant [14, 44]. Objectives of the accident prediction models are [29]:

- To create common understanding of the accident phenomena.
- To prevent personal biases on accident causation.
- To provide opening for wider range of preventive measures.
- To guide investigations regarding data collection and accident analyses.
- To analyze interrelations between factors and conditions of accident causation.
- To identify significant causes of accidents, hence to facilitate preventive measures.

Based on the literature survey, multiple linear regression is the method considered for the analysis. This paper gives details of an accident prediction model on 'Type of LFHS Accidents'. Independent variables considered in the model are number of various 'Types of LFHS Accidents' and dependent variable is mandays loss on account of LFHS accidents.

3 Results

3.1 Development of Accident Prediction Model

Number Industrial Accidents on each type of accident, observed man days loss and estimated man days loss during the period from 2006 to 2015 are given in Table 2.

Following are the variables considered for the regression analysis:

Independent Variables

X51 Fall of objects
X52 Exposure to or contact with electric current
X53 Over exertions or wrong movements, exposure to or contact with harmful substances or extreme temperature, explosions and others
X54 Fall of persons from height
X55 Fall of persons on the same level
X56 Stepping on striking against or struck by object
X57 Caught in or between objects

Dependent Variable

YB5MDL Observed man days loss on account of LFHS accidents

Regression analysis and Analysis of Variance (ANOVA) was carried out with Microsoft Excel and results are given in Table 3. Accident Prediction model developed by the analysis gives following numerical expression:

$$\begin{aligned} \text{YB5MDL} = {} & 20886.47 - 19.41(\text{X51}) - 9718.00(\text{X52}) + 2189.44(\text{X53}) \\ & + 1081.17(\text{X54})^{1.5} + 872.58(\text{X55}) - 6623.14(\text{X56}) + 1365.29(\text{X57}) \end{aligned}$$

Goodness of Fit and Statistical Significance of Accident Prediction Model

Goodness of fit of the model is assessed from Coefficients of Multiple Determination (R-Square and Adjusted R-Square). Statistical significance of the model is determined by conducting F-test, T-test and *p*-values [41–45].

Goodness of Fit

Goodness of fit of multiple regression model is determined from the coefficients of multiple determination; R^2 and Adjusted R^2. R^2 and Adjusted R^2 values of the model are 0.9889 and 0.9502 respectively. R^2 indicates that 98.89% of the total variation of dependent variable is explained by the independent variables. Adjusted

Table 2 Types of LFHS accidents during from 2006 to 2015

Year	Type of accidents (B5)								Observed man days loss	Estimated man days loss
	Fall of objects	Exposure to or contact with electric current	Over exertions or wrong movements, exposure to or contact with harmful substances, extreme temperature, explosions and others	Fall of persons from height	Fall of persons on the same level	Stepping on, striking against or struck by object	Caught in or between objects	Observed number of accidents		
	X51	X52	X53	X54	X55	X56	X57	YB5N	YB5MDL	YB5MDLE
2006	2	0	0	4	0	2	2	10	18,830	18,981
2007	4	1	0	9	0	4	3	21	18,357	17,886
2008	0	2	0	8	0	3	0	13	6312	6045
2009	1	2	0	4	0	1	3	11	7135	7553
2010	0	1	0	7	1	3	1	13	12,355	13,560
2011	3	1	0	1	1	1	0	7	6109	6441
2012	0	1	0	7	1	4	2	15	8620	8303
2013	1	1	2	3	0	1	3	11	18,450	18,619
2014	0	1	1	2	1	0	1	6	18,991	18,654
2015	0	1	0	2	1	1	1	6	10,724	9841
Total	11	11	3	47	5	20	16	113	125,883	125,883

Table 3 Output of regression analysis of types of LFHS accidents during from 2006 to 2015

Summary output						
Regression Statistics						
Multiple R	0.9945					
R^2	0.9889					
Adjusted R^2	0.9502					
Standard error	1240.2657					
Observations	10					
ANOVA						
	df	**SS**	**MS**	**F-Stat**	**Significance F**	
Regression	7	274,903,493.98	39,271,927.71	25.53	0.038	
Residual	2	3,076,518.12	1,538,259.06			
Total	9	277,980,012.10				
	Coefficients	**Standard error**	**t Stat**	***P*-value**	**Lower 95%**	**Upper 95%**
Intercept	20,886.47	2063.56	10.12	0.010	12,007.70	29,765.25
X51	−19.41	349.59	−0.06	0.961	−1523.60	1484.78
X52	−9718.00	995.85	−9.76	0.010	−14,002.79	−5433.21
X53	2189.44	822.53	2.66	0.117	−1349.64	5728.51
X54^1.5	1081.17	158.69	6.81	0.021	398.39	1763.96
X55	872.58	1139.09	0.77	0.524	−4028.52	5773.69
X56	−6623.14	1052.24	−6.29	0.024	−11,150.55	−2095.74
X57	1365.29	466.62	2.93	0.100	−642.40	3372.98

R^2 indicates that 95.02% of the total variation of dependent variable is explained by the independent variables. As R^2 and Adjusted R^2 values are close, goodness of fit of the model is satisfactory and at least one independent variable will contribute information for the prediction of YB5MDL.

Statistical Significance

Statistical significance of the model is assessed with reference to F-Test, T-Test and *p*-values.

F-Test:

F-Stat value is 25.53 and F-Critical value for the model is 19.35. Since F-Stat of the model is more than the F-Critical, the model is statistically significant and at least one independent variable will give significant information for the prediction of YB5MDL.

p-values:

Overall p-value of the model is 0.038 and it is less than the α-value of 0.05. This implies that the model is significant at the α-value of 0.05 and at least one of the regression coefficients is not zero.

The p-values for the estimated coefficients of X52, X54 and X56 are 0.010, 0.021 and 0.024 respectively. These values are less than the α-value of 0.05, which indicates that these variables are significantly related to YB5MDL at the α-value of 0.05.

The p-values for X51, X53, X55 and X57 are 0.961, 0.117, 0.524 and 0.100 respectively. These values are more than the α-value of 0.05.This indicates that these variables are not significantly related to YB5MDL at the α-value of 0.05.

T-Test:

Absolute T-Stat values for the estimated coefficients of X51, X52, X53, X54, X55, X56 and X57 are 0.06, 9.76, 2.66, 6.81, 0.77, 6.29 and 2.93 respectively. T-Critical value for model at the α-value of 0.05 is 4.30.

T-Stat values of X52, X54 and X56 are more than T-Critical. This indicates that these variables are significantly related to YB5MDL at the α-value of 0.05.

The T-Stat values for X51, X53, X55 and X57 are less than T-Critical value. This indicates that these variables are not significantly related to YB5MDL at the α-value of 0.05.

Validation of Accident Prediction Model

Validation of the model was carried out with the independent data for the year 2018. Values of Independent and dependent variables on types of accidents and man-days loss on account of LFHS accidents for the year 2018 are given below:

Independent variables:

$$X51 = 0, X52 = 1, X53 = 0, X54 = 4, \ldots X55 = 1, X56 = 0, X57 = 2$$

Dependent variable:

$$\text{YB5MDL} = 18764$$

Estimated value of dependent variable:

$$\text{YB5MDLE} = 23421$$

Figure 2 shows comparison of observed and estimated man-days loss on account of LFHS accidents during the period from the year 2006 to 2015 and for the year 2018. The figure indicates that estimated man-days loss is in agreement with the observed man-days loss. With independent data for the year 2018, estimated man-days loss is within about 25% of the observed man-days loss. This is a fairly good result and the model is able to reasonably predict the dependent variable.

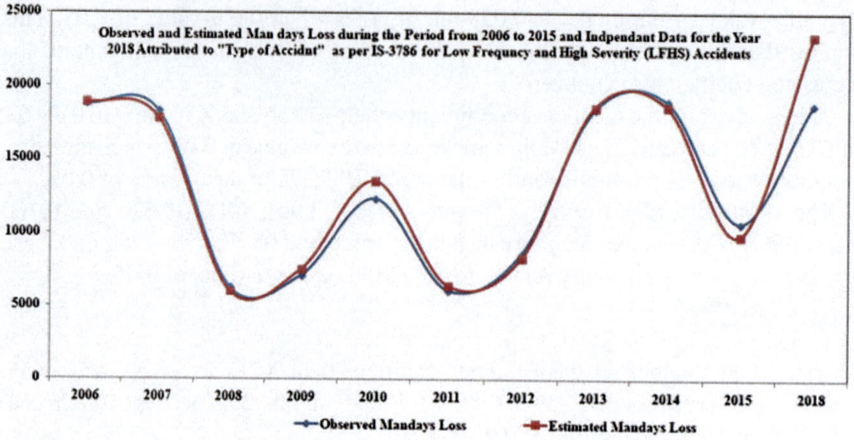

Fig. 2 Observed and estimated man-days loss on account of LFHS accidents

4 Discussion

The paper presents an accident prediction model on 'Types of Accidents' for 113 LFHS accidents comprising of, 95 reportable injuries and 18 fatalities, occurred in the Public Sector Power Company during the period from 2006 to 2015. Based on the statistical significance, following significant types of accidents can be derived from the model:

1. Exposure to or contact with electric current
2. Fall of persons from height
3. Stepping on striking against or struck by object

4.1 Exposure to or Contact with Electric Current

Accidents due to 'Exposure to or Contact with Electric Current' occurred while working on switchgear and electrical equipment.

Causes

- Flash over due to exposure to live circuit breaker components during rack-in or rack-out operations attributed to design deficiency, lack of attention to live electrical circuits, dust accumulation on switchgear due to lack of periodic maintenance; or non-use of arc protection suits and arc rated PPEs.
- Electric shock caused by degradation of electrical cables due to ageing, defective insulation of cables or cable terminals, use of non-standard power plug tops, wrong wiring of electrical extension board, lack of Earth Leakage Circuit Breakers (ELCBs); or inadequate earthing of electrical circuits.

- Insufficient hazard identification and risk assessment, non compliance to guaranteed isolation of power supply or work permit system before starting works on electrical systems.
- Insufficient awareness about the hazards, non use of arc protection suit or inadequate work supervision while working on electrical equipment.

Recommendations

Safety through Design:

- Rectification of design deficiency in the circuit breaker to prevent access to live components during rack-in or rack-out operations and revising technical specifications of circuit breakers.
- Provision of ELCBs and double earthing in electrical circuits and electrical equipment.
- Use of standard extension boards in electrical circuits.

Administrative Controls:

- Conducting Job Hazard Analysis to identify hazards and risks before starting works on electrical systems and performing jobs with safe work procedures.
- Periodic inspection and testing of cables with respect to ageing and replacement of damaged or expired cables.
- Periodic inspection and maintenance of switchgear, electrical equipment and extension boards.
- Independent inspection of cable terminal points during wiring to ensure effective insulation.
- Use of proximity tester while working near electrical systems to identify live electrical components in the vicinity.
- Training of employees on precautions to be taken while working on electrical systems.
- Deploying trained and designated persons for working on electrical systems and wiring of electrical circuits.
- Self checking and peer checking to ensure guaranteed isolation of electrical equipment and compliance to work permit system.
- Performing works on electrical systems under the supervision.
- Training on Cardio-Pulmonary-Resuscitation (CPR) for persons working on electrical systems to rescue victims of electrical shock.

Personal Protective Equipment:

- Conducting Arc Flash Hazard Analysis of the switchgear to determine arc flash incident energy and arc flash protection boundary for identifying required arc protection suit and arc rated PPEs for use while working on electrical switchgear, within arc flash protection boundaries [46, 47].
- Using arc resistant suits and arc rated PPEs while working on electrical switch gear.

4.2 Fall of Persons from Height

Accidents due to 'Fall of Persons from Height' occurred while working on height.

Causes

- Insufficient hazard identification and risk assessment before starting works at height.
- Working at height and on fragile roof without safe scaffoldings, platforms or ladders, safe access, fall protection measures and safety nets; and not complying with safe work procedures and work permit system.
- Lack of guarding or hard barricading with caution signage at locations with fall potential to restrict approach.
- Not wearing or improper wearing of helmet and fall protection safety harness while working at height.
- Inadequate work supervision while working at height.

Recommendations

Safety through Design:

- Erecting scaffoldings and platforms as per standards, using standard materials with safe access.
- Using roof top ladders while working on fragile roofs.
- Restricting approach to floor openings, wall openings and locations with fall potential by guarding or hard barricading, with caution signage.
- Using materials of sufficient strength to cover floor openings, wall openings and manholes.
- Provision of safety net below the locations of works at height.

Administrative Controls:

- Identification of hazards and risks before starting works at height and performing jobs with safe work procedures and work permits.
- Inspection and certification of scaffoldings and platforms by competent agencies before use.
- Prohibiting use of damaged scaffoldings, ladders and platforms, use of scaffoldings, ladders and platforms with expired validity; and inadequate periodic inspection and recertification of scaffolding, ladders and platforms.
- Performing works at height under supervision with approved work permit.
- Enforcement of work permit system for covering floor openings, wall openings and manholes; and periodic inspection of floor openings, wall openings and manholes coverings and rectification of deficiencies if any.
- Periodic illumination survey and ensuring sufficient illumination at work places.
- Self checking and peer checking of the area with respect to fall hazards before starting and while performing works at height.

Personal Protective Equipment:

- Ensuring use of fall protection safety harness by anchoring it to a firm structure or lifeline; and wearing helmet with chinstrap while working at height.

4.3 Stepping on, Striking Against or Struck by Object

Accidents due to 'Stepping on Striking Against or Struck by Object' occurred while working in plant areas.

Causes

- Improper installation of systems, equipment or components, not securing equipment or materials, not providing soft cushions on striking hazards, use of equipment with striking hazards, lack of machine guards or not restricting entry of unauthorized persons to work area.
- Lack of safe work procedures, adopting wrong procedure, lack of attention to striking hazards, non use of mechanical means for material handling, not using leather hand gloves or not wearing helmets in areas with bump hazards.
- Lack of coordination, supervision and signalmen while performing hazardous jobs with striking hazards.

Recommendations

Safety through Design:

- Installing and securing equipment and materials firmly at work place.
- Providing machine guards to prevent approach of workers to hazardous systems, equipment or components.
- Providing soft cushions with caution signage over the bump hazards.

Administrative Controls:

- Inspection of work place before performing job to ensure secured installation of equipment and materials at the workplace.
- Conducting Job Hazard Analysis, implementing hazard control measures; and ensuring use of safe work procedure effective supervision and signalmen while performing operations with striking hazards.
- Restricting entry of unauthorized persons to work area by cordoning with caution signage.
- Self checking and peer checking of work area with respect to striking hazards before starting and while performing jobs.
- Training of persons on use of hazardous equipment components.

Personal Protective Equipment:

- Using leather hand gloves while handling sharp objects and during material handling.
- Enforcing helmets while working in areas with bump hazards.

5 Conclusion

Industrial safety is a multi disciplinary subject involving interaction of human, equipment and work environment. Even though equipment and work environment are under the control organization, human factors are not always under the control organization, as every individual is unique. It is difficult to design equipment and work environment for each individual. Therefore, accident control measures cannot be implemented by 'Safety through Design' alone. In fact 'Safety through Design' should be supplemented by 'Administrative Controls' and 'Personal Protective Equipment'.

Regression analysis is one of the tools for identification of significant contributors by development of accident prediction model. The paper illustrates development of an accident prediction model by multiple linear regression analysis of 113 LFHS accidents occurred in a Public Sector Power Company in India during the period of 10 years from 2006 to 2015. The model gives satisfactory goodness of fit and statistical significance. Validation of the model, using independent data of types of accidents for the year 2018 demonstrates reasonably good prediction capability.

The significant contributors of the LFHS accidents are 'Exposure to or Contact with Electric Current', 'Fall of Persons from Height' and 'Stepping on Striking Against or Struck by Object'. The paper also gives accident control measures to be taken by 'Safety through Design', 'Administrative Controls' and 'Personal Protective Equipment'.

6 Future Research Directions

Extensive literature is available on accident prediction models of traffic accidents and crash modeling. However, accident prediction models on Industrial Accidents are found to be scarce. This gap needs to be addressed through more efforts in this regard.

Acknowledgements The authors acknowledge the support of officials of the Public Sector Power Company for extending technical and administrative supports for carrying out the research.

The article has been written as a voluntary effort and has not of received any funding. Opinions expressed are those of authors and not of their organizations.

Conflicts of Interest
None of the authors have conflicts of interest to declare.

References

1. US Department of Energy, Hand Book on Accident and Operational Safety Analysis: Volume I: Accident Analysis Techniques, USA. 2012.

2. Khanzode, V.V., Maiti, J., Ray, P. K., Occupational Injury and Accident Research: A Comprehensive Review, Safety Science, 2012, Vol. 50, p 1355–1367.
3. Government of India, The Factories Act, 1948 as amended in 1987, Government of India, 2019.
4. Jones, S., Kirchsteiger, C., Bjerke, W., The Importance of Near Miss Reporting to Further Improve Safety Performance, Journal of Loss Prevention in the Process Industries, 1999, Vol. 12, 1999, p 59–67.
5. Bureau of Indian Standards (BIS), Method for Computation of Frequency and Severity Rates for Industrial Injuries and Classification of Industrial Accidents, Indian Standard IS-3786–2014, BIS, 2014.
6. Sonnemans, P. J. M. and Korvers, Patrik M. W., Accidents in the Chemical Industry: Are They Foreseeable?, Journal of Loss Prevention in the Process Industries, 2006, 19, p 1–12.
7. Sonnemans, P. J. M., Körvers, Patrik M. W., Pasman, H. J., Accidents in 'Normal' operation - Can you see them coming?, Journal of Loss Prevention in the Process Industries, 2010, 23, p 351–366.
8. Wu, W., Alistair, G. F., Gibb, Qiming, Li, Accident precursors and near misses on construction sites: An investigative tool to derive information from accident databases, Safety Science, 2010, 48, p 845–858.
9. Zheng, X., Liu, M., An Overview of Accident Forecasting Methodologies, Journal of Loss Prevention in Process Industries, 2009, 22, P 484–491.
10. Mustakim F., Yusof. I, Rahman I., Black spot Study and Accident Prediction Model Using Multiple Liner Regression, First International Conference on Construction in Developing Countries (ICCIDC–I) 'Advancing and Integrating Construction Education, Research & Practice', August 4–5 (Karachi,, Pakistan) 2008, p 121–136.
11. National Safety Council (NSC), Accident Prevention Manual for Business & Industry-Administration and Programs, 12th ed. National Safety Council, USA, 2001.
12. National Safety Council (NSC), Accident Prevention Manual for Business & Industry - Engineering & Technology, 12th ed. National Safety Council, USA, 2001.
13. Bureau of Indian Standards (BIS), Code of Practice on Occupational Safety and Health Audit, Indian Standard IS-14489–2002, BIS, 2002.
14. Bureau of Indian Standards (BIS), Hazard Identification and Risk Analysis-Code of Practice, Indian Standard IS-15656–2006, BIS, 2006.
15. Bureau of Indian Standards (BIS), Occupational Health and Safety Management System-Requirements with Guidance for Use, Indian Standard IS-18001–2007, BIS, 2007.
16. International Organization for Standardization (ISO), Occupational Health and Safety Management Systems-Requirements and Guidance for Use, ISO-45001, ISO,2018.
17. National Institute for Occupational Safety and Health (NIOSH), Hierarchy of Controls, Centre for Disease Control and Prevention, USA, https://www.cdc.gov/niosh/topics/hierarchy/default.html, 2020.
18. Occupational Safety and Health Administration (OSHA), Recommended Practices for Safety and Health Programs, OSHA, USA, https://www.osha.gov/shpguidelines/hazard-prevention.html, 2020.
19. Kinnersley S., Roelen A., The Contribution of Design to Accidents, Safety Science, 2007, Vol. 45, p 31–60.
20. Lees, Frank. P., Loss Prevention in the Process Industries, Vol. 1 to 3, 3rd ed. Butterworth-Heinnemann, Oxford, 2004.
21. National Aeronautics and Space Administration (NASA), Accident Precursor Analysis Handbook, Version 1.0, Office of Safety and Mission Assurance, NASA, 2011.
22. Phimister, James R., Oktem, Uluku, Kleindorfer, Paul R., Kunreuther, Howard, Near Miss Incident Management in Chemical Process Industry, Risk Analysis, 2003, Vol. 23, (3), p 445–459.
23. Saleh Joseph H., Saltmarsh Elizabeth A., Favaro Francesca M., Loıc Brevault, Accident Precursors, Near-Misses and Warning Signs: Critical Review and Formal Definitions within the Frame Work of Discrete Event Systems. Journal of Reliability Engineering and System Safety, 2013, 114, p 148-154.

24. Lucas Deborah A., Organizational Aspects of Near Miss Reporting, Near Miss Reporting as a Safety Tool, 1st ed, Hale A.R, Lucas D.A, Ven der Schaaf T. W, September 1989 (Eindhoven), Commission of European Communities, Butterworth-Heinnemann, Oxford, 1991, p 127–136.
25. Van der Schaaf, Tjerk W, Near Miss Reporting in the Chemical Process Industry, Micro electron. Reliab., 1995, Vol. 35 (9–10), p 1233–1243. 26
26. Seth Ayim Gyekye, Simo Salminen, Anneli Ojajarvi., A Theoretical Model to Ascertain Determinates of Occupational Accidents Among Ghanaian Industrial Workers, International Journal of Industrial Ergonomics, 2012, 42, p 233–240.
27. Taylor Richard H., Van Wijk Lorenzo G.A., May John H.M., Carhart Neil J., A Study of The Precursors Leading to 'Organizational' Accidents in Complex Industrial Settings, Process Safety and Environmental Protection, 2015, 93, p 50–67.
28. Heinrich H. W, Peterson Dan, Roos Nester, Industrial Accident Prevention (A Safety Management Approach), 5th ed, McGraw-Hill Book Company, 1980.
29. Hovden Jan, Albrechtsen Eirik, Herrera Ivonne A., Is there a need for new theories, models and approaches to occupational accident prevention?, Safety Science,2010, 48, p 950–956.
30. Rathnayaka Samith, Khan Faisal, Amyotte Paul., SHIPP Methodology: Predictive Accident Modeling Approach Part I: Methodology and Model Description, Process Safety and Environmental Protection, 2011, 89, p 151–164.
31. Rathnayaka Samith, Khan Faisal, Amyotte Paul., SHIPP Methodology: Predictive Accident Modeling Approach Part II: Validation with Case Study, Process Safety and Environmental Protection, 2011, 89, p 75–88.
32. Ali Al-shanini, Arshad Ahmad Faisal Khan., Accident Modeling and Analysis in Process Industries, Journal of Loss Prevention in the Process Industries, 2014, 32, p 319–334.
33. Tauseef S. M., Abbasi Tasneem, Thiruselvi D. and Abbasi S. A., The risk of domino effect associated with the storage of Liquefied Petroleum Gas (LPG) and the safety codes for accident prevention, International Journal of Engineering, Science and Mathematics, 2017, Vol. 6 (7), p 456–482.
34. Basheer A., Abbasi Tasneem, Tauseef S. M., Abbasi S. A., Designing Risk-Minimized Layout for Hazardous Utilities in Chemical Process Industries Using the New Methodology RIDIMPUS, Journal of Failure Analysis and Prevention, 2019, Vol. 19, p 1127–1134.
35. Abbasi Tasneem, Pompapathi V., Tauseef S. M., and Abbasi S. A., An assessment of the prevailing codes/standards and models for determining safe spacing between two or more hazardous storage tanks, International Journal of Engineering, Science and Mathematics, 2017, Vol. 6 (7), p 255–268.
36. Basheer A., Tauseef S. M., Abbasi Tasneem,, Abbasi S. A., A Template for Quantitative Risk Assessment of Facilities Storing Hazardous Chemicals, International Journal of System Assurance Engineering and Management, 2019, Vol. 10, p 1158–1172.
37. Tauseef S. M., Suganya R., Abbasi Tasneem Abbasi S. A., Chemical Accident Simulation Tool (CAST): A System for Assessing Consequences of Accidents in Chemical Process Industry, Journal of Failure Analysis and Prevention, 2018, Vol. 18, p 101–116.
38. Attwood Daryl, Khan Faisal, Veitch Brian, Occupational Accident Models-Where have we been and where are we going?, Journal of Loss Prevention in the Process Industries, 2006, 19, p 664–682.
39. Kothari C. R, Research Methodology- Methods and Techniques, 2nd ed, New Age International (Pvt.) Limited, New Delhi, India, 1990.
40. Levin Richard I and Rubin David S., Statistics for Management, 6th ed, Printice Hall India (Pvt.) Limited, New Delhi, India, 1996.
41. Jenan Nasha't Sa'eed Kewan, Mohammad Ass'ad and Ali Barakat, Estimation of Multivariate Multiple Linear Regression Models and Applications, National University Nablus Palestine, 2015.
42. Shakil M., A Multiple Linear Regression Model to Predict the Student's Final Grade in a Mathematics Class, Department of Mathematics Miami-Dade College, Hialeah Campus. Florida, USA, 2001.

43. Uyanik Guilden Kaya and Guiler Nese., A Study on Multiple Linear Regression Analysis, Procedia - Social and Behavioral Sciences, 2013, 106, p 234–240.
44. Rokade S., Singh K., Katiyar S. K. and Gupta S., Development of Accident Prediction Model, International Journal of Advanced Engineering Technology, 2010, Vol. I (III), p 25–40.
45. Kraha Amenda, Tumer Heather, Nimon Kim, Zientek Linda Reichwein, Henson Robin K., Tools to Support Interpreting Multiple Regressions in Face of Multicollinearity, Frontiers in Psychology, 2012, Vol. 3, Article 44, p 1–10.
46. Institute of Electrical and Electronics Engineers (IEEE), Guide for Performing Arc-Flash Hazard Calculations, IEEE Standard 1584–2002, The Institute of Electrical and Electronics Engineers, Inc, USA.
47. National Fire Protection Association (NFPA), Electrical Safety in the Workplace, NFPA Standard 70E-2012, NFPA International Codes and Standards Organization, USA, 2012.